Oxy-fuel combustion for power generation and carbon dioxide (CO_2) capture

Related titles:

Developments and innovation in carbon dioxide (CO_2) capture and storage technology, Volume 1: Carbon dioxide (CO_2) capture, transport and industrial applications
(ISBN 978-1-84569-533-0)
Carbon dioxide (CO_2) capture and storage is the one advanced technology that conventional power generation cannot do without. CCS technology reduces the carbon footprint of power plants by capturing and storing the CO_2 emissions from burning fossil-fuels and biomass. Capture technology ranges from post- and pre-combustion capture to combustion-based capture. Storage options range from geological sequestration in deep saline aquifers and utilisation of CO_2 for enhanced oil and gas recovery, to mineral carbonation and biofixation of CO_2. This volume critically reviews carbon capture processes and technology applicable to the conventional power generation sector as well as other high carbon footprint industries.

Developments and innovation in carbon dioxide (CO_2) capture and storage technology, Volume 2: Carbon dioxide (CO_2) storage and utilisation
(ISBN 978-1-84569-797-6)
Carbon dioxide (CO_2) capture and storage is the one advanced technology that conventional power generation cannot do without. CCS technology reduces the carbon footprint of power plants by capturing and storing the CO_2 emissions from burning fossil-fuels and biomass. Capture technology ranges from post- and pre-combustion capture to combustion-based capture. Storage options range from geological sequestration in deep saline aquifers and utilisation of CO_2 for enhanced oil and gas recovery, to mineral carbonation and biofixation of CO_2. This volume critically reviews carbon storage and utilisation, covering all the main geological, terrestrial and ocean sequestration options and their environmental impacts, as well as other advanced concepts such as utilisation and photocatalytic reduction.

Advanced power plant materials, design and technology
(ISBN 978-1-84569-515-6)
Fossil-fuel power plants generate the majority of the world's power, but many plants are ageing and cannot meet rising global energy demands and increasingly stringent emissions criteria. To ensure security and economy of supply, utilities are building a new generation of advanced power plant with increased output and environmental performance. This book initially reviews improved plant designs for efficiency and fuel flexibility, including combined-cycle technology and utilisation of lower-grade feedstocks. Coverage extends to advanced material and component use, and the incorporation of alternative energy conversion technology, such as hydrogen production. Environmental and emissions performance issues round off the book.

Details of these and other Woodhead Publishing books can be obtained by:

- visiting our web site at www.woodheadpublishing.com
- contacting Customer Services (e-mail: sales@woodheadpublishing.com; fax: +44 (0) 1223 832819; tel.: +44 (0) 1223 499140 ext. 130; address: Woodhead Publishing Limited, 80 High Street, Sawston, Cambridge CB22 3HJ, UK)

If you would like to receive information on forthcoming titles, please send your address details to: Francis Dodds (address, tel. and fax as above; e-mail: francis.dodds@woodheadpublishing.com). Please confirm which subject areas you are interested in.

Woodhead Publishing Series in Energy: Number 17

Oxy-fuel combustion for power generation and carbon dioxide (CO_2) capture

Edited by
Ligang Zheng

WP
WOODHEAD
PUBLISHING

Oxford Cambridge Philadelphia New Delhi

Published by Woodhead Publishing Limited,
80 High Street, Sawston, Cambridge CB22 3HJ, UK
www.woodheadpublishing.com

Woodhead Publishing, 1518 Walnut Street, Suite 1100, Philadelphia, PA 19102-3406, USA

Woodhead Publishing India Private Limited, G-2, Vardaan House, 7/28 Ansari Road, Daryaganj, New Delhi – 110002, India
www.woodheadpublishingindia.com

First published 2011, Woodhead Publishing Limited

British Library Cataloguing in Publication Data
A catalogue record for this book is available from the British Library.

ISBN 978-0-08-101719-7 (book)
ISBN 978-0-85709-098-0 (e-book)

The publisher's policy is to use permanent paper from mills that operate a sustainable forestry policy, and which has been manufactured from pulp which is processed using acid-free and elemental chlorine-free practices. Furthermore, the publisher ensures that the text paper and cover board used have met acceptable environmental accreditation standards.

Typeset by RefineCatch Limited, Bungay, Suffolk

Cover image: Schwarze Pumpe 30 MWth oxy-fuel pilot plant, sourced from Vattenfall.

Contents

Contributor contact details

(* = main contact)

Foreword

János M. Beér
Massachusetts Institute of Technology (MIT)
Building 66, Room 301
77 Massachusetts Avenue
Cambridge
MA 02139
USA
E-mail: jmbeer@mit.edu

Editor and Chapter 1

L. Zheng
CanmetENERGY, Natural Resources Canada
1 Haanel Drive
Ottawa
Ontario
Canada
K1A 1M1
E-mail: lzheng@nrcan.gc.ca

Chapter 2

D. Thimsen, J. Wheeldon and D. Dillon*
CoalFleet for Tomorrow Program
Electric Power Research Institute (EPRI)
3420 Hillview Avenue
Palo Alto
CA 94303
USA
E-mail: dthimsen@epri.com
jowheeld@epri.com
ddillon@epri.com

Chapters 3 and 8

Y. Tan
CanmetENERGY, Natural Resources Canada
1 Haanel Drive
Ottawa
Ontario
Canada
K1A 1M1
E-mail: yetan@nrcan.gc.ca

Chapter 4

T. Wall* and R. Stanger
School of Engineering
The University of Newcastle
Callaghan
NSW 2308
Australia
E-mail: terry.wall@newcastle.edu.au
rohan.stanger@newcastle.edu.au

Chapter 5

E. J. Anthony*
CanmetENERGY, Natural Resources Canada
1 Haanel Drive
Ottawa
Ontario
Canada
K1A 1M1
E-mail: banthony@nrcan.gc.ca

H. Hack
Foster Wheeler North America Corporation
12 Peach Tree Hill Road
Livingston
NJ 07039
USA
E-mail: horst_hack@fwc.com

Chapter 6

C. Shaddix*
Sandia National Laboratories
7011 East Avenue
Livermore
CA 94550
USA
E-mail: crshadd@sandia.gov

A. Molina
School of Processes and Energy
Faculty of Mines
National University of Colombia
Car. 80 No. 65-223
Medellín
Colombia
E-mail: amolinao@unal.edu.co

Chapter 7

J. Shan*
Advanced Burner Product Line
SIEMENS Environmental Systems & Services
Siemens Energy
271 Route 202/206
PO Box 410
Pluckemin
NJ 07978
USA
E-mail: jiefeng@wharton.upenn.edu

A. Fry
Senior Engineer
Reaction Engineering International
Suite 210
77 West 200 South
Salt Lake City
UT 84101
USA
E-mail: fry@reaction-eng.com

Chapter 9

Y. Liu*, T. Wall and S. Khare
School of Engineering
The University of Newcastle
Callaghan
NSW 2308
Australia
E-mail: yinghui.liu@newcastle.edu.au
terry.wall@newcastle.edu.au
sameer.khare@csiro.au

R. Gupta
Department of Chemical and Materials Engineering
The University of Alberta
9107 – 116 Street
Edmonton, AB T6G 2V4
Canada
E-mail: rajender.gupta@ualberta.ca

Chapter 10

N. Prosser and M. M. Shah*
Praxair, Inc.
175 East Park Drive
Tonawanda
NY 14150
USA
E-mail: neil_prosser@praxair.com
minish_shah@praxair.com

Chapter 11

M. M. Shah
Praxair, Inc.
175 East Park Drive
Tonawanda
NY 14150
USA
E-mail: minish_shah@praxair.com

Chapter 12

H. Kobayashi* and L. E. Bool
Praxair, Inc.
175 East Park Drive
Tonawanda
NY 14150
USA
E-mail: Sho_Kobayashi@praxair.com,
Lawrence_Bool@praxair.com

Chapter 13

B. Clements*, R. Pomalis, L. Zheng and T. Herage
CanmetENERGY, Natural Resources Canada
1 Haanel Drive
Ottawa
Ontario
Canada
K1A 1M1
E-mail: bruce.clements@nrcan.gc.ca
rpomalis@nrcan.gc.ca
lzheng@nrcan.gc.ca
therage@nrcan-rncan.gc.ca

Chapter 14

H. Jin* and X. Zhang
Institute of Engineering Thermophysics
Chinese Academy of Sciences
P.O. Box 2706
11 Beisihuanxi Road
Beijing 100190
People's Republic of China
E-mail: hgjin@mail.etp.ac.cn
zhangxiaosong@mail.etp.ac.cn

Chapter 15

N. Zhang* and W. Han
Institute of Engineering Thermophysics
Chinese Academy of Sciences
P.O. Box 2706
No. 11, West of North 4th Ring Road
Beijing 100190
People's Republic of China
E-mail: zhangna@mail.etp.ac.cn
hanwei@mail.etp.ac.cn

Woodhead Publishing Series in Energy

1 **Generating power at high efficiency: Combined cycle technology for sustainable energy production**
Eric Jeffs

2 **Advanced separation techniques for nuclear fuel reprocessing and radioactive waste treatment**
Edited by Kenneth L. Nash and Gregg J. Lumetta

3 **Bioalcohol production: Biochemical conversion of lignocellulosic biomass**
Edited by K. W. Waldron

4 **Understanding and mitigating ageing in nuclear power plants: Materials and operational aspects of plant life management (PLiM)**
Edited by Philip G. Tipping

5 **Advanced power plant materials, design and technology**
Edited by Dermot Roddy

6 **Stand-alone and hybrid wind energy systems: Technology, energy storage and applications**
Edited by J. K. Kaldellis

7 **Biodiesel science and technology: From soil to oil**
Jan C. J. Bart, Natale Palmeri and Stefano Cavallaro

8 **Developments and innovation in carbon dioxide (CO_2) capture and storage technology Volume 1: Carbon dioxide (CO_2) capture, transport and industrial applications**
Edited by M. Mercedes Maroto-Valer

9 **Geological repository systems for safe disposal of spent nuclear fuels and radioactive waste**
Edited by Joonhong Ahn and Michael J. Apted

10 **Wind energy systems: Optimising design and construction for safe and reliable operation**
Edited by John D. Sørensen and Jens N. Sørensen

11 **Solid oxide fuel cell technology: Principles, performance and operations**
Kevin Huang and John Bannister Goodenough

12 **Handbook of advanced radioactive waste conditioning technologies**
Edited by Michael I. Ojovan

13 **Nuclear reactor safety systems**
Edited by Dan Gabriel Cacuci

14 **Materials for energy efficiency and thermal comfort in buildings**
Edited by Matthew R. Hall

15 **Handbook of biofuels production: Processes and technologies**
Edited by Rafael Luque, Juan Campelo and James Clark

16 **Developments and innovation in carbon dioxide (CO_2) capture and storage technology Volume 2: Carbon dioxide (CO_2) storage and utilisation**
Edited by M. Mercedes Maroto-Valer

17 **Oxy-fuel combustion for power generation and carbon dioxide (CO_2) capture**
Edited by Ligang Zheng

18 **Small and micro combined heat and power (CHP) systems: Advanced design, performance, materials and applications**
Edited by Robert Beith

19 **Hydrocarbon fuel conversion technology: Advanced processes for clean fuel production**
Edited by M. Rashid Khan

20 **Modern gas turbine systems: High efficiency, low emission, fuel flexible power generation**
Edited by Peter Jansohn

21 **Concentrating solar power (CSP) technology: Developments and applications**
Edited by Keith Lovegrove and Wes Stein

22 **Nuclear corrosion science and engineering**
Edited by Damien Féron

23 **Power plant life management and performance improvement**
Edited by John Oakey

24 **Direct-drive wind and marine energy systems**
Edited by Markus Mueller

25 **Advanced membrane science and technology for sustainable energy and environmental applications**
Edited by Angelo Basile and Suzana Nunes

26 **Irradiation embrittlement of reactor pressure vessels (RPVs) in nuclear power plants**
Edited by Naoki Soneda

27 **High temperature superconductors (HTS) for energy applications**
Edited by Ziad Melhem

28 **Infrastructure and methodologies for the justification of nuclear power programmes**
Edited by Agustín Alonso Santos

29 **Waste to energy (WtE) conversion technology**
Edited by Marco Castaldi

30 **Polymer electrolyte membrane and direct methanol fuel cell technology Volume 1: Fundamentals and performance**
Edited by Christoph Hartnig and Christina Roth

31 **Polymer electrolyte membrane and direct methanol fuel cell technology Volume 2: In situ characterisation techniques**
Edited by Christoph Hartnig and Christina Roth

32 **Combined cycle systems for near-zero emission power generation**
Edited by Ashok Rao

33 **Modern earth buildings: Materials, engineering, construction and applications**
Edited by Matthew R. Hall, Rick Lindsay and Meror Krayenhoff

34 **Handbook of metropolitan sustainability: Understanding and improving the urban environment**
Edited by Frank Zeman

35 **Functional materials for energy applications**
Edited by John Kilner, Stephen Skinner, Stuart Irvine and Peter Edwards

36 **Nuclear decommissioning: Planning, execution and experience**
Edited by Michele Laraia

37 **Nuclear fuel cycle science and engineering**
Edited by Ian Crossland

38 **Electricity transmission, distribution and storage systems**
Edited by Ziad Melhem

39 **Advances in biodiesel preparation: Second generation processes and technologies**
Edited by Rafael Luque and Juan Antonio Melero

40 **Biomass combustion science, technology and engineering**
Edited by Lasse Rosendahl

41 **Ultra-supercritical coal power plant: Materials, technologies and optimisation**
Edited by Dongke Zhang

42 **Radionuclide behaviour in the natural environment: Science, impacts and lessons for the nuclear industry**
Edited by Horst Geckeis and Christophe Poinssot

In collaboration with/En collaboration avec

Canada

Foreword

J. M. Beér, MIT, Cambridge, MA, USA

The provision of electric power is one of the prerequisites of prosperity; there is a strong correlation between electric power generating capacity and per capita GDP. Across the world, economic indicators signal continued growth and increased electricity demand. Twelve hundred gigawatts (GW) of new capacity is projected during the next 15 years to be added to the world's present electric generating capacity of about 4000 GW.

Coal is the primary fuel for electricity generation in the US and many other countries. In the US, about 50% of electricity is generated by the fleet of 335,000 megawatts (MW) capacity pulverized coal (PC) power plants. The advantages of coal are its broad availability with large reserves in several countries around the world, safe and secure supply, low cost, and utilization by mature technologies.

Coal use, however, presents challenges for reducing emissions of air pollutants and carbon dioxide (CO_2). In response to these challenges, a number of technologies that significantly reduce emissions of SO_2, NO_x, particulate matter (PM), and mercury (Hg), have been developed and are in wide commercial use, with further prospective developments toward 'near zero emission' power plants.

CO_2 emissions are gaining significant attention. The most cost-effective and readily available option of mitigating CO_2 emissions is to increase the generating plant's efficiency so that less coal is burned per MWh generated. Compared with the mean efficiency of the existing coal based generating fleet, advanced ultra-supercritical steam plants offer a relative 25% improvement in efficiency and corresponding reduction in all emissions, including CO_2.

Greater reductions of CO_2 emissions can be achieved by CO_2 capture and geological sequestration (CCS). In this process, CO_2 is captured from the cleaned products of combustion or gasification. The CO_2 is compressed to a supercritical fluid, transported by pipeline to a geologic site where it is pumped deep underground for permanent storage.

CCS for base load power generation is likely to become commercially available, i.e. deployable without significant government subsidy, at around 2025, following the construction and operation of several demonstration plants, and also congressional legislation on related legal and liability issues, during the next 15 years.

The most promising coal based technology options for CCS include:

- PC combustion in ultra-supercritical steam (USCPC) cycles, with post-combustion CO_2 capture, because of high generating efficiency;
- integrated gasification combined cycle (IGCC), because of the advantage of pre-combustion CO_2 capture; and
- oxy combustion, because due to the high flue gas CO_2 concentration it does not require CO_2 separation.

The fact that CO_2 separation is not required is an important advantage of oxy combustion because CO_2 capture requires large amounts of steam, leading to significant losses in plant output and efficiency.

When oxygen, instead of air, is used for combustion in a PC boiler, the mass flow rate of combustion products is reduced and the CO_2 concentration correspondingly increased. The clean-up of the four-fold reduced flue gas volume leads to significantly lower equipment and operating costs compared with air combustion. Also, after clean-up and removal of condensables, this oxy-combustion flue gas stream is ready for sequestration, without energy intensive CO_2 separation.

The challenge of oxy combustion is the development of a low energy intensity oxygen production process. Cryogenic air separation presently used by the chemical industry for air separation consumes a significant fraction of the oxy-fired PC plant's output and reduces its efficiency. There is an urgent need for the development and full scale demonstration of novel, membrane type air separation processes, which can lead to major reduction in parasitic energy consumption of a future oxy combustion power plant.

This book is an important and timely addition to the literature of greenhouse gas emissions control. In-depth discussions of oxy combustion technology, current state, and prospective timeline of development, RD&D needs and economics hold valuable information for manufacturers, researchers in industry and universities, and government policy makers.

1

Overview of oxy-fuel combustion technology for carbon dioxide (CO_2) capture

L. ZHENG, CanmetENERGY, Natural Resources Canada, Canada

Abstract: This chapter provides an overview of the concepts, main components, background, advantages, and challenges of oxy-fuel combustion technology for power generation and carbon dioxide (CO_2) capture. Brief descriptions of other carbon capture technologies and their comparisons with oxy-fuel technology are outlined. A concise description of each chapter in this book is included.

Key words: oxy-fuel combustion, carbon capture technologies, clean coal technology.

1.1 Introduction

Oxy-fuel combustion is currently considered to be one of the major technologies for carbon dioxide (CO_2) capture. This book focuses on the development of oxy-fuel combustion technologies using coal as fuel.

1.1.1 Coal as an energy source

Coal plays a very important role in our day-to-day lives. In a comprehensive report published in 2008, the International Energy Agency (IEA) predicted that the demand for coal will surpass oil in absolute terms between 2030 and 2050, and will become the predominant fuel for the world (IEA, 2008–1). Currently, about 40% of the world's electricity is generated with coal (WCI, 2009) making it the largest fuel source for power generation (IEA, 2008–1). In the two largest CO_2 emitting countries, China and the United States of America, more than 77% and 50% of the electricity, respectively, is generated with coal (WCI, 2009). The improvement of global living standards and continuous economic growth will require increased use of energy. From 2000 to 2006, IEA reported that worldwide demand for electricity increased from 12,641 TWh to 15,665 TWh, a stunning rise in demand of nearly 24%. This growth is expected to continue at an average annual rate of 2.5%; coal generation is projected to produce 14,600 TWh of electricity by 2030 – more than double its current contribution of approximately 6300 TWh (IEA, 2008–2).

The attraction of coal as a fuel source is due to several factors. First, it is abundant: even under rapid growth scenarios, known coal reserves can continue to meet our energy needs for at least the next 100 years (Lackner and Sachs, 2005). Indeed, some studies have suggested that there is more than 190 years' worth of coal available, almost four times that of oil and gas combined (WCI, 2009). Unlike oil and gas, coal is well distributed in the world, making it easily accessible and very reliable. Coal is also one of the most affordable energy resources at one to two US dollars per MM Btu; by contrast, oil and gas costs are in the range of 6 to 12 US dollars per MM Btu (MIT, 2007). This combination of attributes makes it very likely that coal will continue to be a critical fuel source well into the future.

1.1.2 Developments in clean coal technology

Coal as an energy source has a number of negative environmental impacts, including (but not limited to) the release of particle matter (PM), oxides of sulphur and nitrogen (SO_x and NO_x), carbon monoxide (CO), and trace metals such as mercury. It is a long procedure from the point of realization of the need to control emissions, to pass appropriate emissions control regulations and standards, to develop control technologies, and to effect their implementation. With the gradual installation of each add-on unit for emissions reduction, the environmental impacts associated with coal combustion have been greatly reduced. Installation of electrostatic precipitators (ESP) and/or baghouses, initiated in the 1970s, allows for flue gas flyash reductions above 95%. Flue gas desulfurization (FGD) technologies, such as wet scrubbers introduced in the 1980s, are capable of 90% SO_x removal. Since the 1990s, low NO_x burners have been employed to reduce nitrogen oxide formation. Combined with selective catalytic reactors (SCRs), it is now possible to reduce NO_x emissions by more than 90% (EPA, 2006). Rapid developments are currently taking place to address the issues of fine particulate matter less than 2.5 μm ($PM_{2.5}$) and mercury emissions. Clean coal combustion technologies have become major business concerns in coal utilization. It has been reported that the capital and operating costs of emission control systems of a typical 500 MW_e coal-fired power plant are roughly 47% and 57% of the total respective costs (Marin *et al*., 2003).

Of increasing concern, coal combustion is also one of the largest sources of anthropogenic CO_2 emissions. In 2006, about 42% of the world energy-related CO_2 emissions were attributable to coal use (IEA, 2008–2). On an annual basis, a typical 500 MW_e coal-fired power plant emits about three million tonnes of CO_2 to the atmosphere (MIT, 2007), the equivalent of the total CO_2 emissions from 374,000 passenger cars (EPA, 2000).

The concern about CO_2 emissions from coal-fired units has prompted intensive research into its control technologies. Some reductions can be achieved by upgrading the coal by washing, drying and briquetting. Optimized operating conditions on excess air and stack temperature reductions lead to better efficiencies and, hence, lower emissions. Furthermore, up to 25% CO_2 emission reduction can be obtained through the utilization of supercritical and ultra supercritical boiler technologies (Beér, 2007).

One of the most feasible options to stabilize CO_2 levels in atmosphere is carbon (CO_2) capture and storage (CCS). CCS is a process in which CO_2 is removed from emission gases, transported, and stored (sequestered) in a location where it is isolated from the atmosphere. The United Nations Intergovernmental Panel on Climate Change (IPCC) has noted that 'CCS has the potential to reduce overall mitigation costs and increase flexibility in achieving greenhouse gas emission reduction' (IPCC, 2005). Due to the large quantity and concentrated nature of CO_2 emissions from coal-fired power generation stations, these emitters have become the focus of CCS development.

1.1.3 Carbon capture technologies

Currently, there are three major CO_2 capture technologies that have reached the level of industrial-scale demonstration. These three technologies are:

- Post-combustion capture: a chemical solvent such as amine or ammonia is used to scrub CO_2 out of the combustion flue gas.
- Pre-combustion capture: solid fuel is gasified with oxygen to produce a gaseous fuel consisting mainly of carbon monoxide (CO) and hydrogen (H_2). A water–gas shift reaction is employed to convert CO and water to H_2 and CO_2 and a physical sorbent is then used to capture CO_2.
- Oxy-fuel combustion: pure oxygen is used for fuel combustion, thereby producing a CO_2-enriched flue gas ready for sequestration once water is condensed from the flue gas and other impurities are removed.

The post-combustion capture approach is the same as the approaches for control of particulate matter, SO_x, and NO_x. This approach involves adding unit operations after combustion, making post-combustion capture an attractive option for retrofitting existing plants or building a CO_2 capture-ready plant. For post-combustion technology, a liquid solvent such as monoethanolamine (MEA) or ammonia is used in an absorption tower to scrub CO_2 from the flue gas. The CO_2-rich solvent is then pumped to a stripper or regeneration tower where heat is used to separate CO_2 from the solvent. The captured CO_2 is then compressed and transported for storage. Chemical solvents are already widely used in refineries, natural gas processing, and petrochemical plants to capture CO_2; demonstrations of solvent applicability for coal-fired power plants are currently underway. Because air is used for combustion, most (70%) of the flue gas is N_2 with comparatively little (< 15%) CO_2, hence large equipment is needed for post-combustion capture processes. Most post-combustion capture technologies are capable of capturing more than 90% of the CO_2 at very high purity levels. A major challenge for post-combustion capture technologies is the intensive energy needed to regenerate solvent or to cool the flue gas when the chilled ammonia process is employed. Furthermore, stringent flue gas limits for SO_x, NO_x, and flyash are necessary to minimize solvent usage when amines are used.

Coal gasification is a central technology for pre-combustion CO_2 capture. In gasification, coal is reacted with oxygen and steam at high temperature and pressure to produce a synthetic gas (syngas). Usually, less than 50% of the oxygen required for complete combustion is used in gasification; thus, the coal is only partially oxidized to CO and H_2, the major syngas components. To capture CO_2 from a gasification process, a water–gas shift reactor is used to convert CO and water to H_2 and CO_2. The CO_2 is then separated with physical sorbent. Gasification technology is widely employed in the chemical industry. However, its application in the power sector is very limited due to the high cost associated with it for electricity generation. At present, there are only four commercial-scale power generation plants in the world that gasify coal (Ratafia-Brown *et al.*, 2002). The process that utilizes coal gasification for power generation is known as the integrated gasification combined cycle (IGCC). In IGCC, once the coal is gasified, up to 99.5% of sulfur in the coal can be recovered as elemental sulfur, making IGCC one of the cleanest coal power generation technologies. The syngas is then combusted in a gas turbine to produce power. Additional power is generated by the gas turbine exhaust in a heat recovery steam generator. IGCC power plants have considerably higher thermal efficiencies and lower CO_2 emission rates than comparable subcritical pulverized coal plants. However, IGCC's high electricity generation cost, plant complexity, and lower availability have limited the technology's uptake in the power industry. Furthermore, gasification of low rank coals, such as lignite, is still in the developmental stage.

One very elegant approach for CO_2 capture uses oxygen instead of air for combustion. By eliminating nitrogen from the oxidant gas stream, it is possible to produce a CO_2-enriched flue gas ready for sequestration after water has been condensed and other impurities have been separated out. This technology is known as oxy-fuel combustion and is the subject of this book.

1.2 Oxy-fuel combustion: concepts and components

Oxy-fuel combustion for CO_2 capture incorporates three main components: the air separation unit (ASU) that provides oxygen for combustion, the furnace and heat exchangers where combustion and heat exchange take place, and the CO_2 capture and compression unit. Due to the large quantity of high purity oxygen typically required in oxy-fuel combustion, cryogenic air separation is currently the technology of choice for oxygen production.

In oxy-fuel combustion (Fig. 1.1), conventional boiler technology is deployed to prepare and combust the fuel, and to transfer the combustion heat from the flue gas to a working fluid (typically steam) to generate electricity. In most implementations, a large portion of the flue gas is recycled back to the furnace to control the flame temperature and to reconstitute the flue gas volume to ensure proper heat transfer (Wall *et al.*, 2005). The resulting flue gas consists mainly of CO_2 and water, as nitrogen has been eliminated from the combustion medium. Flyash in the flue gas is collected in an ESP or baghouse, and flue gas desulfurization is employed to reduce sulfur oxide emissions. Compared with

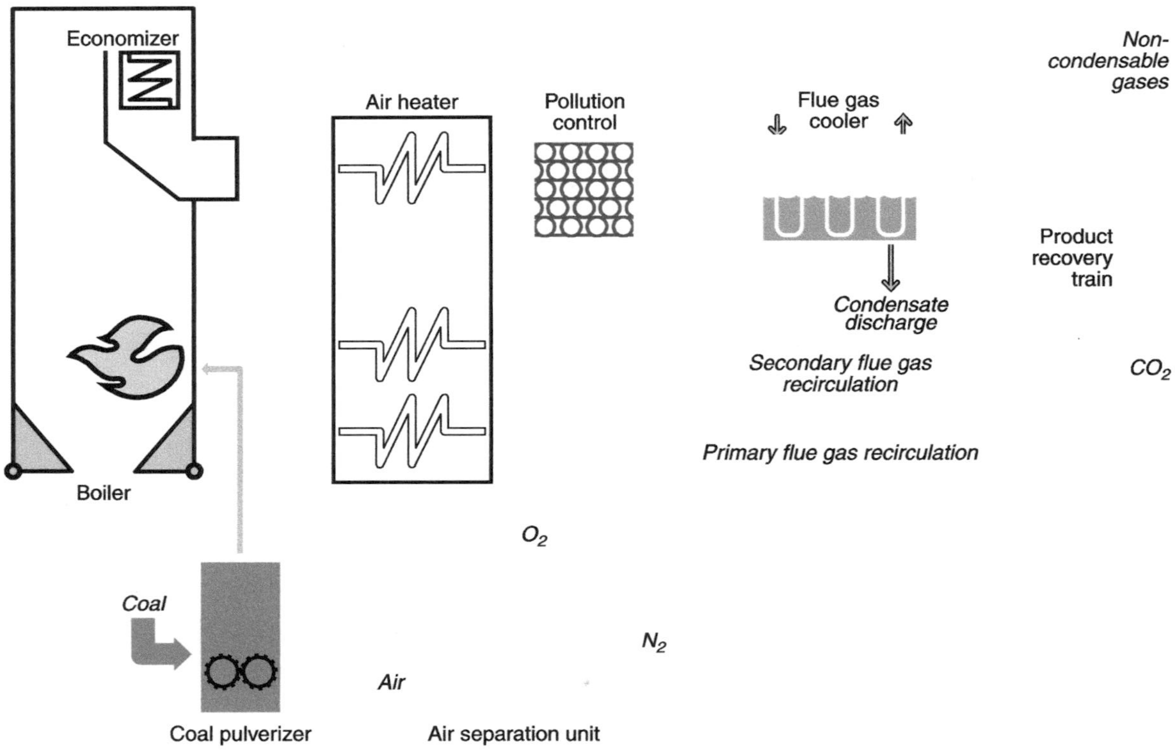

1.1 Oxy-fuel combustion.

most other combustion technologies, oxy-fuel combustion has the inherent advantage of producing low NO_x emissions since oxygen is used for combustion and the re-burning mechanism via flue gas recycle. Consequently, a NO_x control process is typically not required. Recycled flue gas can be drawn at several locations, for example, before or after the flue gas condenser.

Due to the high concentration of CO_2 in the flue gas, no chemical solvent or physical sorbent is required to separate CO_2 from the flue gas. The CO_2 capture and compression unit comprises multi-stage compression and cooling processes in which the flue gas is first dried and the impurities such as oxygen, nitrogen, and argon are separated from the CO_2.

1.3 Oxy-fuel combustion: background and motivation

1.3.1 Pre-carbon capture and storage (CCS) oxy-fuel combustion for enhanced oil recovery and control of nitrogen oxides (NO_x)

For years, the oil and gas industry has been using CO_2 from natural sources and industrial processes (mainly natural gas processing, ammonia and fertilizer manufacturing, and coal gasification plants) to extract oil from depleted oil fields. Operating experience in this process, known as enhanced oil recovery (EOR), is extensive. Many CO_2 pipeline networks have been constructed specifically for EOR operations. Alternatively, in some cases, EOR has employed combustion flue gas directly (Taber, 1985; Taber *et al.*, 1997).

In 1982, Abraham and coworkers (Abraham *et al.*, 1982) proposed the idea of employing oxy-fuel combustion to obtain large quantities of CO_2 for EOR. They suggested that part of the flue gas could be recycled to the furnace to control the combustion temperature. This idea was first tested (Wang *et al.*, 1988) in the 3 MW_{th} pilot-scale test facility at the University of North Dakota's Energy and Environmental Research Center with the Argonne National Laboratory as a research partner.

Industrial furnaces have been using oxy-fuel combustion technology for many years (Dugué, 2000) in the glass, aluminum, cement, steel, and incineration sectors. The technology's primary purposes in these settings are to enhance productivity, reduce fuel consumption, and decrease NO_x emissions (Charon, 2000). These industrial applications are much smaller in scale compared with power generation and usually no flue gas is recycled to the furnace.

1.3.2 Oxy-fuel technology in the context of carbon capture and storage (CCS)

Since the late 1990s, due to concerns about rising atmospheric greenhouse gas emissions levels precipitating climate change, oxy-fuel combustion technology has attracted growing attention and has developed from conceptual exploration and pilot-scale testing to industrial-scale demonstration. This rapid development has arisen primarily from the perceived superiority of the technology on a number of fronts. Compared with other CCS technologies, oxy-fuel is a simple, elegant, and readily available ('off the shelf') technology. Unlike post-combustion capture, there is no need to add a major chemical process to capture CO_2. Furthermore, there is no need for the power generation industry to adopt a new process (such as IGCC) for its core business. As the major components of oxy-fuel combustion, i.e. coal combustion and air separation, are mature technologies that have been extensively employed, the retraining requirements for personnel are minimal. Oxy-fuel also has the advantage of dual firing capability: it is possible to switch to air-firing to meet peak load demand.

The existing fleet of modern pulverized coal-fired power plants represents an opportunity to achieve significant greenhouse gas emissions reductions in the coming years if retrofitted for oxy-fuel operations. Although post-combustion can also be added to an existing plant, it is much more complicated because of the complexity of its process and constraints of space availability.

There are many notable emissions control benefits resulting from oxy-firing. Due to the CO_2-enriched environment and re-burning via flue gas recycle, NO_x formation in oxy-firing is significantly lower than in other combustion methods. Additionally, since oxygen is used for combustion instead of air, the flue gas volume from oxy-firing is only about one-quarter or one-fifth that of air combustion. This in turn results in a need for much smaller flue gas emissions control equipment if the flue gas is recycled upstream of the devices. Low NO_x formation and smaller flue gas volume also leads to substantial capital and annual cost savings. Recent research (White and Fogash, 2009) shows that integrated emissions control of SO_x, NO_x, and mercury might be possible as part of the oxy-fuel flue gas CO_2 capture process, which could further reduce the cost of oxy-fuel combustion technology.

1.4 Existing challenges for oxy-fuel combustion technology

The major challenge for oxy-fuel combustion technology is the high energy costs associated with oxygen production and CO_2 separation. Although the energy efficiency of cryogenic air separation technology has shown steady improvement, it is still an energy intensive process. With costs of 200–220 kWh per tonne of oxygen generated by cryogenic air separation, oxygen production is by far the largest energy expenditure in oxy-fuel technology, resulting in a major system efficiency reduction. A step change technology for oxygen production certainly would be a most desirable development; it would bring the cost of oxy-fuel down considerably and would also benefit other CCS technologies such as pre-combustion.

The US Department of Transportation requires that pipeline CO_2 has a minimum purity of at least 90% (WRI, 2008). Yet, most CO_2 pipeline operators have higher purity standards, usually with a minimum set at 95% for EOR applications (Aspelund and Jordal, 2007). For CO_2 capture in an oxy-fuel process, a multi-stage compression and cooling system is required for purification and compression of the final flue gas from oxy-fuel systems. This process is designed to remove water vapor and other impurities such as oxygen, nitrogen, and argon to produce high purity CO_2 for transportation and storage. Due to the nature of compression and associated cooling, the purification process is also energy intensive, which results in significant cost. In addition, great care must be taken to prevent air infiltration into the oxy-fuel combustion system. Air infiltration would bring additional contaminants into the flue gas with a concomitant energy penalty during CO_2 separation.

While the components of oxy-fuel combustion technology are all mature, their collective operation still must be fully demonstrated.

1.5 Development of oxy-fuel combustion technology

Concentrated and systematic research and development work on oxy-fuel combustion for CO_2 capture has been conducted since the 1990s. Laboratory-scale tests have been conducted in order to understand the fundamental aspects of oxy-fuel combustion issues, including ignition characteristics (Kiga *et al.*, 1997), flame stability (Payne *et al.*, 1989), and rate of char combustion (Shaddix and Murphy, 2003). Many studies have been carried out to investigate heat transfer behaviors under oxy-fuel combustion (Khare *et al.*, 2005) and to evaluate options for retrofitting existing boilers for oxy-firing (Zheng *et al.*, 2002). Pilot-scale oxy-fuel combustion operations have been in service at various installations in Canada, the USA, Europe, and Japan for many years. Through these operations, it has been confirmed that furnaces can be easily switched from air firing to oxy-fuel firing, that air infiltration can be effectively limited, and that a highly enriched CO_2 flue gas can be produced (Tan *et al.*, 2006) for transportation and storage. In addition, oxy-fuel combustion has been shown to significantly reduce NO_x emissions (Croiset and Thambimuthu, 2001). On the evidence, then, it would appear that there are no major technical hurdles in implementing oxy-fuel combustion for CO_2 capture.

Oxy-fuel combustion for CO_2 capture has also been the subject of a number of technical and economic feasibility studies (Dillon *et al.*, 2005; DOE/NETL, 2008; Marin *et al.*, 2003; Singh *et al.*, 2003). These studies provide some basic economic assessments of oxygen generation, CO_2 separation, overall system efficiency, cost of electricity, and cost of CO_2 captured and avoided. Many of these studies also attempted to compare the cost of oxy-fuel with other CO_2 capture technologies, although the lack of commercial experience with all these technologies imparts a high degree of uncertainty to any such comparisons.

Development of oxy-fuel combustion technology has accelerated in recent years due to growing public concern about CO_2 emissions and climate change. Over roughly a two-year period, Vattenfall AB has successfully constructed and operated one of its boilers at Schwarze Pumpe, Germany for oxy-fuel operation.

This is the first comprehensive and large-scale oxy-fuel demonstration plant with CO_2 capture at a capacity of 30 MW_{th}. Since becoming operational in 2008 (Jacoby, 2009), the Vattenfall AB facility has provided a wealth of information on virtually all aspects of oxy-fuel combustion for CO_2 capture. Other industrial-scale projects, such as the Callide 30 MW_e demonstration plant in Australia, are currently underway with anticipated start-up within the next year. In addition, two commercial-scale oxy-fuel plants have been proposed by Vattenfall AB (250 MW_e) and Korea Electric Power Corporation (100 MW_e). In August of 2010, Energy Secretary Steven Chu of the US announced the awarding of \$1 billion under the FutureGen Program to build a 200 MW oxy-fuel coal fired unit in Illinois starting in 2012 (DOE, 2010). This plant is expected to be online by 2016. Interest in the demonstration of oxy-fuel combustion systems has also extended to circulating fluidized bed combustors and pressurized boilers.

Naturally, the development of this technology depends heavily on government policies on CO_2. However, acceptance of oxy-fuel combustion for CO_2 capture is entirely based on whether this technology can be as clean, efficient, and economic as promised.

1.6 About this book

This volume provides a detailed and comprehensive presentation of oxy-fuel technology for power generation with emphasis on its current status as well as its future applications and directions. The book consists of three parts. An overview of oxy-fuel combustion technology is presented in Part I. Part II discusses the fundamental science and engineering aspects of oxy-fuel combustion, and Part III focuses on the advanced development of oxy-fuel combustion technology.

In Part I, a detailed economic analysis of the use of oxy-fuel combustion technology for power generation and CO_2 capture is presented by Drs Thimsen, Wheeldon and Dillon of the Electric Power Research Institute. Comparisons with other CO_2 capture technologies are made. The eventual uptake of this technology will depend on whether it is superior to others, not only on the basis of technological advantages but, far more importantly, also on the economic performance.

Key issues with great implications on the safety, reliability, and smooth operation of the process are discussed in the chapter on operation (Chapter 3). Based on his experience in operating pilot-scale oxy-fuel furnaces, Dr Tan discusses practical operational procedures, such as start-up and shut-down, different flue gas recycle options, effects of air ingress and flue gas leakage, as well as safety considerations.

Oxy-fuel technology demonstration at an industrial level is a critical step in order to scale it up to a commercial level. In Chapter 4, Prof. Wall and Dr Stanger offer an excellent update on the current activities and their objectives on this front with an insightful outline of what is needed for the future.

Circulating fluidized bed combustion has many unique advantages, especially its ability to fire a large variety of fuels including biomass. The development of oxy-fuel combustion technology with circulating fluidized bed is reviewed in depth by Dr Anthony and Mr Hack in Chapter 5.

In Part II, Chapter 6, a comprehensive examination of oxy-fuel combustion characteristics is performed by Drs Shaddix and Molina, where they make some very interesting comparisons with air-fired combustion. They point out that the CO_2-enriched environment might lead to significant changes in flame ignition, flame stability, char combustion, and carbon burnout.

Various design philosophies to establish a stable flame for oxy-fuel burners are discussed by Drs Shan and Fry in Chapter 7. Dual-fire capability considerations and impact on pollutant formations are also explored. Pilot-scale testing results and commercial-scale design criteria are addressed.

Based on extensive experimental data, Dr Tan gives a detailed analysis of oxy-fuel combustion in the context of 'clean coal' technology in Chapter 8. Besides the well-documented lower NO_x emission, oxy-fuel combustion also has effects on other pollutants, such as SO_2, SO_3, and their respective ratio. The discussion on the concept and development of integrated emissions control is especially worth mentioning as it could lead to major reductions in capital and operational costs.

The heat transfer characteristics and its impacts on boiler design are outlined by Drs Liu, Wall, Khare and Gupta in Chapter 9. By studying furnaces of various sizes under oxy-fuel operation, it is concluded that an oxy-fuel-fired furnace can have the same heat transfer rate as an air-fired one with lower furnace exit temperature and higher gas emissivity. Therefore, current boiler design principles and operational practice can be easily adopted for oxy-fuel combustion.

Two chapters on oxygen production and CO_2 separation and compression, the two key components outside the combustion envelope, have been written by industrial experts from Praxair, Inc. In Chapter 10 Drs Prosser and Shah survey the development of oxygen generation technology and identify the current choice for supplying oxygen for oxy-fuel combustion. It is very encouraging to know that even though cryogenic air separation technology is considered a mature technology, there are still opportunities for improvement when it is applied to oxy-fuel combustion. The section on novel oxygen generation using oxygen transport membrane technology has great importance as this could lead to major changes in boiler design and significant reductions in the cost of CO_2 capture. This in-depth analysis is extended to the chapter by Dr Shah on CO_2 separation and compression (Chapter 11). It outlines the important issue of CO_2 purity specification for various applications and related processes and costs. The effects of air infiltration and oxygen purity on the cost of CO_2 capture are clearly demonstrated.

Part III recognizes that current CO_2 capture technologies all require significant energy input and thus result in system energy efficiency reductions. Various efforts are underway to minimize the energy requirement for CO_2 capture. Based on their vast experience on retrofitting industrial furnaces for oxy-fuel operation, Drs Kobayashi and Bool offer many novel ideas in Chapter 12 on minimizing or eliminating the need for recycled flue gas. Many interesting and exciting furnace layouts are presented here that could lead to future R&D efforts.

One of the ideas to increase system efficiency is to operate the boiler at elevated pressures. This might turn the latent heat of the flue gas into a useful heat source within the system. The CanmetENERGY team, led by Mr Clements, review various high pressure system configurations and explore their designs and operation issues in Chapter 13. Other applications of high pressure oxy-fuel combustion in industry, such as direct contact steam generation, are also outlined.

Chemical looping combustion is a special case of the oxy-fuel combustion process, because its approaches to oxygen generation and combustion are completely different from all others. In Chapter 14, Dr Jin, one of the pioneers in the field, with Dr Zhang, gives an in-depth analysis of its principles, advantages, and choices of looping materials, as well as combustion systems that can make use of solid fuels, and future trends.

As noted before, this book is mainly focused on using coal as fuel. Natural gas has many advantages, particularly low emissions of particulates and NO_x; it also has much lower CO_2 emissions. In addition, the natural gas combined cycle is the most efficient system for power generation. However, the situation for natural gas is far more complicated when CO_2 capture is taken into consideration. This has led to the development of several innovative cycle concepts for oxy-fuel natural gas combustion. As such, it was deemed necessary to have a dedicated chapter summarizing the activities in this area. Drs Zhang and Han recap on the key elements of those proposed cycles in Chapter 15, and lay out their thoughts on the advantages and limitations of each cycle.

1.7 Acknowledgements

In 2005, the International Technical Conference on Coal Utilization & Fuel Systems (known as the Clearwater Coal Conference) was the first major international conference to hold a special panel and associated technical sessions devoted to oxy-fuel technology. Since then, oxy-fuel technology has become a major theme of the conference, with more than 30 invited and contributed papers on the subject every year. We would like to express our sincere thanks to the conference organization committee for their support and encouragement. Most importantly, we would like to thank all the conference contributors over the years for their efforts in driving the development of oxy-fuel combustion technology. Last, but not least, we would like to thank our publisher, especially Mr Ian Borthwick, who not only initiated this book but also has always been patient and willing to help.

1.8 References

Abraham B, Asbury J, Lynch E, and Teotia A (1982), 'Coal-oxygen process provides CO_2 for enhanced recovery', *Oil & Gas Journal*, 80(11): 68–70, 75, 15 March, 1982.

Aspelund A and Jordal K (2007), 'Gas conditioning – the interface between CO_2 capture and transport', *International Journal of Greenhouse Gas Control* 1(3): 343–354, July 2007.

Beér J (2007), 'High efficiency electric power generation: The environmental role', *Progress in Energy and Combustion Science* 33: 107–134.

Charon O (2000), *Recent Developments and Future trends in Oxy-Combustion Applications*, International Flame Research Foundation 17th Topic Oriented Technical Meeting, Les Vaux de Cernay, France.

Croiset E and Thambimuthu K (2001), 'NO_x and SO_2 emission from O_2/CO_2 recycled coal combustion', *Fuel* 80: 2117–2121.

Dillon J, White V, Allam R, Wall R, and Gibbins J (2005), *Oxy-combustion Processes for CO_2 Capture from Power Plant*, IEA Greenhouse Gas R&D Programme Report, September 2005.

DOE (Department of Energy) (2010), http://www.energy.gov/news/9309.html and http://www.energy.gov/news/9368.html, accessed on October 21, 2010.

DOE/NETL (Department of Energy, National Energy Technology Laboratory) (2008), '*Pulverized Coal Oxycombustion Power Plants*', DOE/NETL-2007/1291.

Dugué J (2000), *The Use of Oxygen for Industrial Combustion: Summary of the 17th IFRF Topic Oriented Technical Meetings*, International Flame Research Foundation Document No. D121/y/6, 26 September 2000.

EPA (United States Environmental Protection Agency) (2000), *Emission Facts – Average Annual Emissions and Fuel Consumption for Passenger Cars and Light Trucks*, EPA420-F-00-013, April 2000.

EPA (United States Environmental Protection Agency) (2006), *Environmental Footprints and Costs of Coal-Based Integration Gasification Combined Cycle and Pulverized Coal Technologies*, EPA-430/R-06/006, July 2006.

IEA (International Energy Agency) (2008–1), *Energy Technology Perspectives: Scenarios & Strategies to 2050*, Paris, France, 2008.

IEA (International Energy Agency) (2008–2), *World Energy Outlook 2008*, Paris, France, 2008.

IPCC (Intergovernmental Panel on Climate Change) (2005), *Carbon Dioxide Capture and Storage*, IPCC Special Report.

Jacoby J (2009), *First Test Results from Vattenfall's 30 MWth Oxyfuel Pilot Plant in Schwarze Pumpe*, 34th International Conference on Coal Utilization & Fuel System, Clearwater, Florida, 1–4 June 2009.

Khare S, Wall T, Gupta R, Elliott L, and Buhre B (2005), *Retrofitting of Air-fired pf Plant to Oxy-fuel: Heat Transfer Impacts for the Furnace and Convective Pass, and Associated Oxygen Production Requirements*, 5th Asia–Pacific Conference on Combustion, Adelaide, 17–20 July 2005.

Kiga T, Takano S, Kimura N, Omata K, Okawa M, Mori T, *et al.* (1997), 'Characteristics of pulverised-coal combustion in the system of oxygen/recycled flue gas combustion', *Energy Conversion Management* 1997; 38: S129–S134.

Lackner K and Sachs J (2005), 'A robust strategy for sustainable energy', *Brookings Papers on Economic Activity 2005*, (2): 215–284.

Marin O, Chatel-Pelage F, *et al.* (2003), '*Economic Analysis of Oxygen-fired Coal Boilers*', Proceeding of the 28th International Technical Conference on Coal Utilization and Fuel Systems, 10–13 March 2003, Clearwater, Florida.

MIT (Massachusetts Institute of Technology) (2007), *The Future of Coal*, Cambridge, Massachusetts.

Payne R, Chen S, Wolsky A, and Richter W (1989), 'CO_2 recovery via coal combustion in mixtures of oxygen and recycled flue gas', *Combustion Science Technology* 1989; 67:1–16.

Ratafia-Brown J, Manfredo L, Hoffmann J, and Ramezan M (2002), *Major Environmental Aspects of Gasification-Based Power Generation Technologies*, National Energy Technology Laboratory (NETL) Report, December 2002.

Shaddix C and Murphy J (2003), *Coal Char Combustion Reactivity in Oxyfuel Applications*, Twentieth Pittsburgh Coal Conference, 2003.

Singh D, Croiset E, Douglas P, and Douglas M (2003), 'Techno-economic study of CO_2 capture from an existing coal-fired power plant: MEA scrubbing vs. O_2/CO_2 recycle combustion', *Energy Conversion Management* 44: 3073–3091.

Taber J (1985), *Need, Potential and Status of CO_2 for Enhanced Oil Recovery*, Argonne National Laboratory, February 1985.

Taber J, Martin F, and Seright R (1997), 'EOR screening criteria revisited – Part 1: Introduction to screening criteria and enhanced recovery field projects', *SPE Reservoir Engineering*, Vol. 12, No.3, August 1997, pp. 189–198.

Tan Y, Croiset E, Douglas M, and Thambimuthu K (2006), 'Combustion characteristics of coal in a mixture of oxygen and recycled flue gas', *Fuel* 85: 507–512.

Wall T, Gupta R, Buhre B, and Khare S (2005), *Oxy-fuel (O_2/CO_2, O_2/RFG) Technology for Sequestration-ready CO_2 and Emission Compliance*, Proceedings of the 30th International Technical Conference on Coal Utilization and Fuel Systems, Clearwater, Florida, 17–21 April 2005.

Wang C, Berry G, Chang K, and Wolsky A (1988), 'Combustion of pulverized coal using waste carbon dioxide and oxygen', *Combust. Flame*, 72: 301–310.

WCI (World Coal Institute) (2009), *The Coal Resources – A Comprehensive Overview of Coal*, London, UK, 03-06-2009.

White V and Fogash K (2009), 'Purification of Oxyfuel-Derived CO_2: Current Developments and Future Plans', 1st IEA Oxy-fuel Combustion Conference, Cottbus, Germany, 7–10 September 2009.

WRI (World Resources Institute) (2008), *CCS Guidelines – Guidelines for Carbon Dioxide Capture, Transport, and Storage*, Washington, DC, 2008.

Zheng L, Clements B, and Runstedtler A (2002), *A Generic Simulation Method for the Lower and Upper Furnace of Coal-fired Utility Boilers using Both Air Firing and Oxy-fuel Combustion with CO_2 Recirculation*, 27th International Technical Conference on Coal Utilization and Fuel Systems, Clearwater, Florida, 2002.

Part I

Introduction to oxy-fuel combustion

2
Economic comparison of oxy-coal carbon dioxide (CO_2) capture and storage (CCS) with pre- and post-combustion CCS

D. THIMSEN, J. WHEELDON and D. DILLON,
Electric Power Research Institute (EPRI), USA

Abstract: Oxy-coal steam electric plants are a third option for capturing CO_2, an alternative to air-fired combustion with post-combustion CO_2 capture, and gasification/combined cycle power plants with pre-combustion CO_2 capture. The decision on which technology to deploy will largely be made on a comparative cost basis. Comparable costs estimates for the three technology options indicate that oxy-coal with CO_2 capture may out-perform the other two options on capital cost, levelized cost of electricity, and cost of CO_2 emissions avoided. The analysis must be tempered by the fact that costs for oxy-coal steam electric power plants embody significant uncertainties associated with scaling the air separation and CO_2 purification technologies and with design-basis specification for CO_2 purity delivered to CO_2 transport/storage.

Key words: oxy-coal comparative economics, oxy-coal capital cost, oxy-coal cost of electricity, oxy-CO_2 purity, CO_2 avoided cost.

2.1 Introduction

Three options for capturing CO_2 from coal-based power generation plants are under investigation: oxy-coal technology, air-fired combustion with post-combustion CO_2 capture, and gasification/combined cycle power plants with pre-combustion CO_2 capture. The decision to deploy one of these technologies in preference to the others will largely be made on a comparative cost basis, usually the levelized cost of electricity (COE).

There are enormous difficulties in comparing the relevant costs developed by different groups for various purposes. These might have to do with different fuels, different plant design bases, changing unit construction commodity prices, geographical labor rate differences, etc. Hence, to achieve reliable comparison of the technologies they must have the same design basis. Few such studies have been completed to date and not all economic variables of interest have been covered. Of those that have been completed, the most comprehensive are reviewed here.

In addition to the normal challenges of developing meaningful costs for full-scale, coal-fired power plants, inclusion of oxy-coal technology with CO_2 capture is complicated by two additional challenges:

- Oxy-power plants incorporate technologies that are mature at industrial scale but have not been deployed at a scale suitable for the electric power industry.

These are cryogenic air separation units to produce large quantities of purified oxygen, and the purification of the raw CO_2-rich flue gas for geological storage.
- The lack of a unique design basis specification for the product CO_2 purity.

The first challenge introduces technology development risk that is amenable to good management in the development process. The second challenge is more difficult to manage. Product CO_2 purification costs, capital and operating, can be expected to rise along with increasing purity. On the other hand, delivering less pure (and less costly) CO_2 to the power plant boundary will incrementally increase transport and storage costs. Optimizing this cost analysis will be challenging. In addition, the purity of CO_2 that can be stored underground has not yet been specified by regulators. The only reasonable course of action at this point is to clearly indicate the product CO_2 purity associated with the reported costs.

It should also be noted that the costs reported are for oxy-coal flow sheets at an early stage of technology development. As oxy-coal power plant designs advance there is every reason to believe that new flow sheets will emerge with more optimized plant configurations. An example of a near-term improvement is shifting bulk removal of SO_x and NO_x from commonly deployed desulfurization and NO_x reduction processes to the CO_2 purification unit. If these efforts are successful, the costs of the upstream air quality control processes will be greatly reduced, with potential reductions in overall capital cost. An example of a longer-term prospect is development of combustion/heat transfer systems that allow reduced flue gas recycle rates (increasing firing mixture oxygen content) leading to cost reductions in the steam generator island.

2.2 Oxy-coal power plant systems scope

An oxy-coal power plant with CO_2 capture and storage will consist of five major systems:

- Air separation unit (ASU).
- Oxy-coal steam generator island (including air quality control systems).
- Steam turbine cycle island.
- CO_2 purification unit (CPU) (including compression to pipeline pressure).
- Balance of plant (service, electrical, materials handling/storage).

The scope and technical readiness of these systems is reviewed here to frame the cost assessments presented below.

2.2.1 Air separation unit

The air separation unit (ASU) provides purified oxygen to the steam generator island. For the foreseeable future, the only technology sufficiently mature to serve a commercial oxy-coal power plant is cryogenic distillation of oxygen from air. A cryogenic ASU will include the following system components:

- Raw air filtration.
- Air compressors/intercoolers (possibly with heat recovery to the steam turbine cycle).
- Water vapor and trace CO_2 removal.
- Pre-refrigeration.
- The cryogenic cold box including heat exchangers and the oxygen–nitrogen distillation column, and an expansion turbine.

While cryogenic ASUs are mature technology with many years of development supporting modern designs, ASUs deployed to serve an oxy-coal power plant will differ incrementally from those producing industrial oxygen (99.5% purity) in the following ways:

- A full-scale, oxy-coal power plant (800 MW, gross) will require about 20,000 tonnes per day of purified oxygen. This production requirement is four times larger than the largest cryogenic ASU built to date. It is likely that multiple trains will be required for a commercial oxy-coal power plant.
- It is likely that 95%–97% purity oxygen will be acceptable for the oxy-coal application rather than the 99.5% purity commonly produced for industrial gas applications. Purity of 97% can be achieved in a cryogenic ASU without the costly oxygen–argon separation. (Separating argon can be accomplished less expensively in the downstream CO_2 processing unit in an oxy-coal power plant.)
- The oxygen can be delivered to the oxy-coal combustor at relatively low pressure, generally less than 2 bar (14.3 psig). Industrial oxygen is generally delivered at higher pressure. (Note that pressurized oxy-combustion will require delivery of oxygen at a higher pressure.)
- The ASU plant will likely need sufficient operating flexibility to respond to load-following dispatch at least as fast as can the steam generator island. This requirement is not commonly imposed on industrial oxygen plants.

These differences offer ASU system designers several opportunities to increase production efficiency and reduce costs for oxy-coal applications. They also represent scaling and dynamic response design challenges.

It is generally the practice of industrial gas companies supplying ASUs to include contingencies in the cost estimates commensurate with the perceived technical risk. Technical risks will include the scale-up from existing experience (larger diameter distillation columns and the multiplicity of cold box heat exchangers) and guarantees of maximum auxiliary power use to meet the specified delivery, purity, and dynamic performance. These contingencies will generally apply for the first few installations and can be expected to fall as experience is gained. At this point in time, the uncertainty in cost estimates for the ASU is probably greater than the uncertainties in cost estimates for the remaining plant with the exception of the CO_2 purification unit (CPU).

2.2.2 Oxy-coal steam generator island

The first generation of oxy-coal steam generators is likely to be based on the 'synthetic air' approach to combustion/heat transfer. In this approach, relatively cool furnace flue gas (rich in CO_2) is recycled and mixed with the purified oxygen to form a mixture that has combustion/heat transfer properties comparable to air. This allows steam generator vendors to employ well-developed air–coal steam generator designs for oxy-coal service.

Both pulverized coal (PC) and fluidized-bed steam generators have been proposed and are suitable for oxy-coal service. The oxy-coal steam generator system will include the following major components:

- Oxygen–flue gas mixing.
- Fuel metering.
- Furnace/steam generator heat transfer surface.
- Recycle/oxygen heater (comparable to an air heater in conventional air-coal steam generators).
- Particulate control device (bag house or electrostatic precipitator) and ash handling.
- Bulk removal of SO_2 by flue gas desulfurization (FGD) for PC units and in-situ removal by limestone in fluidized-bed units. (Although these stages might be eliminated for low-sulfur coals as discussed below, they will be required for high-sulfur coals to avoid recycling flue gas with high SO_2 content to the steam generator, which would increase corrosion potential on the superheat and reheat tubing.)
- Bulk reduction of NO_x by Selective Non-Catalytic Reduction (SNCR) in fluidized bed units or by Selective Catalytic Reduction (SCR) in a PC unit may be avoided provided the unit is operated on air only during start up.
- Recycle fan, induced draft fan, air-fired stack.

If extended operation with air-firing (other than start up) is specified, flue gas desulfurization and NO_x control will be required, with increased capital and operating cost. If this is not the case, the design basis specification for SO_x and NO_x leaving the steam generator island may be relaxed depending on:

- Technology employed in the CO_2 purification unit where the SO_x would be converted to sulfuric acid and the NO_x to nitric acid (plus absorption of chlorides to hydrochloric acid) with subsequent removal of these water soluble acids in condensate or water wash.
- The amount of SO_2, NO, and NO_2 allowed in the product CO_2 stream sent to geological storage.

The operational approach adopted for the oxy-coal plant will have a significant impact on the environmental controls adopted with corresponding effect on capital and operating costs. This uncertainty results in an increased range covered by cost estimates.

The components in the steam generator island are generally familiar to power plant engineers and designers. The vendors have all conducted burner/pilot plant operations for oxy-coal which should be sufficient to develop oxy-coal steam generator and air quality control system designs. The uncertainty in cost estimates for these components should be no greater than the uncertainty for comparable estimates of components in conventional air/coal-fired steam electric plants.

2.2.3 Steam turbine cycle island

The steam turbine cycle is essentially uncoupled from the details of fuel combustion. There are unlikely to be any significant constraints on delivering steam from an oxy-coal steam generator other than those that constrain delivery of steam from a conventional air-coal steam generator. Thus, oxy-coal steam generators can be designed to deliver supercritical (SC), ultra-supercritical (USC), and advanced ultra-supercritical (AUSC) steam to the steam turbine cycle as the required materials and processes become available for air-coal steam generators. The oxy-coal steam turbine cycle will reap the same efficiency benefits that an air-coal steam cycle does with higher steam temperatures and pressures.

The major components of the steam turbine cycle include:

- High-, intermediate-, and low-pressure (HP, IP, and LP) turbines and controls.
- Condenser.
- Cooling water circulation and cooling towers.
- Condensate and feedwater pumps.
- Feedwater heaters, deaerator.

The costs and performance of steam turbine cycle components deployed in an oxy-coal power plant will have uncertainty comparable to similar components deployed in air-coal power plants.

Unlike post-combustion capture technology, an oxy-coal power plant will employ a standard steam turbine cycle; there is no need to modify the steam turbine cycle by extracting steam to provide the heat required to regenerate CO_2 capture solvent. The net output impact of an oxy-coal power plant results entirely from an increase in auxiliary power required to operate the ASU and CPU.

The steam turbine cycle island components for an oxy-coal plant are identical to those for an air/coal-fired plant and the uncertainties in cost estimates should also be identical.

2.2.4 Carbon dioxide (CO_2) purification unit

The greatest uncertainty in overall cost and performance of an oxy-coal power plant is in design basis specification for the CPU. There is no common specification for CO_2 purity delivered by the oxy-coal power plant to either a pipeline or CO_2 storage site.

The raw flue gas will contain between 75% and 90% CO_2 (dry basis), with nitrogen, oxygen, and argon as the major contaminants and a host of minor contaminants. If the storage geology can accept this quality of CO_2, and the intervening pipeline is not excessively long, the lowest cost option may be to cool and dehydrate the raw flue gas and compress/inject it without further purification. This represents the minimum requirement for the CPU. CO_2 capture for this option is 100% and there are essentially no emissions to atmosphere of any combustion products.

If CO_2 quality comparable to pipeline specifications currently in place for use in enhanced oil recovery is specified, a partial condensation purification process with rectification of the CO_2 is likely to be required, with associated impacts on capital cost and auxiliary power. This represents the maximum requirement for the CPU.

A likely technology option producing CO_2 of intermediate purity is to employ a partial condensation process with a single- or double-stage flash. CO_2 purity in excess of 95% should be achievable with this technology option.

The CO_2 impurities separated from the raw flue gas (primarily oxygen, argon, nitrogen, carbon monoxide, nitrogen oxide) by partial condensation are emitted to atmosphere in a vent gas stream. They will be accompanied by some uncondensed CO_2. The overall CO_2 capture by partial condensation will be near 90%. This may be increased to 98%+ by one of several proposed vent gas recovery CO_2 technologies which are under development.

The major components of a CPU deployed in an oxy-coal power plant will include the following.

Warm side (these components will be included in all oxy-coal power plants):

- Indirect contact flue gas cooler (possible heat recovery to the steam turbine cycle).
- Direct contact flue gas cooler (water wash).
- Compression with inter-cooling (possible heat recovery to the steam turbine cycle).
- Deep drying.

Cold side (these components will be included for production of CO_2 with higher purity than that in the raw flue gas):

- Activated carbon bed to remove mercury and protect the aluminum used in the cold circuit.
- Cold box with heat exchange and partial condensation for CO_2 separation employing either a single-flash (lower CO_2 purity), a dual-flash, or a CO_2 rectifying column (highest CO_2 purity).
- Cold-box vent gas CO_2 recovery (if deployed to achieve 98%+ CO_2 capture).

Supporting components (these components will be included in all oxy-coal power plants):

- Injection pressure compression/cooling (possible heat recovery to the steam turbine cycle).

In addition to the uncertainties in the design basis specifications leading to different CPU flow sheets, the following will also contribute to cost and performance uncertainty (independent of the specification):

- The largest industrial CO_2 purification unit employing partial condensation is approximately 2000 tonnes per day. Full scale (800 MW, gross) plant designs developed to date incorporate multiple CO_2 purification trains of near 4000 tonnes per day capacity.
- Industrial CO_2 purification units generally produce liquid CO_2 and employ an indirect refrigeration cycle using ammonia as the refrigerant. There is general agreement that CPUs for full-scale, oxy-coal applications will employ an 'auto-refrigeration' cycle using the liquid (purified) CO_2 as the refrigerant. Such an 'auto-refrigeration' cycle has yet to be deployed for oxy-coal flue gas purification.

These scaling/performance uncertainties translate into cost uncertainties. As with the ASU, it is generally the practice of industrial gas companies who will supply the cost estimate for the CPU to include contingencies in the cost estimates that are commensurate with the technical risk they perceive. These contingencies will generally apply for the first few installations and can be expected to fall as experience is gained. At this point in time, the uncertainty in cost estimates for a given CPU design basis specification is probably the greatest of the five systems identified here. This uncertainty (and the associated contingencies) could be reduced by:

- Improved vapor equilibrium thermodynamic data for CO_2, O_2, N_2, argon, and the minor contaminants.
- Pilot plant evaluation of the proposed partial condensation/auto-refrigeration purification processes.
- Pilot plant evaluation of the degree to which bulk reduction of SO_x and NO_x can be achieved in the CPU, obviating the need for corresponding air quality control systems to be included in the steam generator island.

2.2.5 Balance of plant

The balance of plant is not affected by the decision to deploy oxy-coal combustion (rather than air-coal combustion) and will include:

- Electric generator.
- Transformers/switchgear.
- Auxiliary power switchgear.
- Water supply/treatment.

- Coal yard/handling.
- Ash handling/disposal.
- Waste water treatment/disposal.

The cost of these system components can be estimated with uncertainty comparable to similar costs for an air/coal-fired steam-electric power plant.

2.2.6 Cost/performance uncertainty summary

The major sources of uncertainty (associated with oxy-coal technology) in estimating full oxy-combustion power plant performance and cost include:

- Scaling the ASU to multiple 5000 tonnes per day trains.
- Maximizing the auxiliary power benefit of relaxing the oxygen purity specification and for delivery at near atmospheric pressure.
- The design basis for specifying air quality control system (FGD and SCR/SNCR) performance.
- The design basis for specifying the final product CO_2 purity.
- Scaling the CPU to multiple 4000 tonnes per day CO_2 trains.
- Performance of the large-scale partial condensation, auto-refrigeration, CO_2 purification process.

2.3 Oxy-coal carbon dioxide (CO_2) capture and storage (CCS) cost estimates and comparisons with post- and pre-combustion CO_2 capture

As the design and cost bases for power plants can vary widely, it is very difficult to compare costs produced from different studies. The most useful cost studies are those that compare technology options with a common baseline. Two such in-depth studies reporting the costs associated with various technology options for low-CO_2 emitting fossil fuel power plants are those published by USDOE/NETL[1] and the Global Carbon Capture and Storage Institute (GCCSI).[2] Costs for both of these studies were developed by Worley Parsons. The study results are reviewed below. Careful attention should be paid to the design basis specifications for the various results as they do impact cost and performance.

2.3.1 United States Department of Energy/National Energy Technology Laboratory (USDOE/NETL) study

An engineering and economic analysis of new-build plants with CO_2 capture and burning US Illinois basin bituminous coal was conducted to estimate the impact of deploying both post-combustion CO_2 capture and oxy combustion with CO_2 capture for disposal in saline geological formations. The design fuel was 2.5%

sulfur, 11.1% moisture and higher heating value of 27,100 kJ/kg (11,700 Btu/lb). A total of ten pertinent cases were evaluated as indicated in Table 2.1. In all cases wall-fired PC steam generators were specified. Two steam cycles were evaluated; a USC cycle with main steam temperature of 1110°F (699°C) and an AUSC cycle with 1350°F (732°C) main steam temperature.

In all cases NO_x control to the same level in the raw flue gas was specified and consisted of low-NO_x burners and overfire air/oxygen. The air-fired cases included SCR. For the oxy-coal cases, it was assumed that the paucity of atmospheric nitrogen entering the furnace, detailed attention to burner design, and high flue gas recycle (to achieving NO_x reduction by reburning) would be sufficient to meet the specified NO_x emissions level without SCR. In all cases wet limestone FGD was specified to achieve the permitted SO_2 emissions leaving the steam generator island. Co-capture in the various air quality control system (AQCS) components was assumed adequate to achieve 90% mercury removal.

In all cases, the plant was sized for a net capacity of 550 MW. In the capture cases, the product CO_2 stream was compressed to pipeline pressure of 154 bar (2215 psig). Oxygen purity of 95% was used for all of the oxy-coal cases save a single case at 99%. (Two reported cases where nascent Ion Transfer Membrane (ITM) technology was used to produce ~100% purity O_2 are not included). Four product CO_2 purity specifications were used, as indicated in Table 2.1. Where CO_2 purification is specified, a dual flash partial condensation, auto-refrigeration process was included. No CO_2 recovery or other treatment of the vent gas was

Table 2.1 USDOE/NETL configuration summary

Case	PC steam generator design	Oxidant	CO_2 purity scenario	CO_2 purity (% mol)
Ultra-supercritical steam conditions 3500 psig/1110°F/1150°F*				
1	Air-fired	Air	No capture	–
3	Air-fired with amine PCC	Air	Regenerated from amine	99+
5	Oxy-coal	95% O_2	Dried flue gas	84
5A	Oxy-coal	99% O_2	Dried flue gas	88
5B	Oxy-coal	95% O_2	Same CO_2 content as 5A‡	88
5C	Oxy-coal	95% O_2	Partial condensation purifier	96
Advanced ultra-supercritical steam conditions 4000 psig/1350°F/1400°F†				
2	Air-fired	Air	No capture	–
4	Air-fired with amine PCC	Air	Regenerated from amine	99+
6	Oxy-coal	95% O_2	Dried flue gas	84
6A	Oxy-coal	95% O_2	Partial condensation purifier	96

* 242 bar/599°C/621°C.
† 277 bar/732°C/760°C.
‡ A slip stream of the flue gas is purified to achieve an overall CO_2 content identical to the unpurified flue gas of case 5A.

included. The post-combustion capture technology specified for cases 3 and 4 was Fluor's Econamine FG Plus.[3]

The estimated performance of the baseline and several design options is listed in Table 2.2 and displayed in Fig. 2.1. The most notable feature of these data is the large auxiliary power load imposed by the ASU, and, to a lesser extent, CO_2 compression. It is notable that with the AUSC the net cycle efficiency expressed on a High Heating Value (HHV) basis is nearly the same for the oxy-coal cases (6 and 6A) and the corresponding post-combustion capture case (4). For the lower efficiency USC cycle the net plant efficiency is one percentage point higher for the four oxy-coal cases (5 and 5A–C) than for the post-combustion case (3).

The stated accuracy in the associated cost study is ±30%. The pertinent aggregate costs are listed in Table 2.3 and displayed in Fig. 2.2. Of note here is that the oxy-coal cases (5, 5A–C, 6) are consistently lower in each cost metric by 5%–7% compared with the post-combustion capture cases (3, 4) for the same steam cycles. The study also looked at two perturbations of the base design, and the effects on the costs metrics are as follows:

- A fuel with sulfur content less than 1% would eliminate the need for wet FGD in the oxy-coal cases (5, 5A–C, 6) but would require additional apparatus to separate sulfur in the CO_2 purification unit. The net effect was a reduction in capital costs and a net reduction in material acquisition/disposal costs (limestone/gypsum) that resulted in a reduction in cost of electricity of approximately 8%.

Table 2.2 USDOE/NETL study performance for 550 MW net PC plant designs

Case	Gross power (MW)	PCC or ASU (MW)	CO_2 compression* (MW)	Balance of auxiliaries (MW)	Net efficiency (HHV) (%)	CO_2 emitted (kg/MWh)	CO_2 capture (%)
Ultra-supercritical steam conditions							
1	580.2	–	–	30.2	39.4	800	–
3	661.1	20.9	44.3	45.9	28.3	111	90
5	785.9	125.7	72.1	38.1	29.3	0	99
5A	787.0	126.1	68.1	37.7	29.5	0	99
5B	785.0	125.5	72.4	38.1	29.3	27	97
5C	785.9	125.7	73.4	38.1	29.2	76	86
Advanced ultra-supercritical steam conditions							
2	576.6	–	–	26.6	44.6	707	–
4	644.4	17.8	37.7	38.9	33.2	95	90
6	759.2	111.4	63.9	32.9	33.0	0	99
6A	753.6	111.4	62.9	34.0	33.0	60	93

* Oxy-coal CO_2 compression power includes pressure loss in the CO_2 purification unit associated with flashing liquid CO_2 to lower pressure to supply the necessary refrigeration.

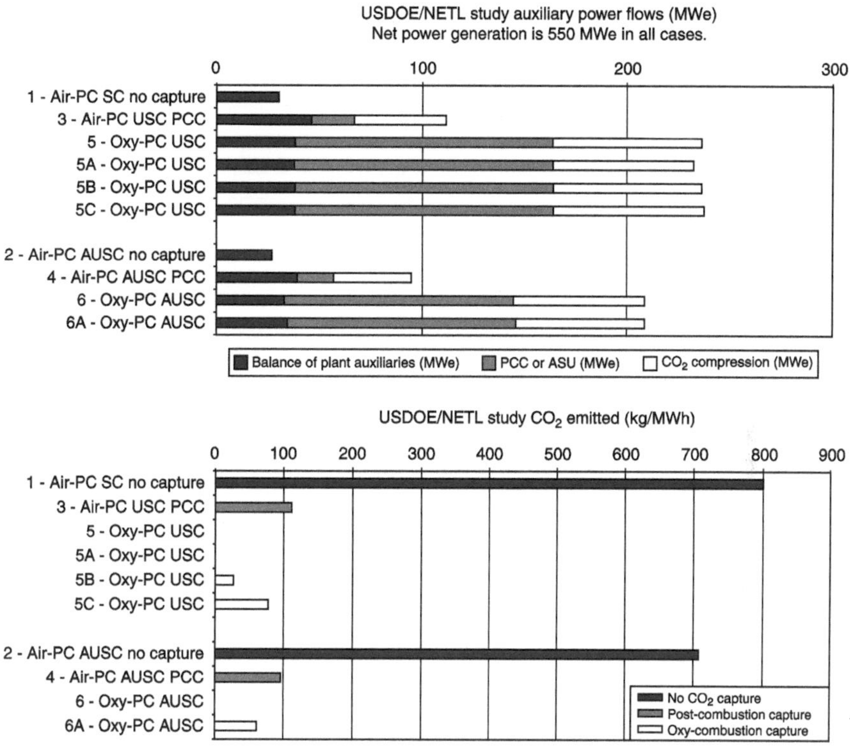

2.1 USDOE/NETL study performance results.

Table 2.3 USDOE/NETL study costs (January 2007 US$)

Case	Total plant cost (US$/kW)	20-year levelized COE (US$/MWh)	Cost of avoided CO_2 emissions (US$/tonne)
Ultra-supercritical steam conditions			
1	1579	63	Baseline
3	2855	109	67
5	2660	101	47
5A	2632	100	46
5B	2687	102	50
5C	2715	103	55
Advanced ultra-supercritical steam conditions			
2	1643	64	12
4	2810	102	56
6	2602	96	41
6A	2683	99	48

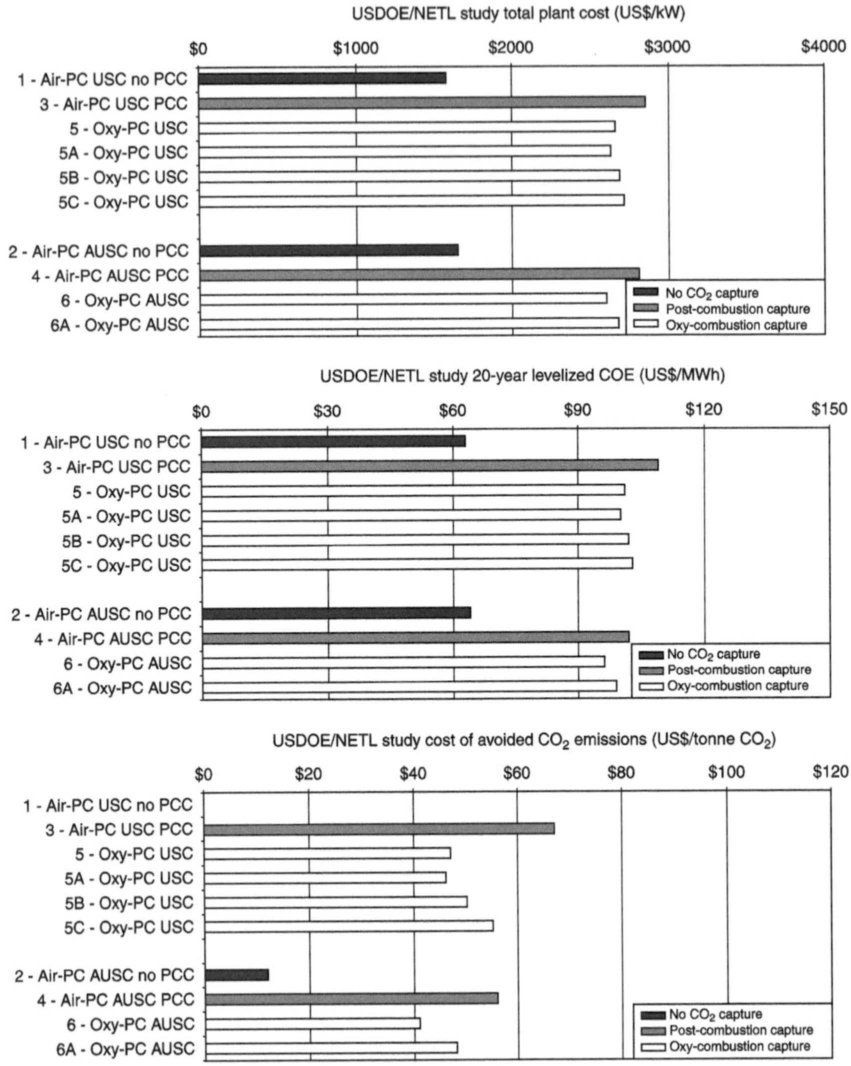

2.2 USDOE/NETL study cost results.

- The study included a 15% contingency for the steam generator in the oxy-coal cases (5, 5A–C, 6) to account for the fact that the first full-scale oxy-coal steam generator has yet to be built. Excluding this contingency reduced the cost of electricity by approximately 2.5%.

2.3.2 Global Carbon Capture and Storage Institute study

The 2009 GCSSI report expanded on the USDOE/NETL study described above. Additional cases of interest included in this study were:

- Integrated gasification/combined cycle with and without pre-combustion CO_2 capture (using bituminous coal).
- Natural gas with and without pre-combustion CO_2 capture.

For this study, costs were inflated to 2009 US$. The overall parameters of the pertinent cases are listed in Table 2.4 and displayed in Fig. 2.3. The performance is shown in Table 2.5 and the capital cost, levelized COE, and cost of avoided CO_2 emissions are listed in Table 2.6 and displayed in Fig. 2.4.

Consistent with the USDOE -NETL study, the oxy-PC cases are less expensive than the pre- and post-combustion capture cases and result in the lowest overall cost of avoided CO_2 emissions. This comparison, however, points out one of the pitfalls of comparing CO_2 capture options. The avoided cost for oxy-PC CO_2 emissions is for a plant delivering scrubbed, dehydrated flue gas to the plant boundary at approximately 84% purity. The comparable costs for the pre- and post-combustion options are for delivery of 99%+ pure CO_2 to the plant boundary. The parties transporting and storing the 84% purity CO_2 will incur incrementally higher costs to handle the impurities, if, indeed, the geology and regulators allow.

2.4 Conclusions

- At the current stage of oxy-coal and post-combustion CO_2 capture technology development, the projected costs have greater uncertainty than costs developed for air/coal-fired steam electric plants without CO_2 capture. As oxy-coal technology evolves the associated costs are expected to fall, which further

Table 2.4 GCCSI study cases

Case	Technology	CO_2 capture scenario
Ultra-supercritical steam conditions 3500 psig/1110°F/1150°F*		
USDOE/NETL Case 1	Air-fired	No capture
USDOE/NETL Case 3	Air-fired	Post-combustion capture
USDOE/NETL Case 5	Oxy-coal	100% Dry flue gas
Advanced ultra-supercritical steam conditions 4000 psig/1350°F/1400°F†		
USDOE/NETL Case 2	Air-fired	No capture
USDOE/NETL Case 4	Air-fired	Post-combustion capture
USDOE/NETL Case 6	Oxy-coal	100% Dry flue gas
IGCC no capture	IGCC	No capture
IGCC capture	IGCC	Pre-combustion capture
NGCC no capture	NGCC	No capture
NGCC capture	NGCC	Post-combustion capture

* 242 bar/599°C/621°C.
† 277 bar/732°C/760°C.
IGCC: Integrated (coal) Gasification Combined Cycle.
NGCC: Natural Gas Combined Cycle.

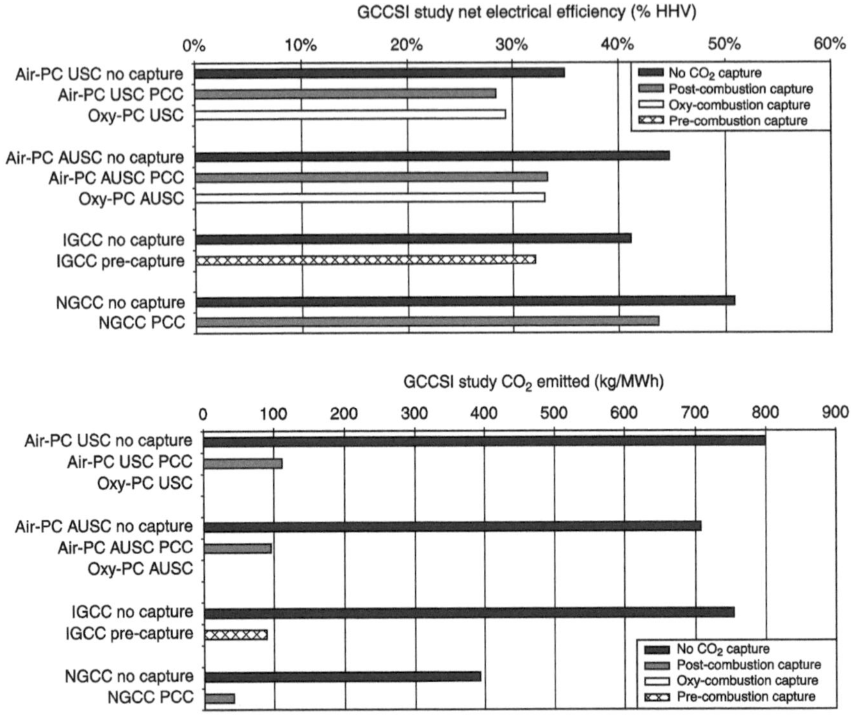

2.3 GCCSI study performance results.

Table 2.5 GCCSI study performance

Case	Gross power (MW)	PCC or ASU (MW)	Net power (MW)	Net efficiency (HHV) (%)	CO_2 emitted (kg/MWh)	CO_2 capture (%)
Ultra-supercritical steam conditions						
USDOE/NETL Case 1	580	30	550	39.4	800	0
USDOE/NETL Case 3	661	111	550	28.3	112	90
USDOE/NETL Case 5	786	236	550	29.3	0	100
Advanced ultra-supercritical steam conditions						
USDOE/NETL Case 2	577	27	550	44.6	707	0
USDOE/NETL Case 4	644	94	550	33.2	95	90
USDOE/NETL Case 6	759	209	550	33.0	0	100
IGCC no capture	748	112	636	41.1	753	0
IGCC capture	694	176	517	32.0	90	90
NGCC no capture	570	10	560	50.8	392	0
NGCC capture	520	38	482	43.7	42	90

Table 2.6 GCCSI study costs (2009 US$)

Case	Total plant cost (US$/kW)	30-year levelized COE Nth of a kind plant (US$/MWh)*	Cost of avoided CO_2 emissions Nth of a kind plant (US$/tonne CO_2)
Ultra-supercritical steam conditions			
USDOE/NETL Case 1	1910	79	baseline for coal cases
USDOE/NETL Case 3	3416	138	88
USDOE/NETL Case 5	3151	126	60
Advanced ultra-supercritical steam conditions			
USDOE/NETL Case 2	2001	76	
USDOE/NETL Case 4	3382	126	84
USDOE/NETL Case 6	3150	120	64
IGCC no capture	2123	96	
IGCC capture	3385	134	78
NGCC no capture	706	78	NGCC baseline
NGCC capture	1435	112	109

* Coal at US$2.91/GJ (US$3.07/million Btu), natural gas at US$6.45/GJ (US$6.80/million Btu).

complicates comparison of the technologies at this stage in time. The uncertainties are due to two broad factors:

(a) Oxy-coal power plants will include technologies that have not been widely deployed in the electric power industry, and neither have they been widely deployed at the required scale. This includes production of purified oxygen by cryogenic air separation and purification of product CO_2 by partial condensation using an auto-refrigeration process. These risks and uncertainties can be reduced by focused management of the technology development process.

(b) While air/coal-fired post-combustion processes produce relatively pure CO_2 (99%+) without extra-ordinary measures, oxy/coal-fired plants will produce raw flue gas that has significant quantities of impurities. The CO_2 can be purified to specified levels at additional cost. The difficulty is specifying the purity required. Three factors are likely to determine the purity requirement, none of which is well defined at the moment:
   - The ability of the pipeline operator and storage geology to accept the various impurities. This is largely a technical/geological question that is amenable to further research and development.
   - Impurity levels that will be tolerated by agencies regulating the transport and geological storage of CO_2. While there will be a technology/safety aspect to this, there are also likely to be unpredictable non-technical influences on regulators.

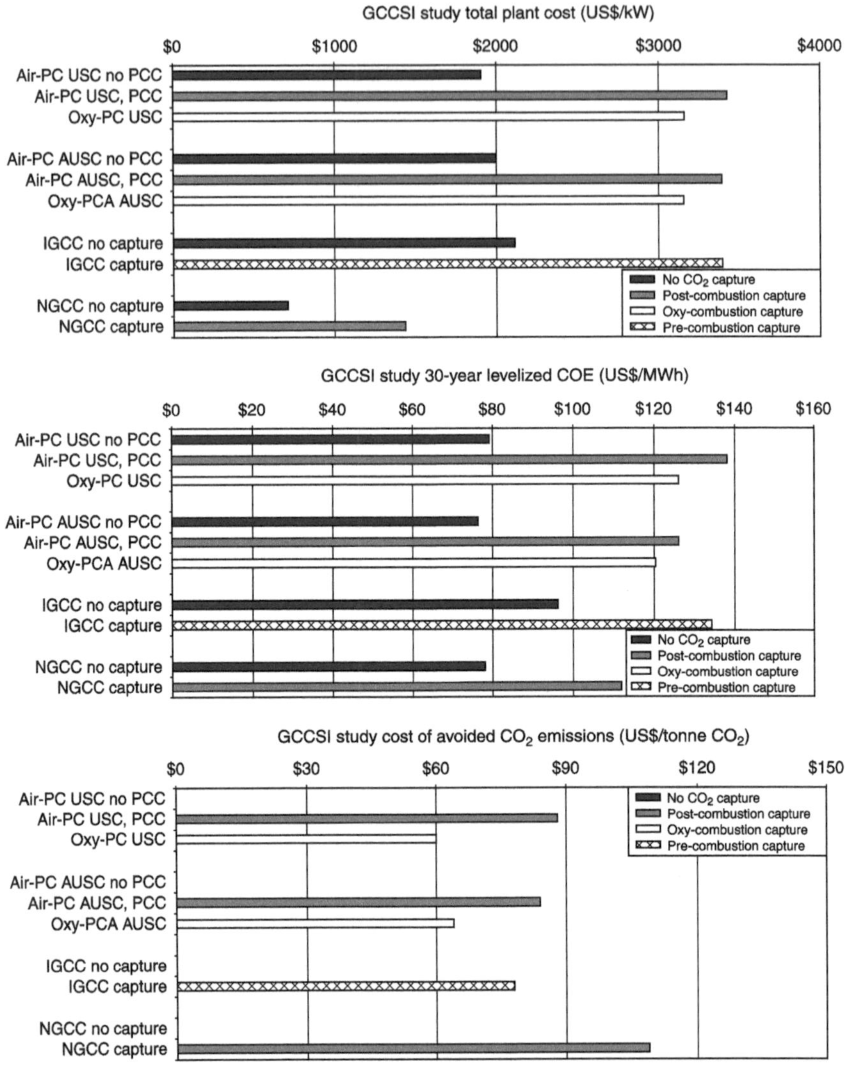

2.4 GCCSI study cost results.

- Economic optimization. The power plant can reduce its costs by limiting the extent of CO_2 purification. However, the pipeline and storage operator will incur higher costs to handle the increased volume associated with the impurities and to accommodate any impact the impurities have on transportation and storage.

- Few detailed studies comparing costs and performance of the three CO_2 capture technology options have been conducted such that the costs and performance can be compared on a relatively consistent basis. The most notable studies are those conducted by Worley Parsons and published by USDOE/NETL comparing oxy-coal technology with post-combustion capture and updated with costs for pre-combustion capture/coal gasification combined cycle costs in a study published by the GCSSI.

- Capital cost estimates (at ±30% accuracy) for the oxy-coal cases (with substantially lower CO_2 emissions) were marginally lower than the capital costs for the post-combustion capture cases. Due to the uncertainties identified above, the cost differences published are indicative, not conclusive.
- Levelized COE for the oxy-coal cases was approximately 7% lower than levelized COE for the air-fired Econamine FG Plus post-combustion capture cases employing the same steam cycles. (This difference is within the uncertainty of the respective cost estimates.) As noted above, the CO_2 purity for the oxy-coal case was very much lower than that of the post-combustion capture case and this may have contributed much of the cost advantage. On the other hand, CO_2 capture for the oxy-coal cases was near 100%, compared with 90% for the post-combustion capture case.
- Levelized COE for the oxy-coal case delivering impure CO_2 to the pipeline was approximately 5% lower than levelized COE for the gasification/combined cycle with pre-combustion capture case delivering relatively pure CO_2 to the pipeline but with a lower overall capture of CO_2. (This difference is within the uncertainty of the respective cost estimates.)
- Capital and levelized COE for the oxy-coal cases with 100% flue gas injection (no CO_2 purification) were uniformly lower than comparable cases employing partial condensation to purify the product CO_2. This suggests that conditions (and costs) under which relatively impure CO_2 can be transported and stored in geological formations should be the subject of further research and development.

While these costs and performance estimates cannot yet be considered conclusive, they do indicate that the oxy-coal technology option is viable and is likely to be competitive with pre- and post-combustion CO_2 capture for new plants. There are at least three development paths that would improve the competitive position of oxy-coal technology:

- Development of bulk SO_x and NO_x removal technologies as part of the CO_2 purification process. If this capability could entirely replace conventional flue gas desulfurization and NO_x control equipment, a commensurate reduction in capital cost is likely.
- Development of low flue gas recycle steam generator designs. Flue gas recycle is employed in current oxy-coal plant designs to produce an oxy–flue gas mixture that simulates air-fired combustion and allows use of well-developed steam generator designs. Reducing recycle to increase the proportion of oxygen (total oxygen flow is not changed) could result in smaller steam generators but would require a significant revision of steam generator architecture. Nonetheless, smaller steam generators are likely to reduce capital

costs. Fluidized-bed steam generators are likely to be most amenable to reduced flue gas recycle operation.

- Reduction of the capital and operating costs of air separation. Two such technologies with longer-term prospects for success are:
  (a) Ion transfer membrane (ITM) oxygen separation. Specific high temperature ceramics allow oxygen ions to migrate through the ceramic material while nitrogen and argon are precluded. This high-temperature, high-pressure technology has been developed at process development unit scale and plans are underway for pilot plant scale development.
  (b) Chemical looping. In this high-temperature, low-pressure process, oxygen is chemically separated from other air constituents by a solid 'carrier' material which is removed from the air stream and mixed with coal in a combustor where the coal–oxygen reactions take place. This process has been developed for a number of carrier materials at bench scale. Process development unit scale investigations are in the planning stage. Dramatic reductions in air separation auxiliary power would likely result from use of this technology.

Deployment of either of these air separation technologies would also involve significant impacts on the steam generator island of an oxy-coal power plant, but probably not result in significant impacts on the steam turbine cycle island, CO_2 purification unit, or balance of plant.

2.5 References

1 *Pulverized Coal Oxycombustion Power plants, Volume 1: Bituminous Coal to Electricity*. Final Report, Rev. 2. DOE/NETL-2007/1291, August 2008.
2 *Strategic Analysis of the Global Status of Carbon Capture and Storage, Report 2: Economic Assessment of Carbon Capture and Storage Technologies*. Global CCS Institute, Canberra, Australia, 2009.
3 *Improvement in Power Generation with Post-Combustion Capture of CO_2*. IEA Greenhouse Gas R&D Programme Report PH4/33, November 2004.

3
Oxy-fuel power plant operation

Y. TAN, CanmetENERGY, Natural Resources Canada, Canada

Abstract: This chapter will provide a brief overview of some of the key issues facing an oxy-fuel power plant, e.g. integration of air separation units and flue gas compression trains, oxygen and recycled flue gas control as well as flue gas recycle strategies. Other important issues that will be discussed include effects of air ingress and flue gas egress, corrosion concerns and power plant maintenance. This chapter will also discuss plant control issues, such as transition between air-blown mode and oxy-fuel mode operations, load changes and plant start-up and shutdown. These issues have significant impact on the performance, safety, reliability and economics of any oxy-fuel power plant.

Key words: oxy-fuel plant corrosion concerns and prevention, oxy-fuel plant start-up, ramping and shutdown, oxy-fuel plant oxygen safety and control.

3.1 Introduction

Oxy-fuel combustion is recognized as an effective means for coal-fired power plants to continue operation in a carbon-constrained world. When air is replaced with oxygen as the combustion gas, an oxy-fuel power plant emits a flue gas stream that is highly enriched in CO_2. Due to its high concentrations (80–90%, dry basis), CO_2 can be recovered relatively easily and economically with the flue gas compression train. Obviously, when oxygen is used as the combustion gas, fuel combustion characteristics change so dramatically that current boiler technologies are not yet ready to cope on a large scale. Therefore, considerable modifications to the operation of a typical power plant will be needed. This chapter addresses the major challenges surrounding the oxy-fuel coal-fired power plants, including safety, plant control, corrosion and maintenance.

Coal-fired power plants can be considered as consisting of several major component blocks: coal preparation; combustion; steam generation; pollution control; and electricity generation. Coal is transported to the furnace with air, combusted in the furnace and the combustion products move in a largely one-directional flow to the stack. In oxy-fuel-fired power plants, additional components are required. These notably include air separation units (ASUs) for oxygen production and flue gas compression trains for CO_2 purification and transport. Coal is now transported by recycled flue gas and the combustion products no longer move in a one-directional flow as a significant proportion is recycled back to the furnace. Apart from these obvious changes, we should also consider that the power plant's control system must also be modified to accommodate these equipment changes and their interoperability. We also need to consider additional health and safety issues related to oxy-fuel firing.

Other chapters address specific issues on the ASU and flue gas compression train. In this chapter, we will look at other issues that have not been addressed, starting with the flue gas recycle system.

3.2 Flue gas recycle system

3.2.1 Reasons for flue gas recycle

At the current state of the art of oxy-fuel power plants, a large portion of flue gas must be recycled to the boiler in order to replicate the combustion temperature and heat transfer characteristics of an air-fired boiler system. More specifically, the recycled flue gas serves several major purposes. One is to temper combustion temperature increase due to the use of oxygen and to ensure proper heat transfer, both in the furnace and in convective passes. Other purposes include conveying coal to the boiler and various auxiliary duties, such as purging, cooling, etc., that are usually accomplished by air in conventional power plants. In short, one of the major differences in oxy-fuel power plants compared with air-blown power plants is that all combustion gases and many auxiliary gases must be switched from air to recycled flue gas, which can sometimes be mixed with high-purity oxygen.

An important related issue concerns the amount of recycled flue gas to extract. The term 'recycle ratio' is often used to denote the percentage of flue gas being recycled over the total amount of flue gas produced by combustion. A typical oxy-fuel plant of current design requires a recycle ratio of 70–80%, depending on the types of coal and flue gas recycle options. As mentioned above, one important role of the recycled flue gas is to control combustion temperature and to ensure proper heat transfer, and the recycle ratio is usually determined by this requirement. As the combustion environment changes from nitrogen-dominant, as in the air-blown case, to CO_2-dominant, as in the oxy-fuel case, the gas thermal properties change considerably. This is reflected in the required higher oxygen concentration in the feed gas for oxy-fuel operation, which averages about 30% v/v, in order to reproduce combustion and heat transfer characteristics similar to air-blown conditions. In a pulverized coal fired (PF) furnace, the bulk of heat transfer is accomplished by radiation, which is mostly affected by the presence of particulate matter in the flue gas as well as that of CO_2 and H_2O. In the oxy-fuel case, the gas emissivity will change due to the considerably higher concentrations of both CO_2 and H_2O. However, since radiation heat transfer in a coal-fired furnace is mostly dominated by particulate matter, it will not be impacted noticeably by the increased concentrations of CO_2 and H_2O. Chapter 9 describes in detail the heat transfer phenomena in oxy-fuel conditions; it is simply emphasized here that, although heat transfer properties change in the oxy-fuel case, the magnitude of these changes can be controlled by recycling the correct amount of flue gas.

One of the most important criteria in furnace temperature control is the furnace exit gas temperature. This temperature must be kept below the ash softening temperature in order to avoid slagging and fouling of high-temperature heat exchange surfaces such as superheaters and reheaters. Other areas that are subject to high-temperature corrosion are waterwalls near the firing zone and low-temperature gas passes and air heaters. These temperatures depend on the properties of specific fuels, and, in oxy-fuel cases, they can be met by recycling an adequate amount of CO_2-rich flue gas. Herein lies an important difference between air-blown and oxy-fuel combustion: in the air-blown mode, the amount of oxidant (air) supplied is dictated by the fuel feed rate and the furnace is built to accommodate such a fuel feed rate. Oxy-fuel combustion provides an additional and effective means to allow the operators to adjust combustion and heat transfer characteristics to a certain extent by varying the amount of flue gas recycled. This being said, it must be acknowledged here that in a retrofit scenario, it would be challenging to find a flue gas recycle ratio that can satisfy all of the above requirements without making modifications to certain heat exchange equipment.

3.2.2 Flue gas recycle options and challenges

Depending on a number of factors, flue gas can be drawn at different locations for recycle. Two main approaches are suggested:

- Wet flue gas recycle: in this case, the flue gas is extracted before it is cooled below its water dew point. In this case, recycled flue gas is usually extracted downstream of particulate removal devices but upstream of any wet scrubbers or condensers;
- Dry flue gas recycle: here the flue gas is extracted downstream of wet scrubbers or flue gas coolers at a temperature that allows most moisture in the flue gas to condense.

There is a third option:

- Extract the flue gas upstream of particulate removal devices such as electrostatic precipitators (ESPs) and bag filters. However, this option will require fans that can resist hot and abrasive flue gas and is likely not practical.

One of the factors in choosing recycle options is to maximize plant efficiency. The other important factor is corrosion concerns. While the wet recycle option may offer slightly higher efficiency compared with the dry recycle option,[1] especially due to the possibility of scaling down the wet scrubbers, it also imposes some constraints. The first limitation is coal conveying – it is not practical to convey coal with a moisture-laden flue gas stream, especially when it has to be maintained above its acid dew point. For example, the sulfuric acid dew point of a flue gas stream with 10% moisture, 80% CO_2, 3% O_2, 0.2% SO_2 and 0.01% SO_3 (assuming 5% of total SO_x is in the form of SO_3) is about 155°C at ambient pressure. While low-sulfur and low-moisture coals may not produce SO_3 and moisture in sufficient amounts to make sulfuric acid dew point a main concern, as discussed in Chapter 8, it is not unusual for a high-sulfur coal to produce SO_2 concentrations well above 1%. In that case, in a wet recycle scenario, the necessity to maintain the flue gas temperature above its sulfuric acid dew point (~ 170°C) will become a serious hurdle to overcome and may well cancel out any other advantages that wet recycle has over the dry recycle option.

While sulfuric acid dew point of the flue gas is a concern for high-sulfur coal, a high-moisture coal can present other challenges with its higher water dew point. Condensation can develop at many places and some of them can lead to equipment malfunction, e.g., moisture in the flue gas can condense on the surfaces of particulate removal equipment and seriously affect their operations. In this case, it is important to recognize these condensation-prone locations and implement measures to either avoid condensation in the first place or constantly drain these locations if, for some reason, condensation cannot be avoided. This is even more important for a coal with both high sulfur and high moisture contents.

For many oxy-fuel power plants, an attractive flue gas recycle option is a combination of wet and dry recycle options. The dry and cleaned recycled flue gas can be used for coal conveying and auxiliary uses that require relatively clean and dry gas and the wet, 'dirty' recycled flue gas can be injected into the furnace for combustion temperature control. A minor disadvantage of such an approach is the need for separate recycle flue gas blowers and piping system, which can introduce more potential for leaks to develop. This is the approach practiced by the Swedish power company Vattenfall in its oxy-fuel demonstration plant in Schwarze Pumpe in Germany as reported by Stromberg *et al.*[2]

As has been discussed previously, with a high-sulfur coal in a wet recycle mode, care must be taken to ensure that the recycled flue gas stream temperature is above the sulfuric acid dew point to minimize piping system and equipment corrosion. This is especially relevant to the minimum cold-end temperature that has to be maintained to avoid corrosion and plugging of the air heater and particulate removal devices. Since this temperature is closely associated with sulfur concentrations in the flue gas (it rapidly increases with increasing sulfur concentrations), it also plays an important role in the choice of recycle options. If the fuel contains exceedingly high sulfur (e.g., 6%) that produces flue gas with correspondingly high SO_2 and SO_3 concentrations, it would be prudent to adopt an exclusively dry recycle option.

3.3 Oxygen (O_2) handling

In an oxy-fuel power plant, issues related to safe oxygen handling are of obvious concern. Being a strong oxidant, oxygen is highly inflammable in the presence of combustible materials. As a result, strict guidelines exist in regulating oxygen handling and various industries have gained vast knowledge in the safe service of oxygen over past decades. We advise readers to refer to their oxygen providers for detailed guidance.

For pure oxygen transport, oxygen-compatible materials must be used. These include copper and its alloys, Inconel and Monel. Stainless steel can be safely used in an oxy-fuel power plant as long as the oxygen flow velocity does not exceed a certain threshold value (below 15 m/s according to Linde Industrial Gases, because of the adiabatic compression effect). In addition, all pipes, gaskets and fittings must be properly cleaned for oxygen services and common precautionary measures taken (e.g., use of check valves wherever needed). For high-velocity operation, other materials mentioned above should be used. Monel is a good choice if the operation requires high-pressure oxygen service.

Pure oxygen in the stainless steel piping system will eventually mix with recycled flue gas, which will contain fine particulate and small amounts of unburned fuel and hydrocarbons. Since oxygen concentration will be lower after it is mixed with the recycled flue gas, the material for the piping system after this mixing point can likely be changed from stainless steel to carbon steel. This is advantageous as stainless steel is several times more expensive than carbon steel. However, carbon steel is prone to trapping particulate and other impurities in the flue gas, which can become a problem over time. Carbon steel also tends to become brittle over time and, finally, the usage of carbon steel also means that any presence of water will lead to corrosion, which may impact on recycle strategy. As a result, the advantages of replacing stainless steel with carbon steel after the mixing point should be carefully weighed against its disadvantages.

Starting at the mixing point, the presence of 'dirty' recycled flue gas and the possible replacement of stainless steel by carbon steel will exert an upper limit on oxygen concentration in the mixed gas stream for safe operation. This additional limit depends to a large degree on oxygen concentration, gas velocity and the nature of the 'dirty' components. If the oxygen concentration in the mixed stream is below 25%, carbon steel may be preferable due to lower cost; stainless steel should be used for higher oxygen concentrations. In order to conform to conventional operation, it may be desirable to maintain oxygen concentration in the mixed stream at about 21% as in the air-blown operation; however, pilot-scale tests have safely used up to 25% as in Farzan *et al.*[3] or 28% as in Tan *et al.*[4] in this mixed stream.

Whether it is 21% or 28% of oxygen in the recycled flue gas, in most cases not enough oxygen is provided through this stream for optimal combustion performance, especially considering that the coal conveying recycled flue gas stream may not contain any additional oxygen, for safety reasons. It has been shown repeatedly in the literature that higher oxygen concentrations, in the range of 28–35% at the burner inlet, are preferred in an oxy-fuel power plant to maintain stable flames as well as furnace temperature and heat transfer characteristics comparable to air-blown conditions. This implies that there must be other means to deliver the balance of the oxygen needed for combustion to the burner, usually in the form of pure oxygen. This may seem to be an additional challenge, but, as discussed in Chapter 8, this pure stream of oxygen actually offers an opportunity to optimize the combustion process so that NO_x emissions can be greatly reduced. An alternative is to design the boiler and the burner system in such a way that lower oxygen concentrations in the feed gas can be effectively employed.

In summary, even though the addition of oxygen adds some complexities to an oxy-fuel power plant, the accumulated experience of oxygen providers and other industry practitioners has shown that, as long as due diligence is observed, there are no particular difficulties in this area.

3.4 Leakages

The purpose of an oxy-fuel power plant is to enrich the flue gas stream with as high a CO_2 concentration as possible so that downstream flue gas cleaning and compression can be facilitated. This poses serious challenges on minimizing air ingress on the one hand and flue gas egress on the other hand. Air ingress will lead to lower flue gas CO_2 concentrations, while flue gas egress can present serious health hazards as CO_2 and other pollutants (SO_x, NO_x, CO and particulate) can escape along with the flue gas, and finally egress of oxygen, which is a strong oxidant, presents obvious safety hazards. This is exacerbated by the fact that many of these gases are heavier than air and thus can accumulate at certain poorly ventilated spots rather than dissipating in air, creating unsafe working conditions. For example, according to the Occupational Safety and Health Administration (OSHA), SO_2 concentration cannot exceed 5 ppm over an eight-hour time-weighted average and the permissible exposure limits are set at 25 ppm for NO, 50 ppm for CO and 5000 ppm for CO_2. Additionally, CO_2 can also displace oxygen in air. A work area is considered oxygen deficient when oxygen level falls below 19.5%. As a result, both ingress and egress are undesirable and must be minimized by sealing the system as much as possible when it is being built and, later on, during operation. It should be noted, however, that gas streams containing high concentrations of sour gas and CO_2 have been successfully managed in many natural gas processing plants. Even though an oxy-fuel power plant will generate a much higher volume of CO_2- and sulfur-containing flue gas, operational experience learned from natural gas processing plants should still be applicable.

In order to minimize air ingress, the easiest and most obvious way is to operate the power plant under a slightly positive draft. On the other hand, a slightly negative operating draft, as is more commonly practiced, will minimize the egress of pollutant-laden flue gas. From a safety and health point of view, it is probably preferable to operate the plant under slightly negative draft. This is especially important for oxy-fuel plants because the recycled flue gas will be enriched in carbon dioxide and, in many cases, high concentrations of sulfur and nitrogen oxides. Any significant egress can ultimately force a plant to shut down and should be avoided.

However, a power plant typically consists of a large number of pieces of equipment and each of these can become a source of leakage. As a result, ensuring and maintaining proper sealing is an important undertaking, which could be facilitated by adopting the wet recycle strategy. In order to maintain CO_2 concentration and safe operation of the power plant, it is important to minimize leakage inside the recycle loop, so the less equipment there is inside the recycle loop and the simpler the duct system, the less likely it is for leaks to arise. Compared with the dry recycle option, which requires the flue gas to pass through condensers or wet scrubbers, the wet recycle option is simpler and leads to less potential for air ingress.

Figure 3.1 illustrates the impact of air ingress on the flue gas quality in terms of CO_2 concentration for a bituminous coal (similar results can be obtained for other coals), assuming that the flue gas contains 3% O_2 v/v on dry basis. Figure 3.1 shows that flue gas CO_2 concentration decreases almost linearly as air ingress increases. In order to maintain a flue gas with >90% CO_2, air ingress should be limited below 4%. To maintain flue gas CO_2 concentration above 80%, air ingress should not exceed 13%.

An additional effect of air ingress is the potentially increased demand on the ASU if the target feed gas oxygen concentration is to be maintained, due to the dilution effect from nitrogen. For example, 4% air ingress would put a 3.3% increased load on the ASU to compensate for the presence of extra nitrogen, while 13% air ingress would impose a 10.8% load increase on the ASU. This obviously will lead to higher oxygen concentration in the flue gas stream, which, if not recovered, would be wasted. On the other hand, if the target flue gas oxygen composition is to be maintained, then oxygen supply from ASU must be reduced to compensate for the presence of oxygen from air ingress. In either case, the flue gas CO_2 quality will be reduced.

In a small-scale oxy-fuel plant, pure CO_2 from the flue gas compression train can be used to help seal specific locations where complete sealing cannot be

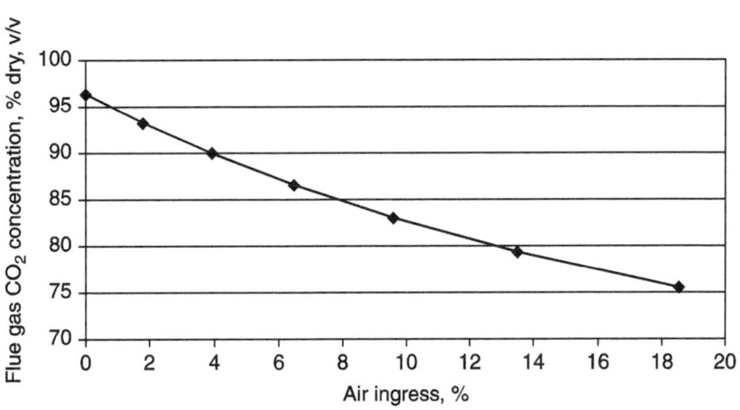

3.1 Impact of air ingress on flue gas CO_2 concentration.

achieved by simply displacing ambient air. This measure, although far from ideal, can help achieve high flue gas CO_2 concentration (e.g., above 90% v/v on dry basis). However, this is far from an ideal solution and is not feasible on large-scale units where leak sources can be more numerous.

Preventing leakage is also important for the vent stream from the flue gas compression train since this stream is highly enriched in CO_2 and CO and probably under positive pressure. As a result, any leaks of this vent stream may cause serious health and safety concerns.

To summarize, preventing air ingress and flue gas egress are important endeavors for an oxy-fuel power plant. To achieve this goal, the first step that should be taken is to ensure excellent sealing throughout the entire plant when it is being built and ensure effective sealing afterward during operation. It is also vitally important that alarm systems interlocking with gas monitors for O_2, CO_2, CO, SO_2 and NO_x be installed throughout the plant area where leaks can occur or accumulate.

3.5 Slagging and ash formation

Ash deposits that form primarily on convection surfaces such as superheater and reheater tubes and cooler regions of furnace walls have been a problem that power plants must solve in order to maintain power plant efficiency and availability. A number of technologies have been developed that greatly reduce problems caused by ash deposits. For example, computer codes to predict the particle-size and composition distribution (PSCD) of the ash produced upon combustion and simplified transport, deposition, and growth programs for specific locations in the boiler have been developed. The validity of these models, which were developed for air-blown conditions, remains to be seen for oxy-fuel conditions. It is pointed out in Chapter 8 that early studies show that ash formation under oxy-fuel conditions can be different from air-blown conditions and this may require modifications to current models being used.

Slag forms mainly on furnace walls and other surfaces exposed to radiant heat or high gas temperatures from fused deposits and resolidified molten material. On cooler surfaces, slag will form a tight bond that can make it very difficult to remove; this bonded slag will then keep growing until fluid temperature is reached and it eventually runs off. In oxy-fuel combustion, higher oxygen concentration may lead to locally high flame temperature and the potential for increased slag formation. It is thus important to ensure proper distribution of oxygen and its mixing prior to entry to the furnace as well as in the furnace.

At this time, there are not a large number of studies dedicated to ash and slag formations under oxy-fuel conditions. As a result, operators will have to pay close attention to these problems by monitoring ash and slag formation and making appropriate operational adjustments as problems occur.

3.6 Flue gas cleaning equipment

Conventional flue gas cleaning equipment, especially electrostatic precipitators (ESPs) or fabric filter and, in many cases, flue gas desulfurization (FGD), will still be needed for oxy-fuel power plants and will operate in an environment that is very different from air-fired plants. These differences are characterized by higher concentrations of CO_2, SO_x, NO_x and H_2O as well as higher flue gas density and, depending on specific plant configurations, lower flue gas volume.

For ESPs and fabric filters, higher moisture content and SO_x concentrations in the flue gas mean a tighter operating temperature range to avoid condensation. For coal-fired power plants, a typical ESP downstream of the air preheater operates at about 160°C. While this temperature may be adequate for oxy-fuel plants with good quality fuels, in some cases (e.g., high-sulfur, high-moisture coals combined with the wet recycle option) the operating temperature needs to be raised. The above discussion also applies to fabric filters, and, in such cases, the filter material needs to be selected according to the higher operating temperature and the harsher operating environment.

For ESP operation, the higher flue gas SO_3 concentration in the oxy-fuel case can be advantageous. As SO_3 is absorbed onto the fly ash, the ash resistivity decreases, thus improving ESP performance. Indeed, in some air-fired plants burning low-sulfur coal, injection of a small amount of SO_3 (20–30 ppmv) is sometimes practiced in order to achieve adequate ash removal rate by ESP.[5] However, it should be noted that the overriding factor to consider here is still the corrosive nature of SO_3; this is true for both the ESP and the fabric filter and regular preventive maintenance should be undertaken to prevent corrosion.

For sulfur removal, wet FGD is one of the most widely used technologies. Using limestone, it is based on the following simplified reaction scheme:

$$SO_2(g) + CaCO_3(s) = CaSO_3(s) + CO_2(g) \quad [3.1]$$

$$CaSO_3(s) + H_2O(l) + \frac{1}{2}\, O_2(g) = CaSO_4(g) + H_2O \quad [3.2]$$

With oxy-fuel combustion, the flue gas CO_2 partial pressure significantly increases compared with air-fired combustion. This increased CO_2 concentration can, in theory, have a detrimental effect on reaction [3.1] above by shifting the equilibrium toward the left-hand side of the reaction and thus negatively impact the performance of the limestone-based sulfur removal process. However, current pilot facilities operated limestone-based FGD with success. For example, operation experiences at Vattenfall's Schwarze Pumpe oxy-fuel pilot plant showed that there was no negative effect on limestone's sulfur-removal performance (>99.5% achieved) due to increased flue gas CO_2 concentration and that the amount of limestone used was the same as under air-fired operation. It remains to be seen if the performance of this process can be maintained on full-scale facilities. In the unlikely event that the performance of the limestone-based process is seriously affected, one alternative approach would be to adopt a lime-based process, based on the following reaction scheme:

$$Ca(OH)_2(s) + SO_2(g) = CaSO_3(s) + H_2O(l) \quad [3.3]$$

$$CaSO_3(s) + H_2O(l) + \tfrac{1}{2} O_2(g) = CaSO_4(g) + H_2O \quad [3.2]$$

This process uses essentially the same type of equipment and as such is familiar to power plant operators.

3.7 Maintenance of oxy-fuel power plants

Because of the increased complexity and harsher operating environment of an oxy-fuel power plant, its maintenance will be more involved. Additional equipment, such as an air separation unit (ASU), recycle flue gas blowers and fans, flue gas compression train and additional piping systems all require regular inspections and maintenance. Obviously, the maintenance schedule should be planned and arranged carefully to ensure as little plant downtime as possible.

One of the major and critically important additional equipments is the ASU. If the entire ASU train goes offline, oxygen supply will stop and an oxy-fuel plant will have to stop operation in oxy-fuel mode and switch to air-blown operation and incur productivity loss. However, modern ASUs have proven over decades that their on-stream reliability can exceed 99.5%, greater than that of typical coal-fired power plants, and major maintenance can be scheduled for intervals of four years.[6] As a result, a properly maintained ASU, especially in multi-train configurations as required for typical oxy-fuel power plants, is expected to have only minor impact on the reliability of an oxygen-fired power plant. Overall ASU availability can be improved by using a multi-train ASU configuration so that the loss of one single ASU train can be compensated to a certain degree.

Another major additional unit in an oxy-fuel power plant is the flue gas compression train. As of this writing, units capable of power plant scales are not available yet so their reliabilities are difficult to evaluate, especially considering that new and innovative designs are still being proposed. It must be said that CO_2 compression units of smaller sizes have been used in the chemical industry and natural gas processing for decades and have proven reliability. One of the main challenges will be to maintain this level of reliability as they are scaled up.

As has been discussed above, leak prevention is crucial in maintaining power plant efficiency, flue gas CO_2 quality and safe operation in an oxy-fuel power plant. Regular leak tests must be performed at various pressure points to minimize any potential leaks, which must be addressed promptly if found. However, some of the repairs may require equipment shutdowns. In this case, a cost/effectiveness analysis should be done to assess the appropriate approach. In the case of air ingress, it may turn out that accepting some minor degradation in flue gas quality can compensate for loss of productivity due to plant shutdown. In the event of minor flue gas egress, though, measures must be taken (such as additional ventilation and personnel protection equipment) to ensure continued safe operation conditions.

A maintenance issue that is present in conventional air-blown power plants but is more pronounced in oxy-fuel power plants is corrosion concerns caused by condensation, especially with high-sulfur, high-moisture fuels, such as some Spanish lignite, which can contain 7% sulfur and 30% moisture. With flue gas recycle and increased SO_x concentrations in the recycled flue gas, acid condensate can lead to rapid corrosion. To minimize this issue, it is very important to maintain the flue gas stream above sulfuric acid dew points whenever and wherever possible. In places where this is not feasible, it is important to ensure that no standing water or condensate exists, that corrosion resistant materials such as stainless steel are used and that regular inspections are being done. The corrosion problem is limited not only to the piping system but also to any equipment in direct contact with flue gases, such as the recycle flue gas blower and flow measuring devices. The potentially severe working conditions that this equipment is subjected to mean that it needs to be serviced on much more frequent schedules, replacement parts must be kept handy and back-up solutions devised. It should be mentioned here that adopting a dry recycle approach significantly reduces corrosion concerns.

Based on the current state of technology, it must be acknowledged that the reliability of an oxy-fuel plant will be lower than its air-fired counterpart. An approximate estimate shows that oxy-fuel firing would reduce the availability for an air-fired unit at 95% to about 91% for an oxy-fuel unit. This is a significant reduction in plant availability; however, as operation experience accumulates, it is expected that the availability of oxy-fuel units will gradually improve and approach those of air-fired units.

3.8 Plant control systems

The plant control system in an oxy-fuel power plant will have to address issues that are not present in conventional air-blown power plants. While most of the core control systems resemble an air-fired plant, the new challenges for an oxy-fuel plant are to integrate the operations of ASU and the CO_2 compression train into the plant control logic.

3.8.1 From start-up to stable operation

An oxy-fuel power plant will most likely start in air-blown mode following typical established start-up procedures for this mode of operation. As stable, low-fire air-blown operation mode has been achieved with coal feed on, the switch to oxy-fuel firing can begin.

The exact procedure for accomplishing the switch-over will depend on individual power plant design. However, the basic principle will involve starting the recycle flue gas blowers to extract flue gases from pre-determined locations and mix part of this flue gas stream with oxygen from the ASU (as secondary and tertiary gas) and route the other part to the coal pulverizer for conveying coal (as primary gas, it can be taken from a different location than for secondary gas). Appropriate control valves (such as linearized proportional valves) can be used to simultaneously reduce air supply and gradually increasing flow rates of the recycled flue gas/oxygen mixture, while maintaining oxygen supply to sustain coal combustion and ensure stable flames during the transition process. As the transition process proceeds and air is gradually replaced by the flue gas/oxygen mixture, pure oxygen supply to the burner system must also come into operation, if implemented, to maintain sufficient oxygen supply for combustion. Adequate primary stream flow rates should be maintained as air is replaced by recycled flue gas to ensure smooth coal delivery to the boilers. Typically, for a roll or race type pulverizer, a primary air-to-coal ratio of 1.8:1 is used. In an oxy-fuel plant, due to higher gas density, this ratio may need to be adjusted to compensate for the lower volumetric flow rate (~ 30% lower) at the same mass flow rate. This also implies that the primary-to-secondary gas ratio will change in oxy-fuel operation compared with air-blown operation with its consequences on flame ignition and flame stability that must be addressed through burner design.

One important aspect during this changeover period is to maintain oxygen levels in a tight range, both for optimal combustion condition and safety purposes. Interlocks and alarms should be put in place so that oxygen concentration in the secondary stream does not exceed the pre-determined threshold level (e.g., 25% on dry basis), which could lead to spontaneous combustion and even explosion. Oxygen level in the flue gas should be prevented from falling too low, as this could result in the production of excessive carbon monoxide and unburned carbons being recycled with oxygen. This oxygen control can only be achieved by carefully controlling and monitoring the flow rates of air, recycled flue gas and oxygen.

Another important aspect worth mentioning concerns the flame monitors. Depending on the oxygen concentrations at the wind box and primary and secondary gas flows, the intensity and shape (flicker frequency) of an oxy-fuel flame can differ from an air-blown flame. This may require some modifications to the flame scanner positioning and sensitivity, especially during the combustion mode transition period. As well, there may be a need to adjust burner swirl settings during the transition period to maintain stable flames and optimal combustion conditions until the fireball condition is established.

The above procedure assumes that the power plant starts on a start-up fuel (such as light oil or natural gas) in air-blown mode and then transfers to air-blown coal combustion and finally switches to oxygen-firing. Another possibility is to start on light oil or natural gas in air-blown mode, transition to oxygen-firing while still using the start-up fuel and then switch fuel to coal.

Both procedures can fulfill the purpose. In the second scenario, the operators will have to deal with oxygen-firing of both the start-up fuel and coal. This can become complicated, especially considering that the burners must be designed to operate in oxy-fuel mode for different fuels, whereas in the first scenario, most of the steps are familiar to existing power plant operators, making it a more appealing choice.

Before the air- to oxy-firing transition starts, the ASU should already be in operation at low load. Modern ASUs have a good degree of load change ability, typically from 60% onward, with multi-train configurations allowing greater scalability, which will allow the ASUs to moderate oxygen production to match fuel input and maximize plant efficiency. At the same time as air is being replaced by recycled flue gas-oxygen mixture, CO_2 concentrations in the exhaust stream will gradually increase. When the CO_2 concentration reaches an acceptable level for the flue gas compression train, this exhaust stream can be sent to the CO_2 compression unit to be processed.

Although the majority of the control system for an oxy-fuel power plant will be similar to that of an air-blown plant, a few notable additional changes must be addressed:

- Oxygen control: oxygen level is crucial in ensuring stable flames and combustion performance. Compared with air-fired boilers that need to control the oxygen level in the flue gas, there will be additional need to control overall oxygen mass flow and concentration at the wind box and local oxygen concentrations in the recycled flue gas/oxygen mixture. While oxygen concentration in the flue gas is controlled by total oxygen flow from the ASU, oxygen concentrations in the recycled flue gas/oxygen mixture and at the wind box must be controlled by flow rates of both the recycled flue gas and oxygen. A simpler control scheme can be achieved if no pure oxygen delivery mechanism is implemented.
- Furnace draft: compared with air-fired operation where the furnace draft is controlled by induced draft (ID) fans, furnace draft will be controlled by both the flue gas compressors and ID fans. In order to minimize air ingress in the recycle loop to maintain flue gas CO_2 concentration, it may be preferable to shift the boiler pressure control point further downstream, especially for a well-sealed new unit and preferably downstream of the point where recycled flue gas is extracted. If a balanced draft is used at this point, the recycle loop will be in slightly positive pressure and air ingress will be minimized. However, this option should be weighed against the increased risks of flue gas egress in practical operation. In addition, induced fans need to operate in accordance with flue gas compressors when the flue gas compression train comes online, as it is important to maintain stable pressure for the flue gas compression train to function optimally.
- Safety measures: additional safety interlocks must be implemented to account for the use of oxygen, so that if oxygen supply is significantly disrupted, appropriate control measures can be taken to avoid either oxygen concentration spikes or severe oxygen starvation in the furnace. Other safety measures should be implemented to account for possible oxygen and flue gas leakage.

After stable oxy-fuel operation is achieved with coal on low-fire, ramping up to full-load operation can start. This step involves ramping up flue gas recycle and oxygen supply according to fuel input increase and transition to full-load operation of the flue gas compression train.

During stable operation, an oxy-fuel power plant operates very similarly to a conventional air-fired one, which is one of the major advantages of the oxy-fuel approach. Still, attention has to be paid to maintaining the oxygen level at the wind box as well as pressure from recycle flue gas blowers and maintaining boiler pressure through the flue gas compressor. Ambient air quality should be continuously measured to ensure safe working conditions and finally, because furnace temperature is very sensitive to oxygen concentration, it is important to closely monitor the furnace temperature profile and furnace exit gas temperature when flow rates of fuel, oxygen and recycled flue gas are adjusted.

3.8.2 Load changes

A power plant's ability to adapt to load change requirement is very important. In a conventional air-blown plant, the load change requirement can be easily met with changes in fuel feed and combustion air flow rates. In an oxy-fuel plant, there is a need for several pieces of equipment to operate in unison to allow smooth load changes.

All load changes will require corresponding changes in fuel feed rates. In an oxy-fuel power plant, the changes in fuel feed rate will have to be matched with changes of flow rates for both oxygen and recycled flue gas, balanced with the ASU and flue gas compressor. As a result, load changes in this scenario are considerably more complicated, but still within the capability of modern power plant control systems. Again, attention must be paid to maintaining oxygen concentrations at the wind box as well as in the flue gas stream as the responses to any such change will be more dynamic than in an air-fired plant.

Load change can also affect flame stability. Compared with conventional air-fired operation where oxygen concentration is fixed, in oxy-fuel conditions flame stability is associated with oxygen flow and oxygen concentration. It is important for operators to pay close attention to maintaining stable flames during load changes, which may require adjusting oxygen flow distribution and burner swirl settings.

Obviously, changes in oxygen flow rates must be met with ASU operation. Modern ASUs allow a good degree of load changes and these load changes can happen relatively promptly at about 3%/min. With more sophisticated ASU multi-train design and control system configurations, a faster load change rate can be achieved, sufficient for normal power plant operations. It is worth noting here that Vattenfall successfully operated its oxy-fuel pilot plant with an ASU load change rate of 1%/min in the 75–100% window, although its ASU design has a load change capacity of 10%/min.

Load change requirement will also impact operations of the flue gas compression train, as flue gas flow rate will change. It is thus important for the design of the flue gas compression train to take into account the rapid power plant load changes and corresponding pressure fluctuations. A multi-train flue gas compression design will help to accommodate the usual power plant load changes. We will once again note that the CO_2 compression plant in Vattenfall's oxy-fuel pilot plant operated with a load change of about 2.5%/min in a capacity window from 60% to 100%.

The above discussion assumes that the oxy-fuel plant is built to accommodate peak load. However, cases can be made to build an oxy-fuel power plant such that the peak load can be met by shutting down the flue gas compression train and venting the flue gas to the stack without CO_2 capture. In this scenario, the size of the ASUs can be reduced, potentially resulting in significant cost savings. With this kind of plant design, the flue gas compression trains can be progressively and selectively shut down to meet the load demand increases beyond the base load. Since a power plant spends most of its operation at base load, this approach can be very attractive. However, depending on future regulations, the power plant owner must take into account the cost of paying for the extra CO_2 emissions during any peak load period.

3.8.3 Plant shutdown

Due to the use of oxygen and recycled flue gas instead of air, plant shutdowns in an oxy-fuel plant must be carefully planned to avoid excessive oxygen concentration fluctuations during the shutdown process. Control logic should be designed so that recycled flue gas/oxygen mixture is gradually replaced with air during the shutdown process and air must be used to purge the system. Obviously, operations of the ASU, flue gas compression train, flue gas recycle and fuel feed should all operate in sync to allow a smooth shutdown.

3.8.4 Component outages

As common practice, plans should be in place in anticipation of equipment malfunctions. One of the scenarios specific to oxy-fuel operation is loss of oxidant, which can be caused by ASU problems; however, as modern cryogenic ASUs are remarkably reliable, it is extremely rare to lose all oxygen supply. It is possible, though, that one of the ASU trains would fail and lead to a decrease in oxygen supply. As this happens, the oxygen supply control system should immediately increase oxygen supply from the other ASU trains if enough reserve capacity is built in. If the working trains do not have enough reserve capacity, then the operators should decrease flow rates of recycled flue gas to maintain appropriate oxygen concentration and decrease fuel feed rate so that it matches the available oxygen supply. It should be noted here that in oxy-fuel combustion with flue gas recycle, a small amount of oxygen is provided by the oxygen in the recycled flue gas. As such, the oxygen concentration control loop must respond swiftly to the dynamic feedback of oxygen flow fluctuations from recycled flue gas and adjust oxygen supply from the working ASU trains accordingly. As the total flue gas flow rates decrease, the flue gas compression train must also adapt its operation in a timely fashion. It is also obvious that operation of other power plant components, such as turbines and steam production, must respond appropriately to the decreased plant load.

As with oxygen supply, a total loss of fuel feed should also be a rare occurrence; however, it is quite possible that some of the coal pulverizers may abruptly go offline causing a sudden decrease in fuel supply. This fuel feed supply disruption will lead to a cascade of effects not dissimilar to those caused by oxygen supply disruption, except in this case the consequences may be more serious if not handled correctly. As fuel feed supply suddenly decreases, there will be a momentary rise in oxygen concentration in the recycle flue gas stream and, potentially, locally in the furnace due to excess oxygen supply in relation to fuel feed. As noted above, if oxygen concentration exceeds a certain limit, explosion hazard will significantly increase. In this scenario, the control system should be able to immediately decrease oxygen flow rates to match the fuel supply and, at the same time, respond promptly to the dynamic feedback from oxygen flow fluctuations in the recycled flue gas stream. Whether it is an oxygen supply or fuel supply disruption, the operators should always make sure that stable flames are maintained when adjustments are made.

One possible scenario that could lead to complete plant shutdown is when the flue gas recycle blowers go offline, because one of the most critical aspects influencing the reliability of an oxy-fuel plant is constantly maintaining a sufficient amount of recycled flue gas at appropriate pressure. The amount of recycled flue gas directly impacts the oxygen flow rate, which dictates the fuel feed rate and power plant output. If flue gas recycled blowers do go offline, oxygen feed must be stopped along with fuel feed unless back-up recycle blowers can be immediately brought online to avoid any oxygen spike in the boiler. If no back-up recycle blowers are available but the cause of the outage can be found and resolved quickly, then the boiler can operate temporarily in air-fired mode at lower load. This will allow a faster transition back to full oxy-fuel operation once the flue gas blowers are back online.

It is important to address the operation of the ASU during a shutdown as a start-up from ambient temperature may take up to 72 hours, sometimes requiring manual operation. Operation experience at Vattenfall's Schwarze Pumpe pilot plant showed that automatic start-up was possible for short stand still periods (< 3 days), but longer standstill (> 5 days) requires emptying liquid from the rectification column, leading to longer start-ups. However, since most ASUs are designed for continuous operation, it is a good practice to maintain ASUs in operation during short shutdown periods when minor maintenance is performed on the power plant.

3.8.5 Switching between air-blown mode and oxy-fuel mode

Smooth transition from air-blown mode to oxy-fuel mode and vice versa is important, not only during the start-up process but also during the shutdown process. Since the configurations of oxy-fuel power plants that are addressed here are fundamentally identical to those of conventional ones, all oxy-fuel power plants are in principle able to function as air-blown ones. This feature may become useful when any of the oxy-fuel-specific systems goes offline because of either oxygen delivery or flue gas recycle system outages as it allows the plant to continue to function. However, an oxy-fuel power plant can only function in air-blown mode in limited capacity due to several constraints.

In an oxy-fuel power plant, because of flue gas recycle, stack flue gas volume is considerably reduced; this reduction can be as high as 75%. This means that certain air pollution control (APC) devices and the stack can be scaled down accordingly to take advantage of this fact to reduce both the capital and operation costs of an oxy-fuel power plant and to improve its efficiency in the meantime. Unfortunately, the scaled-down versions of APC devices and stack also mean that an oxy-fuel power plant cannot operate in full-load in air-blown mode due to the restrictions of the flue gas volume that can pass through the smaller APC devices and the stack. This limitation can be overcome by building stacks with flexible volume (e.g., by combining the plant stack with ASUs for venting nitrogen) and allowing a portion of the flue gas to bypass the APC devices; however, due to emissions regulations, it seems hardly possible to sustain this mode of operation for more than several hours. The lower flue gas volume and difference in flue gas characteristics also mean that heat exchange properties will be considerably different. As a result, it is unrealistic to consider air-blown mode as a viable option for productive back-up operation for an oxy-fuel plant; rather, the air-blown mode can be used as a means to keep the power plant running at low load while problems are being solved, so that the plant can revert to full oxy-fuel operation as quickly as possible.

Smooth transition between air-blown and oxy-fuel operation can be accomplished through appropriate burner design and control system implementations, e.g., linearized proportional valves. Care should be taken to maintain oxygen pressure, oxygen concentrations (in the combustion gas stream as well as in the flue gas stream) and flow rates of both oxygen and recycled flue gas in the right range as combustion gas composition gradually changes. It is important to realize that, depending on their working principles, flow measurement devices must take into account the gas property changes during the transition process.

Since it is important for an oxy-fuel plant to retain the capability for extended air-fired operation, it must be equipped with burner systems that can fire reliably in both air- and oxygen-recycled flue gas mixtures. The burner design should allow its operating settings to be easily adjusted so that optimal flame conditions and combustion performance can be obtained over a wide range of oxygen concentrations. Chapter 7 provides details on burner design issues specifically related to oxy-fuel combustion.

3.9 Conclusion

Oxy-fuel power plants introduce a number of new equipment and operational concerns that need to be taken care of in order to ensure their smooth and safe operation. These include integration of two major new components, the ASUs and flue gas compression train, into the plant operation. Other important features of oxy-fuel power plant consist in the flue gas recycle system and the use of oxygen for combustion. While these additional components undoubtedly increase plant complexity, if properly designed, implemented, managed and maintained, they should have only minor adverse effects on the safety and reliability of oxy-fuel power plants compared with conventional air-blown plants. The areas of concern include leak prevention and detection for both air ingress and flue gas egress; corrosion prevention and detection due to high-moisture and high-sulfur flue gas conditions that can occur with certain fuel feed; oxygen control, especially under system upset conditions; and slag and ash deposit formation that may differ from air-blown combustion due to potentially locally high oxygen concentrations and temperature.

Oxy-fuel power plants also need an improved control system that can react quickly to rapid load changes and system upset so that safe operation can be maintained under any circumstances. These may include outages of air separation units, flue gas compression trains and flue gas recycle systems. In addition, the oxy-fuel plant operation and control system also require extra attention during plant start-up, transition from air- to oxygen-fired operation and shutdown as well as a more flexible burner system. While these demands may seem challenging, they can all be met with existing technologies and, in many cases, with existing experience.

At this time, Vattenfall's 30 MW_{th} Schwarze Pumpe demonstration plant is the only oxy-fuel-fired coal power plant in the world and valuable operating experience is being obtained and learnt from. Apart from Vattenfall, there are plans in both Japan and the United States to build coal-fired oxy-fuel plants in the range of 60 MW_{th}. As these plants start to operate, they will provide more opportunities for gaining operating experience, and it is expected that large-scale units can be built and operated safely and reliably.

3.10 References

1 Zheng, L., Clements, B., Tan, Y. and Pomalis, R., 'Flue Gas Recycle Strategies in Oxy-coal Combustion', *34th International Technical Conference on Clean Coal and Fuel Systems*, Clearwater, Florida, June 2009.

2 Stromberg, L., Lindgren, G., Jacoby, J., Giering, R., Anheden, M., Burchhard, U., Altmann, H., Kluger, F., and Stamatelopoulos, G.-N., 'Update on Vattenfall's 30 MWth Oxyfuel Pilot Plant in Schwarze Pumpe', *Energy Procedia* 1, 2009, 581–589.
3 Farzan, H., Vecci, S. J., Châtel-Pélage, F., Pranda, P., and Bose, A.C., 'Pilot-scale Evaluation of Coal Combustion in an Oxygen-enriched Recycled Flue Gas', *The 30th International Technical Conference on Coal Utilization and Fuel Systems*, Clearwater, Florida, 2005.
4 Tan, Y., Croiset, E., Douglas, M.A., and Thambimuthu, K., 'Combustion Characteristics of Coal in a Mixture of Oxygen and Recycled Flue Gas', *Fuel* 85, 2006, 507–512.
5 Bosch, F., 'Flue gas conditioning – SO_3 Injection Rates for South African Coal Ashes', *9th International Conference on Electrostatic Precipitator* Kruger Gate, Mpumalanga, South Africa, 17–21 May 2004.
6 Castle, W.F., 'Air Separation and Liquefaction: Recent Developments and Prospects for the Beginning of the New Millennium', *International Journal of Refrigeration* 25, 2002, 158–172.

4

Industrial scale oxy-fuel technology demonstration

T. WALL and R. STANGER, The University of Newcastle, Australia

Abstract: As one of the three major carbon capture and storage (CCS) technologies, oxy-fuel technology is currently undergoing rapid development with a number of demonstration projects commencing in the progression of the technology towards commercialisation. An overview of the current pilot plants and demonstration projects is provided (current as of September 2009), together with aspects of a roadmap for the deployment of oxy-fuel CCS technology, with the early commercial phase commencing in 2020 and mature commercial phase in 2030. Industrial scale oxy-fuel technology demonstrations have major significance in defining research needs and, combined with regulations and incentives, are integral in reducing CCS cost and driving efficiency improvements prior to commercialisation.

Key words: oxy-fuel, technology demonstration, carbon capture and storage, commercial deployment.

4.1 Introduction

Reduction of greenhouse gas emission from coal-fired power generation can be achieved by efficiency improvement, switching to lower carbon fuels and CO_2 capture and storage (CCS) (Wall, 2005; 2007). A report released by Massachusetts Institute of Technology (MIT) indicates CO_2 capture and storage is necessary for the future use of coal when carbon costs are established (Katzer, 2007).

There are several options for capture and storage of CO_2 from coal combustion and gasification, including:

- Post-combustion capture (PCC): CO_2 capture from conventional pulverised (pf) coal-firing plant with scrubbing of the flue gas by chemical solvents, solid minerals etc.;
- Pre-combustion capture: integrated gasification combined cycle (IGCC) with a shift reactor to convert steam and CO to make H_2 (a fuel) and CO_2 (that can be stored);
- Oxy-fuel combustion: combustion in oxygen rather than air, with recycled flue gas;
- Emerging options such as chemical looping combustion: oxygen carried by solid oxygen carriers reacts with fuel to produce a high concentration CO_2 stream in the flue gas; oxygen carriers are then regenerated to uptake oxygen from air in a second reactor. This technology is not as advanced in development or scale as the others.

Conventional pf coal-fired boilers, i.e., currently being used in the power industry, use air for combustion in which the nitrogen from the air (approximately 79% by volume) dilutes the CO_2 concentration in the flue gas. During oxy-fuel combustion, a combination of oxygen (typically of greater than 95% purity) and recycled flue gas is used for combustion of the fuel. A gas consisting mainly of CO_2 and water vapour is generated with a concentration of CO_2 that can be purified if required for sequestration. The recycled flue gas is used to control flame temperature and make up the volume of the missing N_2 to ensure proper heat transfer in the boiler.

Figure 4.1 gives the simplified flow sheet and details the unit operations associated with the technology to generate compressed CO_2, showing the additional operations required for a retrofitted standard pf plant, the ASU, recycling flue gas and CO_2 purification and compression. This may be termed the first-generation oxy-fuel plant. Figure 4.1 also indicates that ash removal and sulphur gas removal may be required depending on plant impacts and regulations determining the quality of CO_2 generated.

CO_2 capture and storage CCS by the current technically viable options of post-combustion capture, pre-combustion capture and oxy-fuel combustion will impose a 7%–10% efficiency penalty on the power generation process (Katzer, 2007; Xu *et al.*, 2007; McKinsey, 2008). The major contributors to this efficiency penalty for oxy-fuel and IGCC-CCS technology are oxygen production and CO_2 separation and compression, with regeneration of the solvent required for post-combustion capture particular for that technology. An advantage of oxy-fuel over IGCC-CCS and post-combustion capture is that turbines do not need be developed nor the steam turbine modified for extraction. Since the oxy-fuel process utilises known technologies and existing supply chain the rate of adoption is also expected to move rapidly (McCauley *et al.*, 2009). Furthermore, oxy-fuel technology may be applied to existing power plants as a retrofit. This is an important feature as existing power plant may have a substantial economic lifetime remaining.

The literature contains many reviews of the development of the oxy-fuel technology (Kiga, 2001; Allam *et al.*, 2005; Buhre *et al.*, 2005; Croiset *et al.*, 2005; Wall, 2005; Santos *et al.*, 2006). The state-of-the-art of oxy-fuel technology was reviewed in 2005 (Buhre *et al.*, 2005) and 2007 (Wall, 2007b) respectively. Research has primarily been presented at international conferences and published as journal papers (Wall, 2005; Gupta *et al.*, 2006; Lundström *et al.*, 2006; Rathnam *et al.*, 2006; Yamada *et al.*, 2006; Khare *et al.*, 2007; Rathnam *et al.*, 2007; Spero, 2007; Wall, 2007a, b; Wall *et al.*, 2009). A recent review (Wall *et al.*, 2009) has provided updated progress on oxy-fuel combustion technology development and research.

In terms of technical readiness, the majority of unknowns are based around issues with scale-up and the interaction between unit operations. The state of the art was detailed in the International Energy Agency (IEA) Oxy-fuel Conference 2009 (IEA, 2009) with the need to apply oxy-fuel technology with advanced supercritical

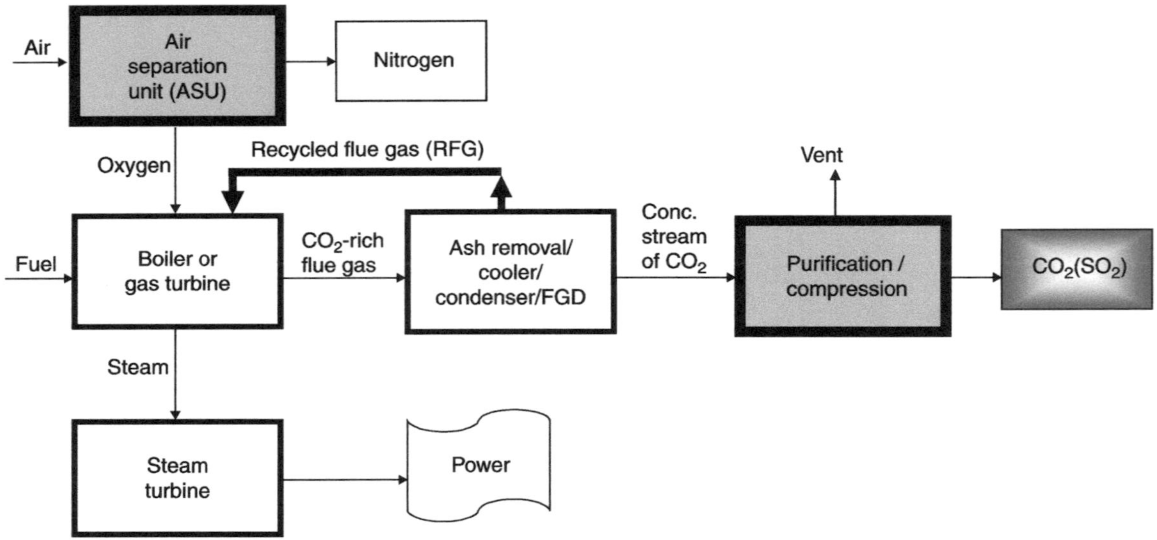

4.1 Simplified flow sheet for oxy-fuel technology, showing in bold the additional operations added to a standard pf plant. FGD: flue gas desulphurisation.

steam cycles. However, while the higher temperature steam cycles result in greater efficiencies, the uncertainty with flame behaviour and radiative heat transfer in larger boilers, pollutant control (i.e., NO_x, SO_x), corrosion of advanced boiler materials and CO_2 cleaning/compression require operational experience with full chain (integrated) pilot plants and demonstration prior to industrial deployment.

4.2 Oxy-fuel demonstrations and large pilot plants

4.2.1 Oxy-fuel carbon capture and storage (CCS) technology development

Currently a number of oxy-fuel demonstrations are being progressed as the technology is developed from pilot scale ($< 5\ MW_t$), with the historical progression detailed in Fig. 4.2.

Projects are classified and detailed in three categories:

- Power plant with CCS, these being full demonstrations which are primarily coal fired, generate electricity, with associated carbon CCS;
- Industrial-scale demonstrations, without carbon storage;
- Pilot-plant, with testing of combustion plant, possibly gas cleaning and carbon capture, but without electricity generation. Such plants may test full-scale burners, and evaluate gas processing and carbon dioxide compression, and storage.

Until recently the technology was developed through pilot plants, these being single burner test furnaces, usually with liquid oxygen supply and recycled flue gas. Buhre *et al.* (2005) detail these developments.

The oxy-fuel demonstrations and large pilot plants of 5–250 MW_e (~ 15–750 MW_t) noted in Fig. 4.2 are listed in Table 4.1. The table gives the scale of plant as MW_e, with $MW_t/3$ for plants without electricity generation. The table includes projects which have commenced operation, through to projects which are at the feasibility study stage, and which await progression if the study is positive. Further information can be found at the IEA Oxy-fuel Network website (IEA, 2009).

The projects listed in Table 4.1 are at different stages of development, with several being at pre-feasibility or feasibility stage. Six of the plants indicated have CCS. Not all are expected to proceed to financial close. Several projects are considered to contribute to particular aspects of oxy-fuel technology development, as follows:

- *The Vattenfall 30 MW_t pilot plant* – this is the first comprehensive project and involves evaluation of burner operation, with key testing of boiler impacts, emissions and impacts on CO_2 compression. The plant also allows evaluation of possible operations such as limestone addition for sulphur capture, and ammonia addition for NO_x reduction. The plant will also demonstrate oxy-fuel combustion with both black coal and pre-dried lignite.

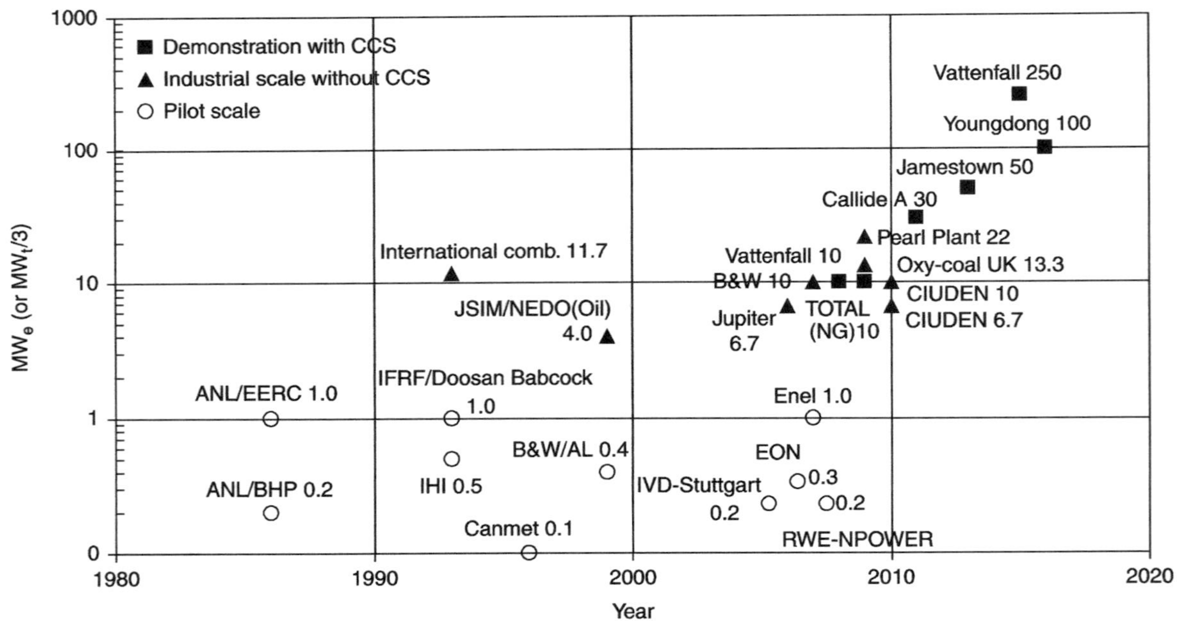

4.2 Historical progression of the scale of oxy-fuel pilot plants and demonstrations.

Table 4.1 List of large demonstration oxy-combustion plants, with some characteristics indicated

No.	Demo/pilot plant name	Scale (demo/pilot plant)	MW_e	New/retro-fit	Start-up/duration	Main fuel	Electricity generation Yes/No	CO_2 com-pression Yes/No	CO_2 use/seq.	CO_2 purity	Gas clean-up	Technology provider: Furnace	O_2	CO_2 compression
1	Vattenfall pilot plant, Germany	P (PC)	10	N	2008	Coal	N	Y	Y (partial by truck)	99.90%	SCR ESP	Alstom	Linde	Linde
2	Callide (CS Energy, Australia)	P	30	R	2010	Coal	Y	Y	N		FF	IHI	Air Liquide	Air Liquide
3	TOTAL, Lacq, France	P (industrial)	10	R	2009	NG	Y	Y	Y	99.90%	FGD	Alstom	Air Liquide	
4	CIUDEN, Spain	P (PC/CFB)	10	N	2010	Coal		Y	N		SCR FF	FW	Praxair	
5	Youngdong, South Korea	D	100	R	2016	Coal	Y		Y					
6	Holland/Praxair Plant, USA	CFB	78	N	2015	Coal	N	N					Praxair	
7	Jupiter Pearl plant, USA	P	22	R	2009	Coal	N	N				Jupiter Oxygen	Jupiter Oxygen	
8	Babcock & Wilcox pilot plant, B&W, USA	P	10	R	2008	Coal	N		N	70% dry	SCR FF	B&W	Air Liquide (American)	
9	Doosan Babcock, USA	P	30	N/A	2008	Coal	N		N					
10	B&W, Black Hills, Wyoming USA	D	100	N	2015	Coal	Y					B&W	Air Liquide	Air Liquide
11	ENEL Oxy-fuel CCS2, Italy	D	16	N	2012	Coal								

PC = pulverised coal; seq. = sequestered; CFB = circulating fluidised bed; ESP = electrostatic precipitator; FGD = flue gas desulphurisation; FF = fabric filter; NG = natural gas; SCR = selective catalytic reduction.

- *The Doosan Babcock Oxy Coal-UK project and B&W USA plants* – these demonstrations have comprehensive burner testing, with burner operational envelopes, stability, turndown, start-up and shut-down, with transition between air and oxy-fuel firing.
- *The Callide 30 MW_e oxy-fuel demonstration project* – will be the first integrated plant, having power generation, carbon capture and CO_2 sequestration. The plant will be the first to demonstrate oxy-fuel combustion of coal in a retrofit scenario, rather than in a new plant, which is significant for existing utilities aiming to convert to CCS.
- *The CIUDEN and Jamestown plants* – these evaluate circulating fluidised bed (CFB) oxy-fuel technology, which is suited to coal/biomass cofiring and to direct sulphur removal using sorbents.
- *The TOTAL, Pearl and Youngdong plants* – evaluate the technology in a commercial context. In particular, the TOTAL plant is unique in using natural gas, rather than coal, and will be the first oxy-fuel plant to transport and inject CO_2. The public engagement required to obtain community support has been found to be a critical issue in this project.

High efficiency supercritical pf plant is only practical at a scale of 250 MW_e or larger and is desirable for practical application of oxy-fuel technology (as well as post-combustion capture (PCC) technology) as the efficiency penalty associated with CCS is a reduced proportion of sent-out electricity compared with sub-critical plant. Only one potential demonstration is of this scale.

Newer projects in Table 4.1, which are in development, have recently been announced, including:

- *B&W Black Hills Oxy-fuel project, Wyoming, USA*. A project has now been submitted to United States Department of Energy (DOE) Restructured FutureGen to build a 100 MW_e oxy-fuel plant with CCS as a greenfield plant for the Black Hills Corporation in Wyoming, with the plant commencing in 2015. Plant simulations for a supercritical unit have included thermal integration to reduce the efficiency penalty for the ASU and CO_2 compression to less than 6% (IEA, 2009). Design options for integration have included the location of the point of recycle extraction (either hot, warm or cold), recycle ratio, heat integration between ASU and CO_2 compression, different emission control units and steam parameters. During pilot scale testing these design options are typically studied separately in order to optimise a particular section of the plant. However, a complete integration of separate units has the potential to optimise the design. Currently, oxy-fuel processes have not been optimised. By comparison, PCC is a relatively mature technology.
- *ENEL Oxy-fuel CCS2 demonstration, Italy*. The project goal of the CCS2 project is to build by 2012 a 50 MW_t zero emission coal-fired power plant based on a pressurised oxy-combustion technology which has been developed

at pilot scale. The process operates at 10 bar pressure, which reduces the furnace size (and therefore CAPEX), increases heat transfer, reduces air ingress, lowers the CO_2 compression energy and allows greater utilisation of energy from water condensation in the flue gas. Specific challenges include ash removal, boiler heat transfer (with scale-up) and coal behaviour at higher pressure.

4.2.2 Pilot plants

- *Vattenfall 30 MW_t Schwarze Pumpe plant* (Anheden, 2008). Vattenfall commenced a research and development (R&D) project on oxy-fuel technology in 2001, leading to a commissioned 30 MW_t oxy-fuel pilot plant in August, 2008. The flow sheet diagram of the plant is shown in Fig. 4.3 (Anheden, 2008) indicating a focus towards combustion and assessing gas treatment options. Pre-crushed pf is fired into the furnace, using either black coal or pre-dried lignite. The lignite is dried at a nearby facility. The flue gas treatment train optionally allows for the removal of NO_x, SO_x, H_2O and fly ash, with provision for direct limestone and ammonia addition. Objectives of the tests are to validate and tune commercially available technologies in oxy-fuel concepts to allow the launch of a demonstration project in commercial scale, including to define optimal operating conditions for oxy-fuel firing in a large-scale facility for the entire process, to identify critical issues for further R&D and to gain operating experience of running oxy-fuel plants. Up to 8 tonnes/day of liquid CO_2 is captured following compression with a 90% capture rate and the purified CO_2 stream (with a food grade quality) is expected to be transported by truck or rail to the storage site.
- *CIUDEN test furnace* (Cortes, 2008). The CIUDEN test facility includes an oxy-fuel 20 MW_t pf and 30 MW_t CFB. Plans include provision for limestone preparation/feed system and optional selective catalytic reduction (SCR), fabric filter (FF) and wet flue gas desulphurisation (FGD). For air fired experiments, CO_2 capture will occur through an absorption tower and be fed into a compression/cooling unit, which can be fed directly during oxy-fuel mode.
- *Babcock and Wilcox (B&W) USA project* (McDonald *et al.*, 2007; McDonald *et al.*, 2008). B&W's 30 MW_t Test Facility is located in Alliance, Ohio (McDonald *et al.*, 2007; McDonald *et al.*, 2008). During 2007 and early 2008, B&W's existing 30 MW_t Clean Environment Development Facility (CEDF) was modified to operate in the oxy-coal combustion mode. The plant includes removal of fly ash, SOx and water prior to CO_2 cleaning and compression.
- *OxyCoal-UK project.* Currently Doosan Babcock are retrofitting a 90 MW_t full-scale burner test facility with an oxy-fuel combustion firing system; operation with a 40 MW_t, oxy-fuel burner commenced in mid-2009. The

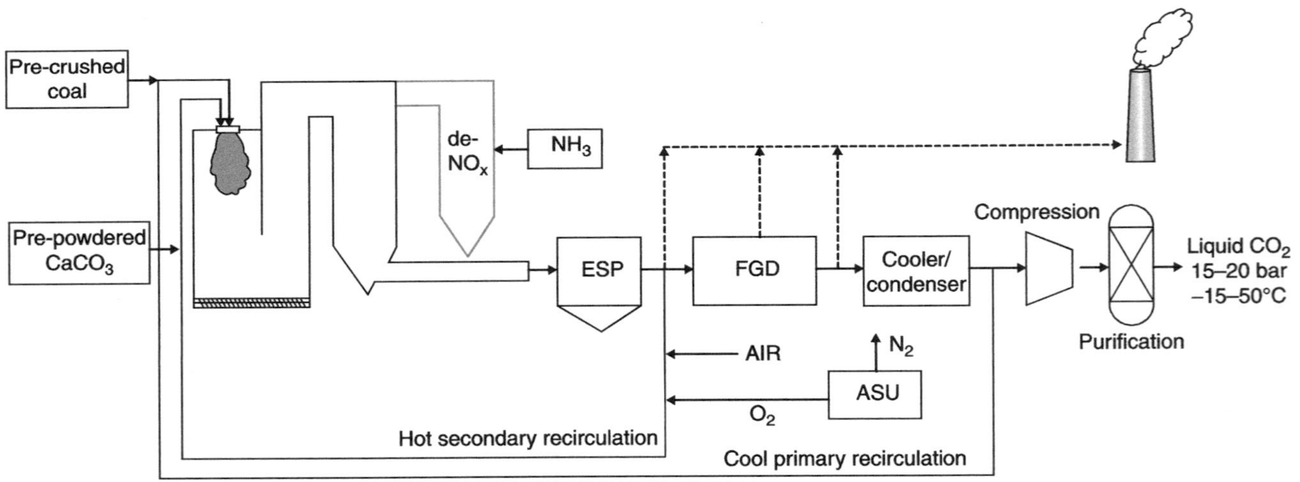

4.3 Flow sheet diagram of Vattenfall's oxy-fuel pilot plant, showing options for the removal of NO_x, SO_x, H_2O and fly ash, with provision for direct limestone and ammonia addition (Anheden, 2008).

project, OxyCoal-UK 2, is sponsored by the UK Government. The full-scale burner demonstration follows on from an earlier project which started in 2007 and involved ignition testing (in an ignition bomb apparatus), drop-tube furnace tests, computerised fluid dynamics (CFD) modelling, pilot-scale testing at 160 kW_t and 1 MW_t, corrosion tests and engineering studies. Testing of the 40 MW_t oxy-fuel burner, the largest burner planned to be tested to date, was completed at the end of 2009.

- *Pearl Power Station (Jupiter Oxygen).* A 22 MW_e four-burner oxy-fuel combustion system is proposed by Prairie Power Inc. at Pearl Power Station in the USA and burner testing started in 2008. The Jupiter Oxy-fuel combustion plant would include both power generation and a CO_2 compression train. Mercury is removed before CO_2 capture. In the proposal, the captured CO_2 will be transported by pipeline.

4.2.3 Demonstrations of power plants with CCS

- *The Callide oxy-fuel demonstration project* (Spero, 2005, 2007). The Callide oxy-coal demonstration project was initiated by a feasibility study in 2004 and is managed by CS Energy, an Australian utility. The concept and flow sheet of the Callide A oxy-coal firing demonstration project is illustrated in Fig. 4.4. The focus is on demonstrating a retrofit with electricity generation during oxy-fuel firing and storage of CO_2. The coal is crushed on-line and therefore partial drying of the recycled flue gas is required. The plant design includes: 2×330 tonnes/day ASU with 98% O_2 purity, four-year operation, 40% flue gas recirculation, slip stream compression – drying + cryogenic purification unit (CPU) with two-stage compression, Hg removal and ~99% CO_2 product quality. Stack emission modelling for the lower stack velocities encountered in oxy-firing mode was also undertaken, giving ground level concentrations for SO_2. The design liquid CO_2 production rate is 75 tonnes/day with the target geosequestration rate being 60 tonnes CO_2/day, over three years. Transport by truck to the storage site is expected.
- *TOTAL Lacq project.* The TOTAL Lacq CCS pilot-scale oxy-fuel project uses natural gas as fuel and aims to transport the CO_2 via a 27-km pipeline to the sequestration site. The unit has a capacity of 30 MW_t. The plant produces 92% purity CO_2 stream and the well reservoir to be monitored in the existing gas field has a 4500 m depth. This is a two-year project with commissioning beginning in 2009. The project will be the first large scale oxy-fuel demonstration of a retrofitted natural gas-fired process with pipeline CCS.
- *Praxair Jamestown CFB oxy-coal project.* Praxair has announced a near-zero emissions flue gas purification project for existing fluidised bed power plants retrofitted with oxy-fuel combustion technology project in Jamestown, New York. Goals of this project are to cost-effectively capture more than 95% of

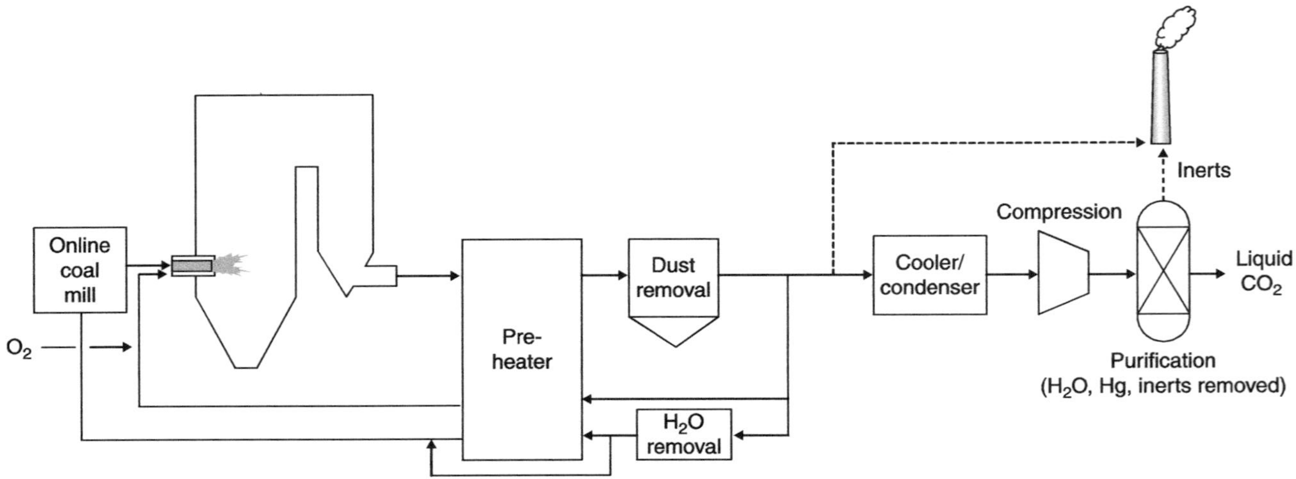

4.4 Flow sheet of the retrofitted oxy-coal firing power plant of the Callide A demonstration project (Spero, 2005, 2007).

CO_2 emissions from a CFB boiler. The plan is to capture up to 98% of the CO_2, transport by pipeline and inject into a sequestration well.

- *Youngdong oxy-coal demonstration project.* A re-powered 100 MW_e unit is planned, replacing the Youngdong unit #1 boiler in Korea, which currently fires domestic anthracite. The oxy-coal plant will be designed by 2013 and is to be constructed by 2015. High-volatile bituminous coal is the design coal, but sub-bituminous coal or lignite, possibly imported from Indonesia, may be used in the plant. The research phase, led by the KEPRI group, has been approved with the conceptual design over three years. The storage site of CO_2 produced from the demonstration plant is yet to be decided and is the element of greatest uncertainty.
- *Vattenfall 250 MW_e oxy-fuel demonstration plant.* In May 2008, Vattenfall announced its plans to build a demonstration plant for CCS technologies at one of the 500 MW blocks of the conventional lignite power plant in Jänschwalde in the State of Brandenburg, Germany. The investment for the demonstration is estimated to be one billion euros. The Jänschwalde lignite power plant consists of six 500 MW blocks. For the demonstration plant, one of the blocks consisting of two boilers will be equipped with CCS facility. One boiler will be a new plant with oxy-fuel technology and the other will be retrofitted with a post-combustion technology.

4.2.4 International vendors and demonstrations

Oxy-fuel technology uses standard operations familiar to power station vendors, and therefore most vendors of pf plant are also involved in the developments. Two categories of international vendors are associated with the development of oxy-fuel technology: traditional power station designers and manufacturers; suppliers of ASUs, who are commonly also developers of CO_2 compression technology.

Table 4.1 indicates some of the associations for the demonstrations, with vendors such as IHI, Alstom, Doosan Babcock, Foster Wheeler, Jupiter Oxygen, Praxair, Air Products and Air Liquide (IEA, 2009).

4.3 Demonstrations and progress towards commercial deployment

Developments in the field of oxy-fuel combustion have progressed through laboratory-scale and pilot-scale demonstrations and are approaching pre-commercial demonstration (or industrial scale) and commercial scale. The Vattenfall project is the most advanced with construction of its pilot plant completed in 2008. A number of research campaigns have been designed to investigate mode switching (e.g., air to oxy-fuel) with constant heat flux, burner configuration, ash mineralogy, air ingress minimisation, flame management with flue gas recycle and O_2, corrosion of the boiler, fuel qualities (moisture, sulphur), de-NO_x with SCR for lignite, alternative CO_2 compression plant configurations and biomass blending. The knowledge gained from these test campaigns will feed directly into the design of the industrial demonstration plant (300–700 MW_e) planned for 2013–2015.

Currently the oxy-fuel concept has been proven to be sound; however, many of the unknowns in designing a commercial-scale oxy-fuel power plant are directly related to scale and operability. New additions such as recycled flue gas, ASU and CO_2 compression plant, must operate in conjunction with current power plant systems. Currently, these systems have developed as stand alone operations. Issues such as corrosion must be tested over longer time periods to determine which materials are suitable for larger-scale construction. Material choice is particularly important if supercritical steam cycles (higher efficiency) are to be realised. This is because the higher sulphur levels caused by recycling the flue gas may increase corrosion of materials currently under consideration for supercritical systems.

Environmental pollutant behaviour and control is not adequately understood in oxy-fuel systems. Lower NO_x levels and higher SO_x emission levels (as mass/energy) have been observed in laboratory and pilot systems; however, the optimal choice of control options is far from certain. Typical NO_x control can involve a low NO_x burner installation coupled with either selective or non-selective catalytic reduction. However, because of the lower NO_x levels in oxy-fuel processes, these control systems may not be necessary. Furthermore, the standard SOx control unit, the wet flue gas desulphurisation unit (wet-FGD), may not be necessary with emerging options as part of the CO_2 compression circuit (IEA, 2009). Such systems (such as Air Product's Sour Gas Compression) are still being trialled at pilot scale. Mercury behaviour and capture in oxy-fuel systems is of significance due to its corrosion potential in the cryogenic inerts removal (O_2, N_2, argon) section of CO_2 compression.

The anticipated cost of CCS technologies as they are researched, demonstrated and deployed is illustrated in Fig. 4.5 (Dalton, 2009). Prior to demonstration, costs are underestimated. The first-of-a-kind plant generally has a lower expected cost than those following, as some design and operational issues are not foreseen. After a small number of plants are operating, the experience gained reduces the cost, i.e., costs reduce as the technology is deployed and matures. The cost reduction will be associated with a lower capital cost of plant, better design and economies of scale as well as competition due to the availability of technology suppliers.

In Fig. 4.5 oxy-fuel technology is seen to be in development and entering the demonstration phase, whereas other CO_2 capture technologies (PCC and IGCC-CCS) are being demonstrated. CO_2 storage of the scale required for CCS is also under development. Advanced ultra supercritical (USC) pf plant straddles several phases depending on the steam conditions. This is relevant to oxy-fuel technology as the efficiency gain from higher temperature steam cycles will reduce the impact of energy penalties associated with O_2 supply and CO_2 compression.

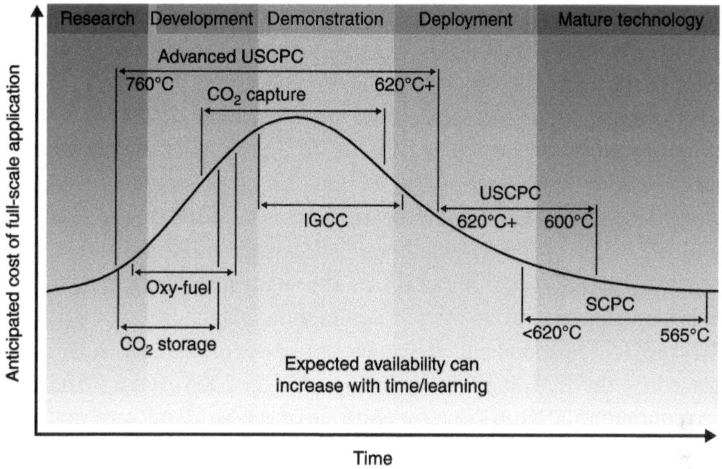

4.5 Anticipated cost of technologies as they are developed and applied, with current status of technologies indicated.

Operating experience and optimisation of the demonstration plants are crucial for commercial oxy-coal combustion technology. Inputs of the demonstration plants through to the commercial plants provide guidelines not only to the operation of full-scale power plants, but also to the CO_2 handling (such as compressing and transportation) and storage (such as injection and monitoring). Examples from the Vattenfall pilot plant (detailed in IEA, 2009; Strömberg *et al.*, 2009b; Strömberg *et al.*, 2009a) show the value of experience gained at pilot scale. Examples include the ASU requiring 60 hours to cool leading to a decision to leave it running when operating in air mode. The FGD unit was also designed for oxy-fuel mode (~1/3 of volumetric flow of air mode), which required either limestone added to the furnace or treating 1/3 of the total flow when in air mode. Furthermore, it was found that when switching from air to oxy-fuel mode, the slurry temperature in the FGD took ~5 hours to reach a new equilibrium. This process was thought to be the slowest in the system, requiring specific consideration in plant dynamics. Safety aspects of working with O_2, CO_2 and NH_3 (for the compression plant) required specific gas monitoring and shut-down procedures with systematic purging to prevent condensation and corrosion by sulphur components in the flue gas.

Three phases – 1: concept, 2: front end engineering design (FEED), and 3: construction/commissioning – are associated with the development of a coal-fired power plant with geosequestration. A typical time from concept development through construction to handing over the plant by the contractor is 6–9 years for a 500 MW_e plant. After each phase the project might not proceed, but one critical requirement is that a feasible CO_2 storage site must be validated at the end of phase 2, typically after 3–5 years. For projects requiring exploration of a number of uncertain storage sites, the cost of establishing a suitable storage site can exceed the FEED cost (Wall, 2009).

A sequence of the development of oxy-fuel technology for first-generation plant can be proposed, which is expected to use an ASU for O_2 supply, standard furnace designs with externally recirculated flue gas, and limited thermal integration of the ASU and compression plant with the power plant. This is the technology of the currently announced pilot-scale and industrial-scale plant and demonstrations with and without CCS. The suggested deployment of commercial plant is expected to overlap the pre-commercial plant.

The project development sequence for the commercial scale CCS plants suggested by the IEA and G8 Workshop (CSLF, 2008) states, 'The G8 must act now to commit by 2010, to a diverse portfolio of at least 20 fully integrated industrial-scale demonstration projects (>1 Mtpa), with the expectation of supporting technology learning and cost reduction, for the broad deployment of CCS by 2020'. For the IEA plants, the project development should commence prior to 2012, requiring that storage sites be proven prior to 2016. The EPRI roadmap of Wheeldon and Dillon (Dillon *et al*., 2005; Novak, 2007) prepared in 2007 has the pre-commercial demonstrations running from 2012 to 2022.

Figure 4.6 gives the phases planned for design, construction and operation of the current demonstrations with CCS. Several projects have a commercial phase, while most are designed for limited operation. Vattenfall's Schwarze Pumpe demonstration is the most progressed, having been in operation since September 2008 (Strömberg *et al*., 2009a), TOTAL's Lacq project was inaugurated in January 2010 and CS Energy's Callide project is planned to commence operation in 2011 (IEA, 2009).

Following industrial-scale deployment and experience gained from demonstration projects, technology costs are expected to reduce, with the early commercialisation phase from 2020 indicated in Fig. 4.6. Commercial deployment will follow when the CO_2 emission cost exceeds the CO_2 avoidance cost of the technology, as shown in Fig. 4.7. Figure 4.7 is based on a report by McKinsey & Company published in 2008 (McKinsey, 2008), in which CCS based capture CO_2 costs were estimated, these being found to be independent of the capture technology for the methodology used in the study. The figure indicates that during the demonstration phase, the shortfall between carbon price and CCS technology cost must be provided – by government support, technology vendors and electricity providers or the coal industry. The early commercial phase will depend on technology costs reducing, and CO_2 emission costs established by a trading scheme or a tax must match the technology costs.

Figure 4.8 gives a suggested sequence of development to commercial deployment, with components of R&D, technology demonstration, cost and regulation targets to address barriers, together with nominated periods for development and completion. Each category is discussed here.

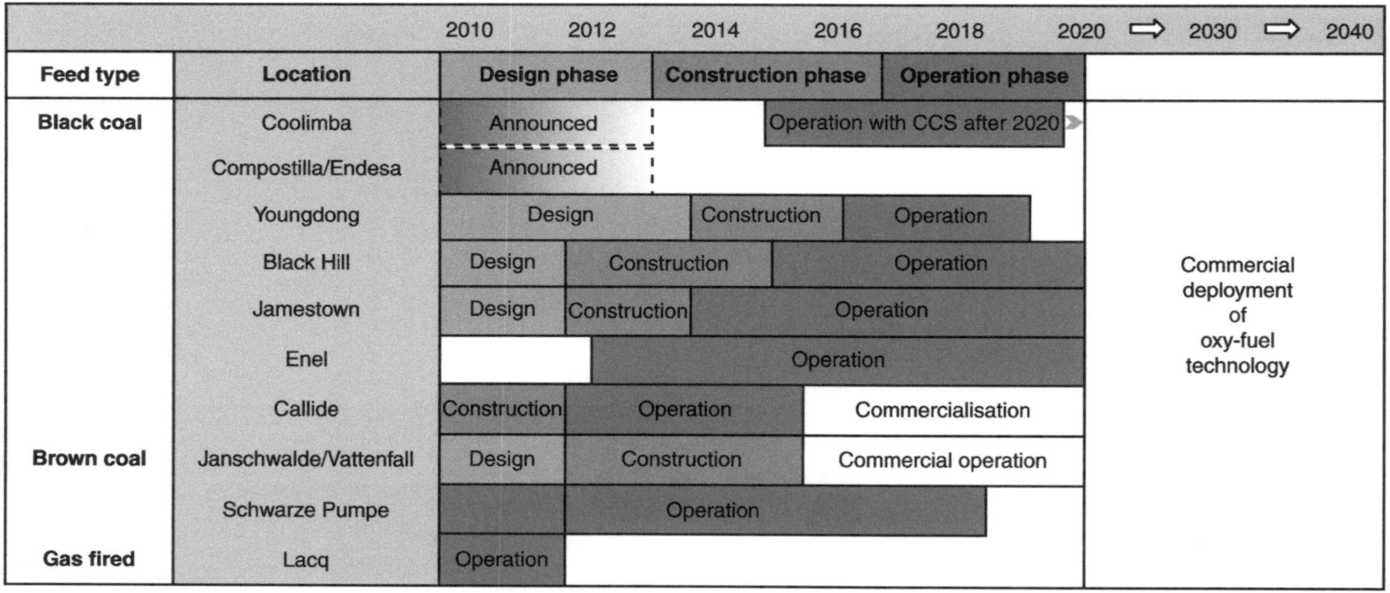

4.6 Nominated phases of oxy-fuel demonstration projects with CCS.

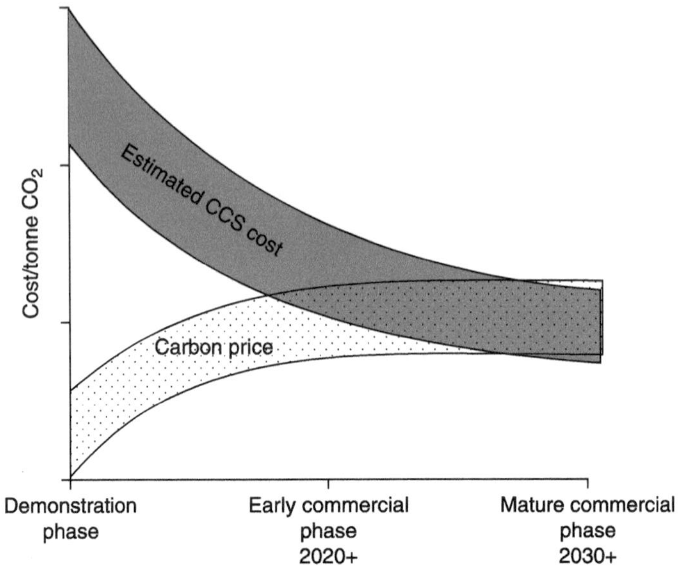

4.7 Representative diagram of CCS cost and the emerging carbon market.

R&D targets

Research and development targets are nominated in Fig. 4.8, with elaboration elsewhere (Wall *et al.*, 2009). Comparative techno-economic and process modelling of CCS technologies provide comparisons and benchmarks for design and affect the technology choices and process configurations that are adopted. These assessments require thermal, physical, chemical and often empirical information such as overall process or unit performance. Such models must continue to be developed and revised as they are improved following information from the early demonstrations. For oxy-fuel processes, the CO_2/H_2O atmosphere combined with higher levels of impurities influences heat transfer, flame ignition, coal reactivity, minor species formation and the overall separation efficiency of flue gas cleaning and purification units. The gas quality and concentrations of impurities can have a major impact on energy usage and purity of the final CO_2 product.

Laboratory-scale measurements of coal reactivity in oxy-fuel environments and vapour/liquid equilibrium (VLE) data for supercritical CO_2 with impurities are needed for design optimisation. Research on minor species formation and impacts, particularly NO_x, SO_x, soot and Hg is necessary. Corrosion studies should be undertaken on materials for both furnace and transport equipment and validated by the demonstrations. Development of materials for higher temperature USC steam (+700°C) plant in oxy-fuel gas conditions is necessary.

Options for O_2 separation options with lower energy penalty require development – such as membrane and chemical looping technologies. For

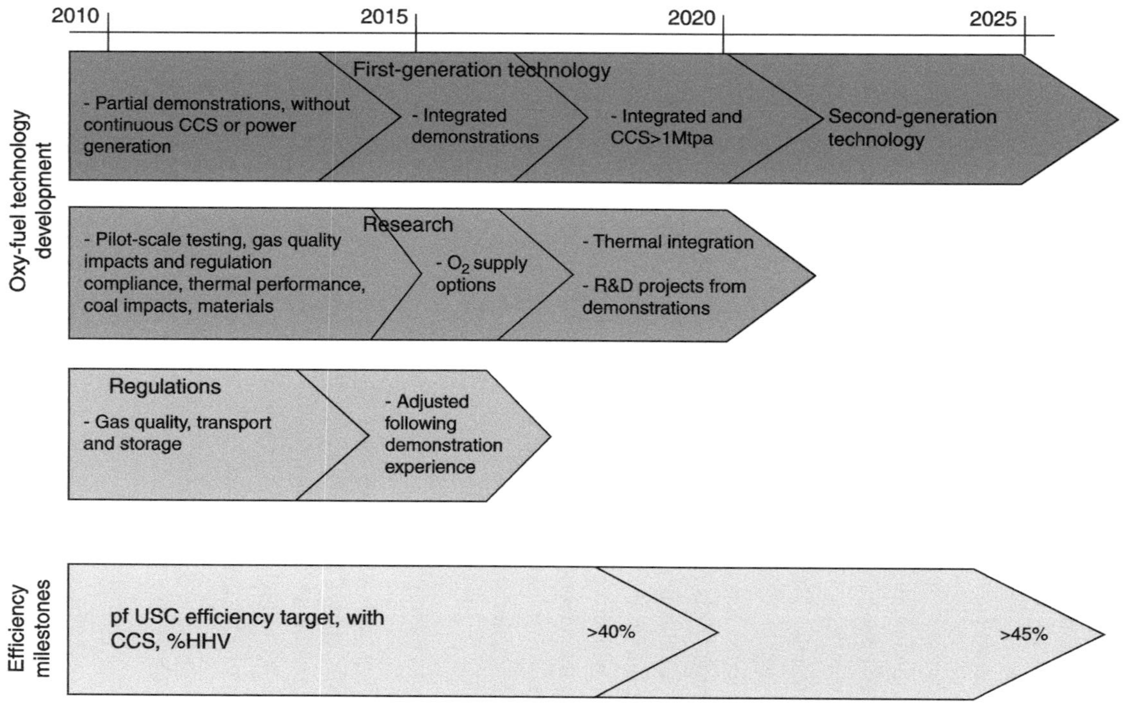

4.8 Phases and sequences of the projected development of oxy-fuel technology.

membrane systems, the integration of heat from the combustion side will play a role in determining the energy penalty. For chemical looping systems, oxygen transfer materials, particle degradation and coal behaviour are all areas in which future research can make significant reductions in energy penalty.

Experience from the operation of demonstrations will continuously feed back into the R&D effort.

Demonstration and deployment targets

Pilot scale facilities such as Endesa's CIUDEN project (IEA, 2009) provide design information of coal and process behaviour at industrial scale. This information is important, particularly for oxy-fuel processes, as the treatment of flue gas prior to recycle can alter the concentration of impurities in the furnace and in compression. Therefore, large pilot plants must be available for testing coal resources to support commercial deployment.

For a reference scenario, the IEA estimates that 1680 GW_e of new electrical generation capacity is required between 2005 and 2030 (IEA, 2007) with coal-fired power stations making up the dominant share, rising from 40% to 45% of total generation. Furthermore, the IEA predicts that the deployment of clean coal technologies will drive up the average thermal efficiencies in coal-fired power stations with the adoption of best available technology (BAT) and the use of SC plant. SC oxy-fuel plant will emerge following demonstration at subcritical conditions at smaller scale.

Cost and capture targets

Future development and deployment of SC and USC systems will improve generation efficiency. Coupled with expected efficiency increases in gas treatment, thermal integration and compression, the overall plant efficiency is expected to rise to above 43% (higher heating value, HHV, including CCS) by 2030, with a stretch target of 45% given in Fig. 4.8. Targets for oxy-fuel CO_2 capture penalties will reduce.

Regulations

An immediate regulatory issue is the requirement for 'capture ready' features to be built into new power stations. These regulations are currently evolving. The UK has recently announced (Milliband, 2009) that new power plant (>300 MW) will require sufficient space available to retrofit CCS, suitable potential sites to store carbon dioxide, a feasible potential transport route from the power station to the storage area and no foreseeable barriers to retrofitting CCS.

For oxy-fuel technology, the site must allow for an ASU, a compression train and recycle transfer lines. Further capture-ready requirements may also include specifying a minimum demonstrated capture level (e.g., 25%) with the ability to add further capture capacity.

Gas quality regulations for transport and storage have an impact on oxy-fuel technology as the concentrations of impurities are significantly higher than in other technologies. Of particular concern are the H_2O and SO_2 contents allowed in the final CO_2 product as they represent a possible corrosion issue when combined. The CO_2 product specification will affect technology choices in gas treatment. Therefore, impurity levels for transport options (pipeline, truck, ship) and final storage sites should be defined.

Incentives for demonstrations are also needed. Initial financial incentives for early movers of oxy-fuel demonstration leverage private sector investment. After construction and commissioning, even with electricity being generated, early demonstrations of oxy-fuel may require financial support during operation to compensate for the costs associated with CCS. This can be in the form of CO_2 credits or $/tonne of CO_2 sequestered. A commercially viable technology that is competitive with a carbon cost or market will develop.

4.4 Conclusions

The current status of first-generation oxy-fuel combustion demonstration projects is presented indicating that the technology is entering the pre-commercial demonstration stage with a number of demonstration plants having commenced or being under construction.

Milestones and dates of a roadmap for deployment are suggested, requiring projects leading to integrated commercial plants larger than 500 MW targeted for completion by 2020 with CCS at scales exceeding 1 MTpa.

Aspects of the technology development are outlined which are related to the demonstrations, including R&D, regulations and incentives for early movers and efficiency targets for pf USC plant.

4.5 Update

As stated in section 4.2.1 (detailing technology development), not all oxy-fuel projects are expected to go to completion, and other projects will emerge. This section is provided as a project update prior to printing (current as of 12 April 2010), rather than altering the original article (current as of September 2009).

The US oxy-fuel projects (B&W's Black Hills, Praxair's Holland and New York State's Jamestown projects) did not receive funding from the third funding round of the US DOE Clean Coal Power Initiative (US DOE, 2009). As such, it is unlikely that these projects will be proceeding through to the design phase without an alternative funding mechanism. At the time of writing the original manuscript, some confusion surrounded the Holland and Jamestown CFB projects. Praxair was originally associated with the Jamestown (50 MW CFB) proposal, but joined the Holland (78 MW CFB) project DOE submission. The Jamestown proposal was submitted by the Jamestown Board of Public Utilities, despite local opposition (Bertola, 2009). In Europe, Vattenfall's 250 MW pulverised coal Jänschwalde project and Endesa's 300 MW CFB Compostilla project have received funding under the EU's European Energy Programme for Recovery and will proceed into the design phase (EUROPA, 2009). The Compostilla project will be based on data generated as part of the CIUDEN 30 MW CFB pilot project. TOTAL'S oxy-natural gas pilot plant at Lacq was inaugurated in January 2010 (TOTAL, 2010) with monitoring of the injection site to proceed for three years after the two-year injection period.

The most recent oxy-fuel project to attract funding (announced August 2010) has been the re-vamped FutureGen2.0 project in the United States, which will re-power a 200 MWe unit of the Meredosia plant, Illinois, with pf oxy-fuel technology. The project will have a storage component, which has yet to be determined (FutureGen2.0, 2010).

4.6 Acknowledgements

Some of the material presented in this chapter was established from meetings and courses of the Asia Pacific Partnership (APP) Oxy-fuel Working Group (OFWG), as detailed on the OFWG website at http://www.newcastle.edu.au/project/oxy-fuel-working-group/. We acknowledge the information provided by the proponents of oxy-fuel demonstrations.

4.7 References

Allam, R. J., V. White, *et al.* (2005). *Optimising the design of an oxyfuel-fired advanced supercritical PF boiler*. The Proceedings of the 30th International Technical Conference on Coal Utilization & Fuel Systems. Coal Technology: Yesterday – Today – Tomorrow. Clearwater, Florida, USA; Coal Technology Association.

Anheden, M. (2008). *Vattenfall's Schwarze Pumpe oxyfuel pilot – an update*. 3rd IEA GHG Oxyfuel Workshop, Yokohama, Japan.

Bertola, D. (2009). 'Jamestown Power Project Left in the Cold', article for *Buffalo Business First*, accessed 12 April 2010, http://buffalo.bizjournals.com/buffalo/stories/2009/12/14/story8.html

Buhre, B. J. P., L. K. Elliott, *et al.* (2005). Oxy-fuel combustion technology for coal-fired power generation. *Progress in Energy and Combustion Science* 31(4): 283–307.

Cortes, V. J. (2008). *Test facility for advanced technologies for CO_2 capture in coal power generation update and upgrade (CIUDEN, Spain)*. 3rd IEA Oxy-fuel Combustion Workshop, Japan.

Croiset, E., P.L. Douglas, *et al.* (2005). *Coal oxyfuel combustion: a review*. The Proceedings of the 30th International Technical Conference on Coal Utilization and Fuel Systems – Clearwater Coal Conference, Clearwater, Florida, USA; Coal Technology Association.

CSLF (2008). 'G8-IEA Calgary Workshop on Near Term Opportunities for Carbon Capture and Storage', accessed 30 July 2009, http://www.cslforum.org/publications/documents/pgtg_ResultsG8-IEA-CSLFWorkshop0408.pdf

Dalton, S. (2009). EPRI (personal communication).

Dillon, D. J., R. S. Panesar, *et al.* (2005). Oxy-combustion processes for CO_2 capture from advanced supercritical PF and NGCC power plant. *Greenhouse Gas Control Technologies 7*. Oxford: Elsevier Science Ltd, pp. 211–220.

EUROPA (2009). 'List of 15 Energy Projects for European Economic Recovery', Official website of the European Union, accessed 12 April 2010, http://europa.eu/rapid/pressReleasesAction.do?reference=MEMO/09/542&format=HTML&aged=0&language=EN&guiLanguage=en

FutureGen2.0 (2010). Company website, accessed 22 November 2010, http://www.futuregenalliance.org/

Gupta, R., S. Khare, *et al.* (2006). *Adaptation of gas emissivity models for CFD based radiative transfer in large air-fired and oxy-fired furnaces*. The Proceedings of the 31st International Technical Conference on Coal Utilization and Fuel Systems, Sheraton Sand Key, Clearwater, Florida, USA; Coal Technology Association.

IEA (2007). 'World Energy Outlook', accessed 31 August 2010, http://www.worldenergyoutlook.org/2007.asp.

IEA (2009). 'Oxy-fuel Network', accessed 24 July 2009, http://www.co2captureandstorage.info/networks/oxyfuelmeetings.htm.

Katzer, J. (2007). 'The future of coal: an interdisciplinary MIT study', accessed 31 August 2010, http//web.mit.edu/coal/

Khare, S.P., A.Z. Farida, *et al.* (2007). *Factors influencing the ignition of flames from air fired swirl PF burners retrofitted to oxy-fuel*. The Proceedings of the 32nd International Technical Conference on Coal Utilization and Fuel Systems, Clearwater, Florida, USA; Coal Technology Association.

Kiga, T. (2001). O_2/RFG combustion-applied pulverised coal fired plant for CO_2 recovery. In: *Advanced Coal Combustion*. T. Miura (ed.). Huntington, New York: Nova Science Publishers Inc., pp. 185–241.

Lundström, D., J. Eriksson, *et al.* (2006). *The use of CFD modeling to compare air and oxy-firing of a retrofitted pulverized fuel boiler*. The Proceedings of the 31st International Technical Conference on Coal Utilization and Fuel Systems, Sheraton Sand Key, Clearwater, Florida, USA; Coal Technology Association.

McCauley, K.J., H. Farzan *et al.* (2009). *Commercialisation of oxy-coal combustion: applying results of a large 30MWth pilot project*. Greenhouse Gas Technologies 9 Conference, sourced from *Energy Procedia* 1, pp. 439–446.

McDonald, D., D. DeVault, *et al.* (2007). *Oxy-combustion in pulverized coal power plants for carbon dioxide concentration*. 2007 Electric Power Conference. Chicago, Illinois, USA; Babcock & Wilcox Power Generation Group.

McDonald, D.K., T.J. Flynn, *et al.* (2008). 30 MWt clean environment development oxy-coal combustion test program. The Proceedings of the 33rd International Technical Conference on Coal Utilization and Fuel Systems. Clearwater, Florida, USA; Babcock & Wilcox Power Generation Group.

McKinsey & Company (2008) 'Carbon Capture & Storage: Assessing the Economics', accessed 31 August 2010, http://www.mckinsey.com/clientservice/ccsi/pdf/ccs_assessing_the_economics.pdf

Milliband, E. (2009). 'Press Release – No New Coal without CCS', Department of Energy and Climate Change, United Kingdom, accessed 23 April 2009, http://www.decc.gov.uk/en/content/cms/news/pn050/pn050.aspx

Novak, J. (2007). *Electric Power Sector Perspectives on CO_2 Capture and Storage (CCS): Achieving the Promise of CCS for Electric Power Generation*. EPRI, USA.

Rathnam, R. K., L. Elliott, *et al.* (2006). *Differences in coal reactivity in air and oxy-fuel conditions and implications for coal burnout.* The Proceedings of the 31st International Technical Conference on Coal Utilization and Fuel Systems, Sheraton Sand Key, Clearwater, Florida, USA; Coal Technology Association.

Rathnam, R. K., B. Moghtaderi, *et al.* (2007). *Differences in pulverised coal pyrolysis and char reactivity in air (O_2/N_2) and oxy-fuel (O_2/CO_2) conditions.* The Proceedings of the 32nd International Technical Conference on Coal Utilization and Fuel Systems, Clearwater, Florida, USA; Coal Technology Association.

Santos, S., M. Haines, *et al.* (2006). *Challenges in the development of oxy-combustion technology for coal fired power plant.* The Proceedings of the 31st International Technical Conference on Coal Utilization and Fuel Systems, Sheraton Sand Key, Clearwater, Florida, USA; Coal Technology Association.

Spero, C. (2005). *Australian Japanese co-operation on oxyfuel pilot project for plant retrofit – Callide-A project.* Inaugural Workshop of the Oxy-fuel Combustion Network, Cottbus, Germany.

Spero, C. (2007). *Status of Callide (30MWe) oxyfuel project.* 2nd Workshop of the Oxy-fuel Combustion Network, Hilton Garden Inn, Windsor, Connecticut, USA.

Strömberg, L., G. Lindgren, *et al.* (2009a). Update on Vattenfall's 30 MWth oxyfuel pilot plant in Schwarze Pumpe, *Energy Procedia* 1(1): 581–589.

Strömberg, L., G. Lindgren, *et al.* (2009b). *First results from Vattenfall's 30MWth oxyfuel pilot plant in Schwarze Pumpe.* Clearwater Coal Conference, 34th International Technical Conference on Clean Coal and Fuel Systems, Sheraton Sand Key, Clearwater, Florida, USA; Coal Technology Association.

TOTAL (2010). Company press release, accessed 22 November 2010, http://www.total.com/en/press/press-releases/consultation-200524.html&idActu=2265

US DOE (2009). 'US DOE Clean Coal Technology & The Clean Coal Power Initiative', accessed 12 April 2010, http://www.fossil.energy.gov/programs/powersystems/cleancoal/index.html

Wall, T. (2005). *Fundamentals of oxy-fuel combustion.* Inaugural Workshop of the Oxy-fuel Combustion Network, Cottbus, Germany.

Wall, T. (2007a). *Performance of PF burners retrofitted to oxy-firing.* 2nd Workshop of the Oxy-fuel Combustion Network, Hilton Garden Inn, Windsor, Connecticut, USA.

Wall, T. F. (2007b). *Combustion processes for carbon capture.* Proceedings of the Combustion Institute 2007, Pittsburgh, USA.

Wall, T. (2009). 'Coal Based Oxy-combustion for Carbon Capture and Storage: Status, Prospects, Research Needs and Roadmap to Commercialisation', 28 March 2009, Invited Lecture, Purdue University Energy Centre, accessed 31 August 2010, http://www.purdue.edu/dp/energy/pdfs/presentations/WALL_PURDUE_oxyfuel_2009.pdf

Wall, T., Y. Liu, *et al.* (2009). 'An overview on oxyfuel coal combustion-state of the art research and technology development.' *Chemical Engineering Research & Design* 87(8): 1003–1016.

Yamada, T., M. Tamura, *et al.* (2006). *Comparison of combustion characteristics of between oxy-fuel and air combustion.* The Proceedings of the 31st International Technical Conference on Coal Utilization and Fuel Systems, Sheraton Sand Key, Clearwater, Florida, USA; Coal Technology Association.

Xu, B., R. A. Stobbs, *et al.* (2007). 'Future CO_2 Capture Technology Options for the Canadian Market', accessed 18 November 2009, www.berr.gov.uk/files/file42874.pdf

5
Oxy-fuel combustion on circulating fluidized bed (CFB)

E. J. ANTHONY, CanmetENERGY, Natural Resources Canada, Canada and H. HACK, Foster Wheeler North America Corporation, USA

Abstract: Oxy-fuel fluidized bed combustion can burn fuels in an energy-efficient manner, while offering a pure stream of CO_2 suitable for sequestration. Pilot-scale test units demonstrate all of the well-established advantages of air-blown circulating fluidized bed combustion (CFBC) technology in terms of low emissions and fuel flexibility when operating with pure oxygen. Research is still required to study the sulphation process and determine the potential for fouling due to carbonation if the unit is operated above 900°C. Currently, efforts to commercialize this technology are making rapid strides and it is anticipated that there will be full-scale demonstrations within the next several years.

Key words: oxy-fuel combustion, fluidized bed combustion, emissions issues, demonstration projects.

5.1 Introduction

Anthropogenic CO_2 production is primarily driven by fossil fuel combustion and the current energy situation gives no indication that fossil fuel combustion demand will change in the near future. Consequently, it is increasingly necessary to find ways to reduce CO_2 emissions when fossil fuel is used. Of the various potential reduction options, CO_2 capture and storage (CCS) appears to be among the most promising for large stationary power plants. All of the CCS technologies for power plants involve producing an almost pure stream of CO_2 either by concentrating it in some manner from flue gases, or by using effectively pure oxygen as the combustion gas (Buhre *et al.*, 2005). The latter option, referred to as oxy-fuel combustion, has been well studied for pulverized coal combustion (Toftegaard *et al.*, 2010), but to date has received relatively little attention for oxy-fuel circulating fluidized bed combustion (CFBC), although the concept was examined over 20 years ago for bubbling FBC (Yaverbaum, 1977). More recently, the boiler companies Alstom and Foster Wheeler have explored the oxy-fuel CFBC concept using pilot-scale testing (Eriksson *et al.*, 2007; Stamatelopoulos and Darling, 2008). Alstom's work included tests in a unit of up to 3 MWt in size, but did not involve recycle of flue gas (Liljedahl *et al.*, 2006). Foster Wheeler's work (Eriksson *et al.*, 2007) also involved pilot-scale testing, using a small pilot-scale (30–100 kW) CFBC unit owned and operated by VTT (Technical Research Centre of Finland) and this work along with CanmetENERGY's work with its own 100 kW CFBC, appears to be the first in which units were operated with oxy-fuel combustion using flue gas recycle.

The advantages of CFBC are already well known in terms of its ability to burn a wide range of fuels, both individually and co-fired, to achieve relatively low NO_x emissions, and accomplish SO_2 removal by limestone (Grace *et al.*, 1997). Another advantage of CFBC technology, in the context of oxy-fuel firing, is the relatively low heat flux in the furnace. This low heat flux may allow either a significant reduction of the amount of recycled flue gas or, alternatively, permit the use of a much higher oxygen concentration in the combustor. Both of these will improve the economics of oxy-fired CFBC relative to pulverized coal (PC) or stoker firing by reducing the size of the CFBC boiler island by as much as 50% (Liljedahl *et al.*, 2006). In considering the scale-up of CFBC units above 300 MWe, both Foster Wheeler and Alstom are now offering much larger units and Foster Wheeler has in operation a 460 MWe supercritical CFBC boiler (Stamatelopoulos and Darling, 2008; Hotta *et al.*, 2008).

More-difficult-to-quantify advantages for the technology relate to the possibility of co-firing biomass, so that with CCS, the overall combustion process may potentially result in a net reduction of anthropogenic CO_2, and the potential for this technology to be used with more marginal fuels, as premium fossil fuels become in short supply. The co-firing option offers a potentially interesting advantage of CFBC technology since it is well established that CFBC can burn biomass and fossil fuels at any given ratio in a range of 0–100%, thus offering the possibility of using local and seasonally available biomass fuels in a CO_2 'negative' manner.

The ultimate availability of premium coal for a period of hundreds of years has also recently been called into doubt with suggestions that coal production may peak well before the end of this century. Thus, Mohr and Evans (2009), for example, have developed a model which indicates that coal production will peak between 2010 and 2048 on a mass basis and between 2011 and 2047 on an energy basis, with a best-guess scenario of peaks in 2034 on a mass basis and 2026 on an energy basis. In the event of such solid fuel shortages, fluidized bed combustion is ideally suited to exploit the many marginal coals and hydrocarbon-based waste streams available worldwide.

Currently R&D on oxy-fired CFBC technology is being explored in numerous countries, including Canada, Finland, Poland, China and the United States among others. However, to date most test work has been done at small scale (in the <100 kW range), and/or using bottled gases to simulate recycled flue gas to achieve the necessary gas velocity and solid circulation rate in terms of heat transfer requirement. CFBC systems can potentially operate with less flue gas recycle because the hot solids that are recycled in a circulating fluidized bed can also be used to produce steam. This cools the fluidized bed, and hence less flue gas needs to be recycled, compared with a suspension-fired boiler. However, the minimum amount of flue gas recycling is governed by maintaining sufficient fluid velocity in the fluidized bed while at the same time providing enough water/steam-cooled surface area to provide adequate internal and external solids heat transfer. Probably the most important advanced test programmes in oxy-CFBC are the development of a number of large pilot-scale units. However, first an exploration of the early work will be presented, followed by a discussion of results from the various pilot-scale tests done so far, finishing with a look at the major demonstration projects now underway.

5.2 Early work

One of the earliest detailed studies was presented by Alstom in 2004 for the US Department of Energy (Nsakala *et al.*, 2004; Liljedahl *et al.*, 2006), which identified oxygen-fired CFBC as having near-term development potential. The Phase II report was devoted to a pilot plant study on a 3 MWt test facility which was operated with O_2/CO_2 mixtures of up to 50% O_2 by volume, with local oxygen concentration in the lower part of the combustor of up to 70% O_2 by volume. Tests were carried out with one petroleum coke and two US bituminous coals. Because of concerns about recarbonation of the lime-based sorbents, the tests were carried out at 900°C, at which temperature indirect sulphation occurs via the global reaction:

$$CaO + SO_2 + \frac{1}{2}O_2 = CaSO_4 \qquad [5.1]$$

as opposed to direct sulphation, which can be described by the following reaction:

$$CaCO_3 + SO_2 + \frac{1}{2}O_2 = CaSO_4 + CO_2 \qquad [5.2]$$

For these tests the unit was operated with mixtures of O_2/CO_2, supplied from bottled gases, to achieve the gas velocity and composition for oxy-CFB operation. Other tests were also done with a bench-scale unit. The test campaigns carried out by Alstom are summarized in Table 5.1.

Under these conditions, the studies indicated that CO levels were higher, and the NO_x levels were lower than for air firing. Alstom also noticed that, for the bituminous coal, SO_2 levels were higher than expected, which they attributed to the temperature of operation, but SO_2 levels were lower for petroleum coke. Their

Table 5.1 Test campaigns

April 2004	June 2004	June 2005
Bituminous coal Limestone 0.6–1.4 MWt Air-fired 20–30% O_2 mixtures	Bituminous, petroleum coke Aragonite 1.2–2.3 MWt O_2-fired 40–50% O_2 mixtures	Bituminous, petroleum coke Lime, limestone, aragonite 3 MWt Air-fired, O_2-fired 30% O_2-fired

work also suggested no particular problem with either unburned hydrocarbons or N_2O emissions, the latter being an unsurprising result given the high temperatures required to ensure direct sulphation.

Also, Alstom reported some evidence of recarbonation of the lime-based sorbent in the cooler regions of the boiler. This last finding is interesting because the fouling by carbonation is a well-known phenomenon for lime kilns and has also been reported and discussed by other workers exploring fouling in petroleum coke boilers; this subject will be discussed later in more detail (Anthony *et al.*, 2001).

The issue of recarbonation was also of concern to the Alstom engineers for, among other things, its potential to affect the external fluidized bed heat exchanger where solids are typically cooled to below ~680°C. In particular, there was concern that, since some flue gas would be entrained with the solids if the heat exchanger was fluidized with air, there would be air leakage into the combustor, thus causing a reduction of the purity of the CO_2 leaving the boiler. Alternatively, if the unit was fluidized with gases containing CO_2, in particular recycled flue gases, there was concern that the solids might recarbonate, thereby consuming CO_2 and reducing the amount of gas available to fluidize the unit. It was observed that when the unit was fluidized with CO_2, the seal pot ceased to operate because most of the CO_2 was adsorbed. In practice this meant that they chose to fluidize the external fluid bed heat exchanger with air, but they did note that in a full-scale unit the temperature would be such that carbonation would not occur.

In general, it is not clear what might happen if the temperature falls below the calcination temperature as two competing reactions may be involved. However, experience with Ca looping cycles (Blamey *et al.*, 2010) shows that sulphation of lime-based particles significantly reduces carbonation (Sun *et al.*, 2007), while the addition of steam, even at temperatures well above those at which $Ca(OH)_2$ is stable, significantly enhances carbonation (Symonds *et al.*, 2009). Alstom engineers also proposed that they might avoid the entire problem by developing a moving bed heat exchanger which requires no fluidizing gases. If, however, it is decided to proceed with fluidized bed heat exchangers, there is no doubt that the engineering issues described here must be resolved.

One of the more important general conclusions from this work was that there appeared to be no major barrier to oxy-firing in CFBC boilers at a commercial scale. Also, as expected, there was no significant difference in the heat transfer to the walls and surfaces because in the CFBC environment heat transfer by solids is the dominant factor, unlike in suspension or PC combustion. Alstom also carried out techno-economic evaluations and came to the conclusion that the mitigation cost of about US$37/ton of CO_2 avoided was achievable based on their work.

Another major company working in the area of CFBC oxy-fuel combustion is Foster Wheeler (FW) (Eriksson *et al.*, 2007). FW has been working on a range of CCS solutions, and has chosen to use its Lagisza 460 MWe plant as the basis of an oxy-fuel combustion study. The company also carried out extensive pilot plant work using pilot-scale facilities at VTT in Finland.

VTT has a pilot-scale (30–100 kW) CFBC reactor that can be operated either in air- or oxygen-firing mode, as well as a smaller fluidized bed or bench-scale facility. The pilot-scale CFBC has an inner diameter of 167 mm, and a height of 8 m, while the bench-scale unit has a diameter of 32.8 mm and a height of 610 mm. Unlike the Alstom facility, the CFBC unit was capable of operation with flue gas recycle. In their 2007 study they found that SO_2 and NO_x levels were similar to air firing when using a bituminous coal in their bubbling bed unit. Tests in their pilot-scale facility were rather similar to those for the bench-scale facility for NO_x, and showed that emissions were quite similar to air firing, but one-third of those seen in the bench-scale unit, presumably reflecting the effects of better staging that can occur in a larger unit.

Table 5.2 gives results for four tests reported by FW. What is surprising about these results is the extremely good sulphur capture, which, unlike the bench-scale tests, is better than most of the literature would suggest for typical CFBC operation (Anthony and Granatstein, 2001). These results also seem not to be dependent on flue gas recycle rates, or in one case without recycle at all, which makes a very significant difference to the H_2O content of the gas. As can be seen, all other emissions, namely CO and N_2O, are quite typical of the levels that one might expect to see from small fluidized bed facilities, and suggest that for such emissions at least there are no major changes from air to oxygen firing. These results are also consistent with those obtained in Alstom's pilot plant facilities, as will be discussed below. A possible slight difference from the Alstom work, besides the superior sulphur performance, which cannot be compared because the Alstom tests were all done at conditions where indirect sulphation occurred, is that the CO emissions do not appear elevated for this series of tests, and are quite typical of what one would expect from small pilot-scale units.

Table 5.2 Test matrix of VTT's CFB pilot plant oxygen combustion test facilities

Fuel input (kW)	46.2	97.4	88.2	88.9
Ca/S molar ratio	1.5	1	1	1
Feed gas (FG) oxygen (%)	21.0	30.7	28.6	39.8
FG recycle ratio (from total) (%)	0	68	70	57
Bed temperature (°C)	799	787	762	875
Grid velocity (m/s)	1.43	1.74	2.03	1.52
Gas emissions and sulphur capture				
O_2 % (dry)	5.2	5.4	5.9	4.9
CO_2 % (dry)	91.8	94.5	92.5	91
CO ppm (dry)	147	82	189	157
NO ppm (dry)	75	126	93	213
N_2O ppm (dry)	189	131	367	50
SO_2 ppm (dry)	32	214	301	136
H_2O % (dry)	5.8	18.9	18.6	18.4
Sulphur capture (%)	94.1	75.9	66.1	87.1

Subsequently, FW had a second series of tests carried out, and this work produced fairly similar results (Table 5.3). Interestingly, at least for the South African coal, the sulphur utilizations were significantly lower.

5.3 Other test facilities

In addition to FW and Alstom, other laboratories have also carried out tests on oxy-fired CFBC. However, with the exception of CanmetENERGY's, all of those tests have been carried out without flue gas recycle. Zhejiang University utilized a 77 mm diameter, 2.8 m high unit, and conducted tests on O_2/CO_2 mixtures with O_2 levels ranging from 21 to 35%. Air-fired baseline tests were also conducted, with an excess O_2 level of about 5%, using a bituminous coal and a 93% $CaCO_3$ calcitic limestone (Fang *et al.*, 2007). Tests appear to have been carried out at bed temperatures in the range of about 830–970°C, which would mean that sulphur capture occurred via direct sulphation at the lowest temperature and indirect sulphation at the highest temperature of operation.

These workers reported very little difficulty in operating their unit, and reported a slight increase in combustion efficiency with increasing oxygen concentration. NO_x levels also increased with increasing O_2 in the base of the combustor, but again as with other workers they found that NO_x levels were, if anything, lower for oxy-fuel combustion. Unlike the VTT results, however, they found that sulphur capture was slightly lower for oxy-fired combustion, ranging from about 88 to 90% for a Ca/S molar ratio of 2.8, i.e., with Ca utilizations in the range of about 30% compared with 35%. Somewhat surprisingly these workers observed that sulphur

Table 5.3 Comparative sulphur capture

Test parameter				
Foster Wheeler (Polish coal) – Test Series 1				
Bed temp. (°C)	760	787	799	875
CO_2 (% dry)	92.5	94.5	91.8	91
Ca util. (%)	66.1	75.9	62.7	87.1
Alstom (coal tests at 900°C)				
Ca util. (%)		36–48		
Alstom (petroleum coke tests at 900°C)				
Ca util. (%)		47–49		
Foster Wheeler (Polish coal) – Test Series 2				
Bed temp. (°C)	762	787	799*	875
Ca util. (%)	87	91	60	94
Foster Wheeler (South African coal) – Test Series 2				
Bed temp. (°C)	760	–	819	–
Ca util. (%)	52		53	

* Tests carried out without flue gas recycle.

capture improved with increasing bed temperature for both air and oxygen firing, the former result being atypical of usual experience with CFBC combustion.

These workers reference other, earlier, work on oxy-fired CFBC, and incorrectly identify the CanmetENERGY unit as a batch reactor and note that the earliest work that they are aware of is on a bubbling bed unit firing O_2/CO_2 mixtures reported in 1998 (Hirma *et al.*, 1998), in which very dramatic decreases in NO_x levels were found in tests, but such results are evidently not directly applicable to CFBC application. Overall, with the exception of the results on sulphur capture, these workers appear to be in broad agreement with work done elsewhere.

Finally, Polish workers have also done limited oxy-fuel tests using a 50 mm diameter, 1.7 m high CFB reactor (Czakiert *et al.*, 2006). However, these experiments were done with gas mixtures supplied by bottled gases, and were also performed in batch mode, which means that the data in this paper cannot easily be compared with the other studies discussed here.

5.4 CanmetENERGY tests

The CanmetENERGY test results will be described at length here, as results for these tests have been presented in much greater detail in the open literature than for other facilities, operated with flue gas recycle. This unit was first described in 2006, for the tests carried out by Hughes *et al.* (2006). However, that initial work was marred by the fact that the unit still had significant leaks and high CO_2 levels were not produced in the off-gases. Completely successful results from the rig were first presented in 2007 (Jia *et al.*, 2007).

CanmetENERGY's mini-CFBC contains a 0.1 m inside diameter stainless steel riser (Fig. 5.1) covered with 100 mm of insulation. Independent feed augers can supply multiple fuel types and a sorbent, with solid fuel feed rates of up to 15 kg/h. Oxygen, CO_2 and recycled flue gas flow rates are controlled by a combination of mass flow controllers and rotameters. Bed temperature is in the range of 750–950°C. Superficial gas velocity is up to 8 m/s, although the unit is more normally operated at around 4 m/s. The mini-CFBC has been extensively modified for oxy-fuel CFB combustion. Modifications include addition of an oxygen supply line and flue gas recycle train.

Initially, Eastern bituminous (EB) coal and Highvale coal (a sub-bituminous coal) were used. Table 5.4 gives the analysis of the fuels, which were crushed to –5 mm. For sulphur capture tests, Havelock limestone was used when EB coal was fired in the initial test series. Table 5.5 gives limestone analysis. The particle size of the limestone was in the range of 0–0.5 mm. In all cases, sand of size 0.15–0.35 mm was used as initial bed material. Subsequently, the work was extended to use both petroleum coke and Kentucky bituminous coal, and a Polish limestone was also used to confirm this initial work (Jia *et al.*, 2010).

The mini-CFBC was started in air-firing mode. When the bed temperature reached the desired level the air supply was shut off, the flue gas recycle blower

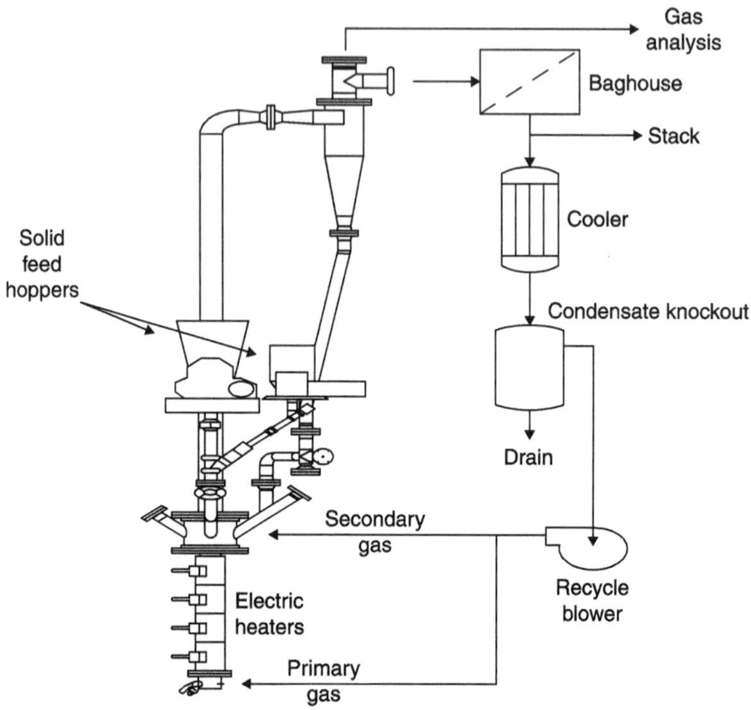

5.1 CanmetENERGY's minibed oxy-fired CFBC.

Table 5.4 Analysis of fuels

	Eastern bituminous	Kentucky coal	Highvale	Petroleum coke
Proximate analysis, wt% (dry)				
Moisture, wt% (as analyzed)	1.08	2.01	10.39	0.66
Ash	8.86	11.31	19.17	1.00
Volatile matter	35.78	37.35	33.76	11.46
Fixed carbon	55.56	51.34	47.076	86.97
Ultimate analysis, wt% (dry)				
Carbon	77.81	74.05	59.78	86.91
Hydrogen	5.05	5.06	3.49	3.22
Nitrogen	1.49	1.62	0.79	1.83
Sulphur	0.95	1.56	0.22	5.88
Ash	8.86	11.31	19.17	1.00
Oxygen (by difference)	6.04	6.40	16.58	1.16
Heating value (MJ/kg)	32.51	30.93	23.27	34.71

Table 5.5 Analysis of limestones (wt%)

	Havelock	Czatkowice
CaO	53.99	54.32
MgO	0.59	0.48
SiO_2	1.23	
Al_2O_3	<0.38	0.04–0.09
Fe_2O_3	<0.55	0.08–0.11
Na_2O	<0.17	0.01–0.03
K_2O	<0.08	0.02–0.04
MnO	0.08	
TiO_2	<0.04	
Cr_2O_3	<0.01	
P_2O_5	<0.02	
SO_3	0.20	0.08
V_2O_5	<0.02	
SrO	0.02	
BaO	0.02	
NiO	0.01	
Loss on fusion	43.34	42.70
Sum	99.48	

was turned on and oxygen was then supplied to the mini-CFBC. The transition from the air-firing mode to the oxy-fuel-firing mode occurred easily and took only a few minutes. The mini-CFBC was operated at nominal bed temperature of 850°C. Superficial gas velocity was about 4 m/s. Global oxygen concentration during oxy-fuel combustion periods was ~28–30%.

The first fully successful combustion tests carried out used Highvale coal, since it is a low-sulphur fuel (0.22%) and does not require limestone addition. Experimental results during periods of stable operation under air and oxy-fuel firing are given in Table 5.6. Table 5.6 shows that, during oxy-fuel firing, temperatures across the mini-CFBC were very similar to those seen during air firing. Gas velocities in the dense bed zone and in the riser were slightly lower under oxy-fuel firing conditions. Oxygen concentrations in the primary gas and secondary gas were 34.0% and 67.1% by volume. Global oxygen concentration was 28.8%. Primary and secondary combustion gas fractions were 0.684 and 0.056. The rest of the gas was supplied through the return leg of the CFBC and the coal feed port. Errors associated with these estimates were about 5–10%. The flue gas recirculation ratio was estimated at 58.03 ± 3.3%. One of the reasons for the low recycle ratio was the fact that in the mini-CFBC the solid particles were cooled in the return leg (all insulation material was removed from the return leg for these runs). The temperature of the solids entering the combustor from the return leg was 514 ± 4°C. If an external solid heat exchanger were available to cool the solids to a lower temperature, then flue gas recirculation

Table 5.6 Oxy-fuel mini-CFB combustion tests with Highvale and Eastern bituminous coals

	Highvale coal		Eastern bituminous coal	
	Air firing	Oxy-fuel	Air firing	Oxy-fuel
O_2 (%)	4.95	6.67	3.66	4.55
CO_2 (%)	13.4	80.5	13.9	83.5
CO (ppm)	53	39.4	180	195
SO_2 (ppm)	169	100	295	1262
NO_x (ppm)	123	2571	188	214
Fuel mixture feed rate (kg/h)	4.0	4.0	3.5	4.0 4
Gas velocity in the bed (m/s)	2.94	2.60	2.74	2.33
Gas velocity in the riser (m/s)	3.72	3.56	3.72	3.37
Average bed temperature (°C)	849	845	848	851
Average riser temperature (°C)	824	823	827	868
Average cyclone temperature (°C)	651	722	756	770
Average baghouse inlet temperature (°C)	227	219	258	216
Ca/S molar ratio	–	–	2.0	2.0
Sulphur capture efficiency (%)	–	–	68.4	40.1
Fuel N to NO_x (%)	6.61	6.64	5.61	2.83
Recycle ratio (%)	–	58.03	–	53.05

ratio could be further reduced. During stable periods of oxy-fuel CFBC firing with flue gas recirculation, the average CO_2 concentration in the flue gas was 80.5 ± 1.3%.

The combustion of Highvale coal was excellent under oxy-fuel firing conditions and a change from air-fired mode to oxy-fired mode was accomplished in 20 min or less as shown in Fig. 5.2. The CO concentration of 39 ppm in the flue gas also provides a good indication of the excellent combustion performance, and was slightly lower than that achieved by air firing. Since cyclone temperature has a powerful effect on CO concentration in CanmetENERGY's mini-CFBC, as shown in Fig. 5.3, higher average cyclone temperatures in the oxy-fuel firing period are most likely responsible for the observed decrease. Flue gas recirculation had no effect on CO concentrations under oxy-fuel CFB combustion conditions.

NO_x concentration in the flue gas during the oxy-fuel firing period was twice as high as under the air-firing mode for Highvale, but only slightly higher for the EB coal. Since temperatures across the mini-CFBC were very similar, the increase appears to be the result of flue gas recirculation. There are no data on oxy-fuel CFB combustion with flue gas recirculation in the open literature. As a result, it is not possible to evaluate the CanmetENERGY results with comparable test data reported elsewhere, but there are some studies without flue gas recycle, as noted above. Thus, Nsakala *et al.* (2004) conducted CFB combustion tests

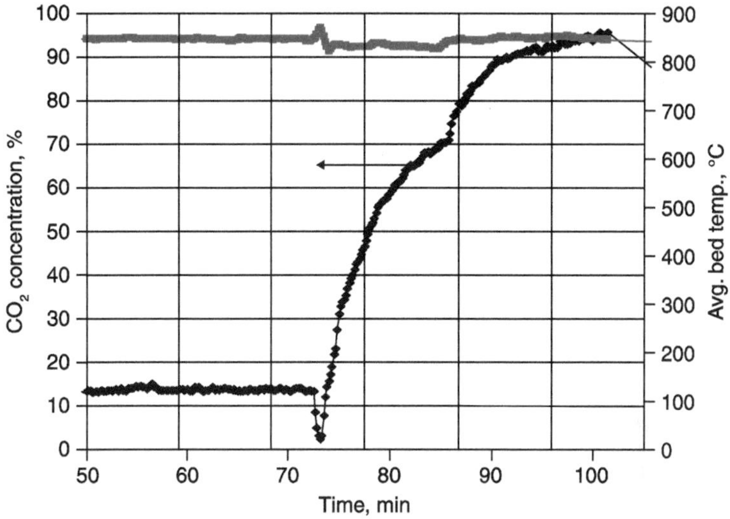

5.2 Transition from air firing to oxy-fuel firing in CanmetENERGY's mini-CFBC during second Highvale coal test.

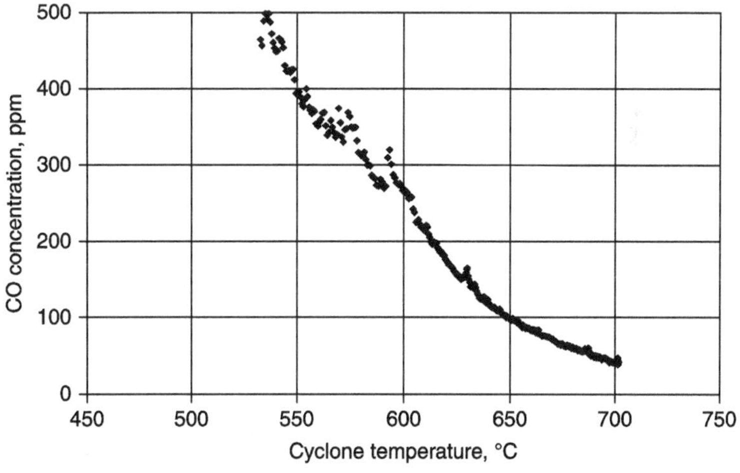

5.3 Effect of cyclone temperature on CO concentration, Highvale coal.

with O_2 and CO_2 mixtures in a 1.02 m diameter CFBC firing medium volatile bituminous coal. They noted that at about 30% O_2 in the combustion gas, NO_x concentrations in the flue gas were 55 to 72 ppm, which was significantly lower than for air firing of the same coal in their tests (220 ppm). These NO_x results with O_2/CO_2 mixtures were lower than those from the CanmetENERGY study. Since these measurements were made without flue gas recycle, this suggests that flue gas recirculation resulted in higher NO_x levels. In a small-scale CFBC study Czakiert *et al.* (2006) have also suggested that oxy-fuel CFB combustion results in a significant reduction of fuel nitrogen conversion to NO_x. However, this study, which was carried out with brown coal, again did not use recirculated flue gas mixtures.

Tan *et al.* (2006) reported oxy-fuel pulverized fuel combustion with flue gas recirculation in a 0.3 MWt test facility. For Highvale coal firing (the same coal as used in the CFBC work), results indicated that NO_x concentration increased to 1183 ppm from 707 ppm in the air-firing mode. The magnitude of increase in NO_x concentration was lower than in the current study (67% increase compared with 200%), albeit the NO_x level from the PC combustor was much higher than for CFBC operation. To verify the oxy-fuel firing results, a repeat test was conducted with Highvale coal. The average CO_2 concentration in the flue gas during the stable oxy-fuel firing period was slightly higher at 88.7 ± 4.7%. All operating conditions and concentrations of gaseous species were similar to the initial oxy-fuel CFBC test. However, it should be noted that, although NO_x concentration was higher, the total amount of NO_x emitted from the combustor was about the same because the amount of flue gas discharged from the oxy-fuel-fired unit is less than that for air firing. Thus fuel nitrogen-to-NO_x ratio remained essentially unchanged (6.61% for air firing and 6.64% for oxy-fuel firing with flue gas recirculation). And it seems safe to conclude that oxy-fuel CFBC firing is comparable to or better than air firing for NO_x emissions.

To assess sulphur capture performance under oxy-fuel CFBC firing conditions, EB coal was burned with Havelock limestone addition at a Ca/S molar ratio of 2. Table 5.6 gives the experimental results for these tests. Temperatures in the riser were higher during oxy-fuel firing compared with air firing by about 40°C. However, the average bed temperature was essentially the same for the two modes of combustion. Superficial gas velocities were slightly lower during the oxy-fuel firing period (~10%). The oxygen concentrations in primary and secondary combustion gases were 40.1% and 73.1%, respectively. Primary and secondary gas fractions were 0.588 and 0.062. The global oxygen concentration was 32.9%, higher than in the tests performed with Highvale coal, which may explain the higher combustor temperatures measured in the EB coal tests.

The flue gas recycle ratio was very similar to that of the Highvale coal tests, at 55.03%. The average temperature of the solids entering the CFBC riser from the return leg was 439 ± 15°C. These results confirmed that the recycle ratio for oxy-fuel CFBC can be much lower than in oxy-fuel pulverized fuel combustion, which represents a major advantage of CFBC, allowing one to reduce operational cost, improve cycle efficiency and reduce the amount of CO_2 generated per unit electric power produced.

Concentrations of CO were basically the same during air firing and oxy-fuel firing. The trend was the same as in the Highvale coal tests, although absolute concentrations were higher at 180 to 195 ppm. NO_x concentration was only slightly higher during the oxy-fuel firing period. The different behaviour in NO_x concentration variation compared with Highvale coal tests (Table 5.6) was likely the result of the characteristics of the coal. In this case, the ratio of fuel nitrogen converted to NO_x was decreased by almost 50%.

Sulphur capture efficiency was 68.4% during air firing. During the oxy-fuel firing period, the sulphur capture efficiency decreased to 40.1%. These values were lower than expected as up to 90% sulphur capture is typically achievable in utility/industrial CFBC units with Ca/S ratios of 2 to 2.5 (Anthony and Granatstein, 2001). However, the current tests were short runs (7–8 h), and steady state of ash composition in the mini-CFBC had not been achieved. These results were problematic since VTT and Foster Wheeler workers found that very high Ca utilizations could be achieved in their CFBC with full flue gas recycle (Eriksson *et al.*, 2007, Hotta *et al.*, 2008). Alstom workers, who ran their equipment with bottled gases, also found that, with a reactive limestone, sulphur capture ranged from 72 to 96% (or 36 to 48% Ca utilization) compared with an expected 95% with air firing for bituminous coal, while with petroleum coke captures were in the range of 94 to 99%, which for a Ca/S molar ratio of 2 represents good performance.

To check that the limestone remained fully calcined, CFBC bed ash generated during the oxy-fuel firing period was subjected to thermogravimetric analysis (TGA) (Fig. 5.4). It is clear from the TGA results that added limestone did not calcine in the CFBC. Therefore, sulphur capture was via direct sulphation of $CaCO_3$.

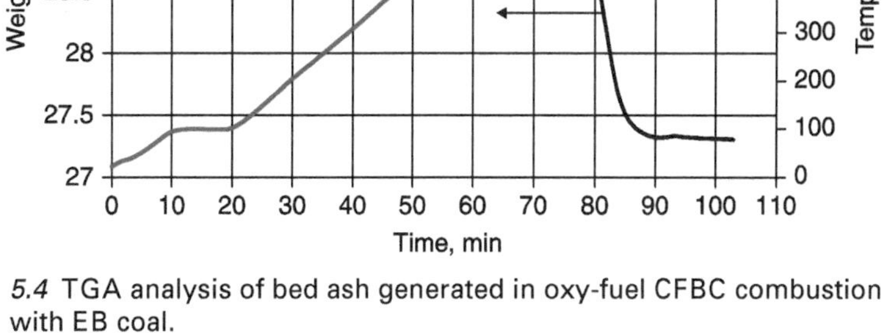

5.4 TGA analysis of bed ash generated in oxy-fuel CFBC combustion with EB coal.

5.5 Longer duration sulphation tests

To attempt to resolve these discrepancies on sulphur capture in particular it was decided to do longer duration runs with an additional coal (a Kentucky bituminous coal) and a petroleum coke, with tests lasting 12 h or more. For these runs Ca/S molar ratios were maintained at a value of 3, and the temperature range studied was from 850°C to 950°C, i.e., to allow a change from direct to indirect sulphation. The test results are presented in Tables 5.7 and 5.8 and show essentially the same features as seen before. NO_x levels were similar to air firing for the coal runs, and the sulphur capture was worse than for air firing.

Very similar results were obtained with Kentucky coal (Table 5.8), although these results showed that NO_x emissions were sensitive to combustion temperatures. However, the results are still well within the range of typical CFBC operation and the lower fuel NO_x conversions may represent the benefits of the longer runs. The performance for petroleum coke over time is shown graphically in Fig. 5.5.

Interestingly, the concentration of SO_2 decreased significantly when the bed temperature increased from 850°C to 950°C. This can be seen more clearly in Fig. 5.6, which shows that as the bed temperature increased, SO_2 concentration quickly decreased from about 6000 ppm at 850°C to about 2000 ppm at 950°C, a reduction of more than 65%. For the Kentucky coal, the test results are shown in Fig. 5.7.

We can see that, as bed temperature started to increase from 850°C to 950°C (at about 300 min elapsed time), the major species exhibited significant fluctuations in concentrations. Although similar fluctuations were observed for EB coal, in this case the fluctuations were very pronounced, especially after 350 min when bed temperature rose above 930°C. Fluctuations in both the bed temperature and SO_2 concentration were too large to draw any conclusion concerning the effect of increasing temperature on SO_2 emissions. During the coal tests, it was observed that these fluctuations seemed to be closely correlated to the limestone calcination temperature. As bed temperature approached 930°C, there was a sudden drop in the lower bed pressure accompanied by poor solid recirculation, which led to a rapid rise in bed temperature. In the case of petroleum coke, this effect was not as obvious.

In terms of actual sulphur capture for petroleum coke the sulphur capture was about 55% at less than 900°C, but increased to 85% as the bed temperature increased above 930°C. For the coal the sulphur capture was around 50–60% regardless of temperature.

5.5.1 Sulphation issues

At this point there is clearly considerable disagreement in the literature as to sulphation behaviour under oxy-firing conditions. For direct sulphation conditions at least, the literature on pressurized fluidized bed (PFB) combustion, where sulphation normally occurs with $CaCO_3$, should provide some guidelines, although

Table 5.7 Experimental results for Eastern bituminous coal and petroleum coke

	Eastern bituminous coal			Petroleum coke		
	Air firing Czatkowice (7 Nov.)	Oxy-fuel Czatkowice (7 Nov.)	Oxy-fuel Havelock (14 Nov.)	Oxy-fuel Czatkowice (1 Oct.)	Oxy-fuel Czatkowice (1 Oct.)	Oxy-fuel Havelock (25 June)
O_2 (%)	3.04	3.72	3.47	4.74	3.64	3.28
CO_2 (%)	14.7	91.1	92.9	84.40	86.3	92.3
CO (ppm)	47.1	186	83.0	94.7	25.1	30.5
SO_2 (ppm)	281	1165	704	5866	2142	2386
NO_x (ppm)	245	210	170	394	443	387
Fuel mixture feed rate (kg/h)	3.5	4.5	4.5	6.25	6.5	6.0
Gas velocity in the bed (m/s)	1.70	2.25	2.59	2.25	3.25	2.73
Gas velocity in the riser (m/s)	3.23	3	3.64	3.17	4.20	3.73
Average bed temperature (°C)	836	838	828	842	930	961
Average riser temperature (°C)	902	860	853	843	946	950
Average cyclone temperature (°C)	854	808	842	813	884	896
Average return leg temperature (°C)	280	268	320	612	471	612
Average baghouse inlet temperature (°C)	374	330	449	256	314	288
Ca/S molar ratio	3	3	3	3	3	2
Sulphur capture efficiency (%)	74.4	65.0	77.1	64.5	87.3	85.6
Fuel N to NO_x (%)	6.22	1.76	1.54	3.35	3.69	3.28
Recycle ratio (%)	–	59.7	61.0	54.6	62.5	54.28
O_2 in primary gas (%)	20.9	50.8	44.0	58.3	44.1	60.5
O_2 in secondary gas (%)	20.9	3.72	60.9	69.1	75.0	–
Global O_2 (%)	20.9	34.9	34.2	44.2	36.2	43.7

Table 5.8 Experimental results for Kentucky coal

	Kentucky coal and Czatkowice limestone		Kentucky coal and Havelock limestone	
	Air firing (15 Oct.)	Oxy-fuel (15 Oct.)	Air firing (1 June)	Oxy-fuel (10 Oct.)
O_2 (%)	2.88	3.78	4.05	4.49
CO_2 (%)	14.8	87.5	13.5	84.6
CO (ppm)	93.3	93.7	215	56.1
SO_2 (ppm)	268	1176	418	1504
NO_x (ppm)	184	156	206	466
Fuel mixture feed rate (kg/h)	4.75	5.00	4.0	6.0
Gas velocity in the bed (m/s)	2.23	2.46	2.14	2.56
Gas velocity in the riser (m/s)	3.75	3.64	3.30	3.81
Average bed temperature (°C)	820	838	832	921
Average riser temperature (°C)	842	852	846	952
Average cyclone temperature (°C)	834 ±	824	785	897
Average baghouse inlet temperature (°C)	257	277	522	557
Average return leg temperature (°C)	575	584 ± 7	236	302
Ca/S molar ratio	3	3	3	3
Sulphur capture efficiency (%)	86.6	67.8	78.2	72.0
Fuel N to NO_x (%)	3.78	1.14	4.53	3.51
Recycle ratio (%)	–	60.4	–	49.11
O_2 in primary gas (%)	20.9	47.4	20.9	66.06
O_2 in secondary gas (%)	20.9	66.1	20.9	32.4
Global O_2 (%)	20.9	34.1	20.9	45.00

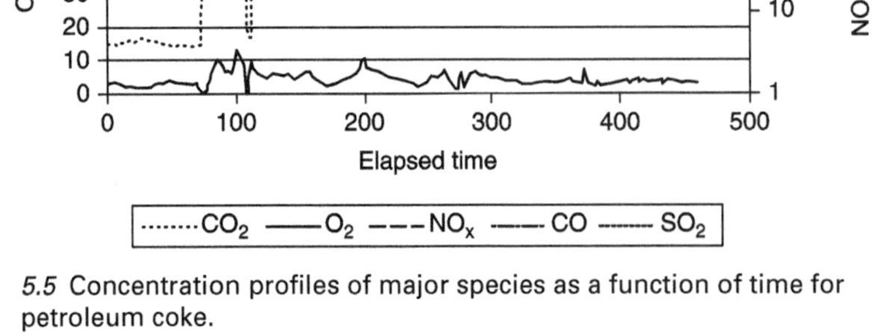

5.5 Concentration profiles of major species as a function of time for petroleum coke.

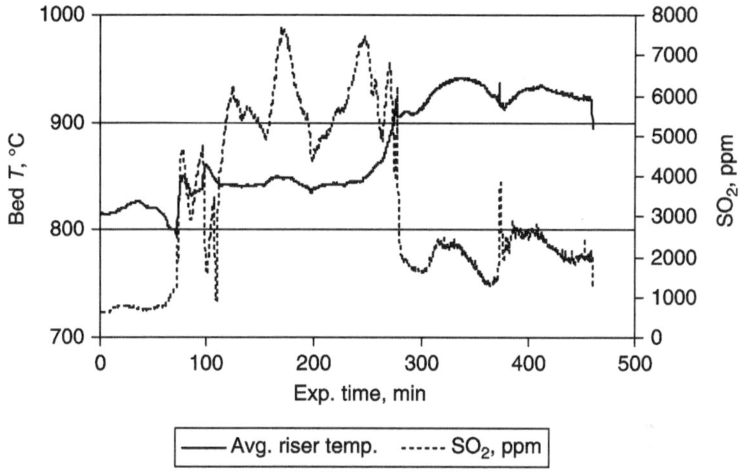

5.6 Profiles of SO_2 concentration and average bed temperature for petroleum coke.

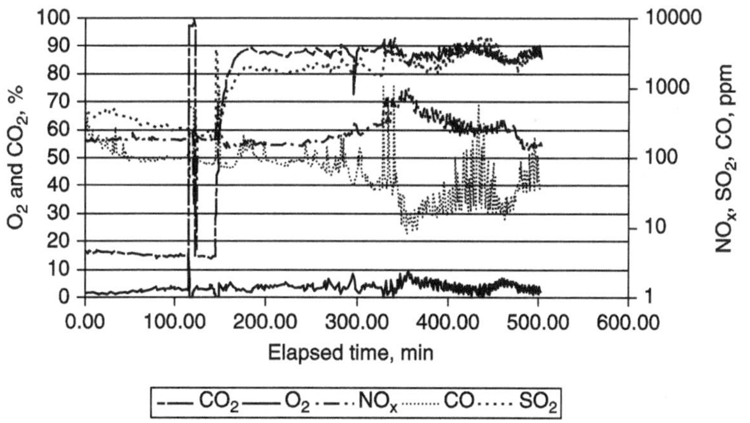

5.7 Concentration profiles of major species as a function of time for Kentucky coal.

even there it is clear that the mechanisms for direct sulphation are still not completely understood (Hu *et al.*, 2007). Very early work suggested that direct sulphation might be more effective than indirect sulphation (Hajaligol *et al.*, 1988; Snow *et al.*, 1988) and an explanation was given that counter-diffusion of CO_2 allowed improved sorbent performance. Similar results were produced much later by Liu *et al.* (2000), who studied limestone sulphation using two limestones at atmospheric pressures with different concentrations of CO_2 in the sulphation gases in a fixed bed reactor. Interesting findings from that study are that indirect sulphation is faster at low degrees of Ca conversion (<0.3), but thereafter direct sulphation becomes faster; sintering was much reduced in direct sulphation so that at higher conversions direct sulphation was faster than for indirect sulphation. Very similar arguments have been put forward recently by Chen *et al.* (2009), who again note the earlier literature concerning almost complete conversion of $CaCO_3$, and mention the back-diffusion by CO_2 argument. Like Liu and his co-workers they find that direct sulphation becomes faster at fractional conversions of 0.3 for limestone samples sulphated in a TGA. In this context it is worth noting that while Hu and his colleagues agree with the observation that the sulphate layer is porous, they argue that this is due not to back-diffusion of CO_2, but instead to the way the product crystal grains form on the calcite surface and to local porosity (Hu *et al.*, 2007, 2008). Finally, to add a note of caution, it is worth remarking that results from pilot and commercial PFBC units tend to show comparable performance to air-fired FBC (Cuenca and Anthony, 1995) and not the superior performance reported from fixed beds and TGA tests for which most of the studies have been done.

The work of Hu *et al.* (2006, 2007, 2008) notes that both high partial pressures of CO_2 and H_2O might cause significant sintering of the $CaCO_3$, thus impairing overall sulphation performance. In a TGA study done in a fixed bed laboratory reactor at temperatures ranging from 500 to 700°C, a concentration of 7.5% H_2O produced an enhancement of sulphation behaviour. In practice one would always expect H_2O to be present in a pilot-scale combustor at levels greater than that so one might expect that this would explain enhancement of sulphation, but such an enhancement is not apparently evident in Table 5.2, where water concentrations vary from 5.8 to almost 19%, nor indeed reflected in the lower sulphation levels seen for the CanmetENERGY work, where sulphur capture is better for air-fired combustion. However, it may be that there is a threshold value of water in all coal combustion systems such that they are, in reality, insensitive to the changes in water concentrations produced by flue gas recycle. Whatever the explanation it is clear that the sulphation process in fluidized beds requires much further work.

5.5.2 Carbonation issues

As noted earlier there are concerns about carbonation of CaO for high-temperature oxy-fuel CFBC processes, with the possibility of carbonate fouling occurring. To study this issue a TGA study was performed at temperatures ranging from 250 to 800°C, with CO_2 concentrations set at 80% and H_2O concentrations of 0, 8 and 15% (Wang *et al.*, 2008). This study noted that no noticeable carbonation could occur without water when the temperature was below 400°C, but was able to occur down to a temperature of 250°C if water was present. Much more surprising was the fact that H_2O made a significant difference in the carbonation even at temperatures as high as 700°C, well above the temperature for which $Ca(OH)_2$ would be expected to be thermodynamically stable, with that effect becoming more important as the temperature fell.

5.6 Large pilot-scale and demonstration projects

Currently, CanmetENERGY is conducting oxy-fuel tests for Foster Wheeler to generate data for a large-scale Foster Wheeler demonstration project. Figure 5.8 shows the CO_2 levels produced for two days of operation in oxy-fuel mode as part of a week-long test using CanmetENERGY's 0.8 MWt CFBC unit. While results from these tests are not yet available in the open literature, they have so far indicated that operation in the oxy-fuel mode for extended times appears to pose few or no problems.

Foster Wheeler is using this and other test work carried out by its partners, including VTT and Lappeenranta University of Technology, Finland, to develop Flexi-Burn™ CFBC technology which would permit a boiler to operate in either normal air-firing mode or oxy-fuel mode for carbon capture and storage. In addition, Foster Wheeler will also offer the possibility of retrofitting existing plants to operate as oxy-fired CFBC boilers. This is a potentially important development since it offers the opportunity of meeting the anticipated market for carbon capture-ready plants, which are a requirement likely to be increasingly demanded worldwide (International Energy Agency, 2007).

Foster Wheeler is the first company to commercialize supercritical CFBC technology (Lagisza power plant, Poland). Using this technology as the basis, it is now working with Spain's largest utility power company, Endesa Generación, on the development of a 300–500 MWe supercritical Flexi-Burn™ Oxy-CFBC design for CCS (Hack *et al.*, 2009). This technology would have a capture rate of 90% of CO_2 emissions and it anticipates it could be available by 2015 (Eriksson *et al.*, 2009). The expected configuration of such a power plant is given in Fig. 5.9. Ultimately, Foster Wheeler believes that it could offer such technology at the 600–800 MWe size with 600°C steam temperature.

Following extensive testing to be completed in Canada at CanmetENERGY, demonstration tests are scheduled to start in 2011 at the 30 MWt CIUDEN pilot

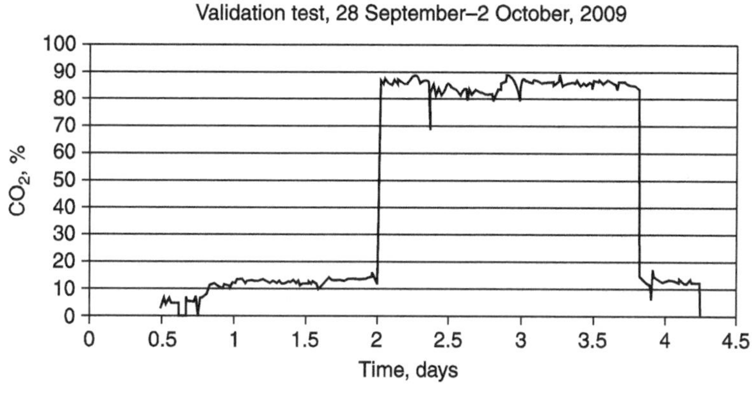

5.8 Tests on oxy-fired combustion in CanmetENERGY's 0.8 MWt CFBC.

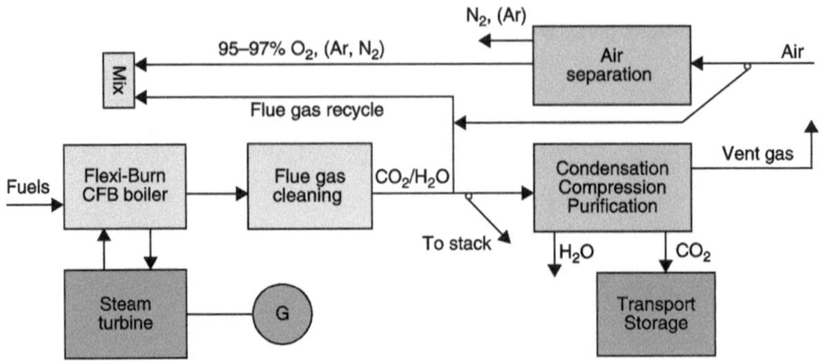

5.9 Schematic of a Flexi-Burn™ CFB power plant.

CFBC facilities in Spain, which will provide a full experimental CCS platform for the demonstration and validation of the flexible air/oxy-fuel CFBC combustion. Finally, it should be noted that Alstom has also announced its intention of carrying out a 100 MWe oxy-fuel CFBC demonstration, although at the time of writing no further information appears to be available in the open literature (Suraniti *et al.*, 2009). Overall, it is apparent that oxy-fired CFBC technology is making major strides to enter the commercial arena, and there seems little doubt that before the end of the decade it will be available as a commercial and competitive CCS technology.

5.7 References

Anthony, E.J., Granatstein, D.L. (2001) Sulfation Phenomena in Fluidized Bed Combustion Systems, *Progress in Energy and Combustion Science* **27**, 215–236.

Anthony, E.J., Jia, L., Laursen, K. (2001) Agglomeration of High-sulfur Fuels, *Canadian Journal of Chemical Engineering* **79**, 356–366.

Blamey, J., Anthony, E.J., Wang, J., Fennell, P. (2010) The Calcium Looping Cycle for Large-scale CO_2 Capture, *Progress in Energy and Combustion Science* **36**, 260–279.

Buhre, B.J.P., Elliot, L.K., Sheng, C.D, Gupta, R.P., Wall T.F. (2005) Oxy-fuel Combustion Technology for Coal-fired Power Generation, *Progress in Energy and Combustion Science* **31**, 283–307.

Chen, C., Zhao, C., Liu, S., Wang, C. (2009) Direct Sulphation of Limestone Based on Oxy-fuel Combustion Technology, *Environmental Engineering* **22**, 1481–1488.

Cuenca, M.A., Anthony, E.J. (eds.) (1995) *Pressurized Fluidized Beds*, Blackie Academic and Professional, London, UK.

Czakiert, T., Bis, Z., Muskala, W., Nowak, W. (2006) Fuel Conversion from Oxy-fuel Combustion in a Circulating Fluidized Bed, *Fuel Processing Technology* **87**, 531–538.

Eriksson, T., Sippu, O., Hotta, A., Myohanen, K., Hyppanen, T., Pikkarainend T. (2007) *Oxyfuel CFB Boiler as a Route to Near Zero CO_2 Emission Coal Firing*, Power Generation Europe, Madrid, Spain, 26–28 June 2007.

Eriksson, T., Sippu, O., Hotta, A., Fan, Z., Ruis, J.A., Sacristán, A.S.-B., Jubitero, J.M., Ballesteros, J.C., Shah, M., Prosser, N., Haley, J., Guidici, R. (2009) *Development of*

Flexi-Burn™ CFB Technology Aiming at Fully Integrated CCS Demonstration, Power Generation Europe, Cologne, Germany, 26–28 May 2009.

Fang, M., Yang, L., Mao, Y, Luo, Z., Cen, K. (2007) *Experimental Study on O_2/CO_2 Combustion in a CFB Test Facility*, 2007 International Conference on Coal Science and Technology, University of Nottingham, UK, 28–31 August 2007.

Grace, J.R., Avidan, A.A., Knowlton, T.M. (eds.) (1997) *Circulating Fluidized Beds*, Blackie Academic and Professional, London, UK.

Hack, H., Fan, Z., Seltzer, A., Robertson, A. (2009) *Pathway to Supercritical Flexi-Burn™ CFB Power Plant to Address the Challenge of Climate Change*, 2009 International Pittsburgh Coal Conference, Pittsburgh, PA, USA, 21–24 September 2009.

Hajaligol, M.R., Longwell, J.P., Sarofim, A.F. (1988) Analysis and modeling of the direct sulfation of $CaCO_3$, *Industrial and Engineering Chemistry Research* **27**(12), 2203–2210.

Hirma, T., Hosoda, H., Azuma, N., *et al.* (1998) *Drastic Reduction of NOx and N_2O Emissions from BFBC of Coal by Means of CO_2/O_2 Combustion: Effects of Fuel Gas Recycle and Coal Type (C)*, International Symposium of Engineering Foundation Fluidization IX, USA, 1998.

Hotta, A., Nuorimo, K., Eriksson, T., Palonen, J., Kokki, S. (2008) *CFB Technology Provides Solutions to Combat Climate Change*, Proceedings of the 9th International Conference on Circulating Fluidized Beds, in conjunction with the 45th International VGB Workshop, 'Operating Experience with Fluidized Bed Systems', Werther, J., Nowak, W., Wirth, K-E., Hartge, E-U. (eds.), Hamburg, Germany, 13–16 May 2008, pp. 11–17.

Hu, G., Dam-Johansen, K., Wedel, S., Hansen, J.P. (2006) Review of the Direct Sulfation of Limestone, *Progress in Energy and Combustion Science* **32**, 386–407.

Hu, G., Dam-Johansen, K., Wedel, S. (2007) Direct Sulphation of Limestone, *AIChE Journal* **53**, 948–960.

Hu, G., Shang, L, Dam-Johansen, K., Wedel, S., Hansen, J.P. (2008) Indirect Kinetics of the Direct Sulphation of Limestone, *AIChE Journal* **54**, 2663–2673.

Hughes, R., Jia, L., Tan Y., Anthony, E.J. (2006) *Oxy-Fuel Combustion of Coal in a Circulating Fluidized Bed Combustor*, Proceedings of the 19th International Conference on FBC, Vienna, Austria, 2006.

International Energy Agency (2007) *CO_2 Capture Ready Plants*, Technical Study, Report No. 2007/4, May 2007.

Jia, L., Tan, Y., Wang, C., Anthony, E.J. (2007) Experimental Study of Oxy-Fuel Combustion and Sulphur Capture in a Mini CFBC, *Energy and Fuels* **21**, 3160–3164.

Jia, L., Tan, Y., Anthony, E.J. (2010) Emissions of SO_2 and NOx during Oxy-fuel CFB Combustion Tests in a Mini-CFBC, *Energy and Fuels* **24**, 910–915.

Liljedahl, G.N., Turek, D.G., Nsakala, N.Y., Mohn, N.C, Fout T.E. (2006) *Alstom's Oxygen-fired CFB Technology Development Status for CO_2 Mitigation*, 31st International Technical Conference on Coal Utilization and Fuel Systems, Clearwater, Florida, USA, 21–25 May 2006.

Liu, H., Katagiri, S., Kaneko, U., Okazaki, K. (2000) Sulfation Behaviour of Limestone under High CO_2 Concentrations in O_2/CO_2 Coal Combustion, *Fuel* **79**, 945–953.

Mohr, S.H., Evans, G.M. (2009) Forecasting Coal Production until 2100, *Fuel* **88**, 2059–2067.

Nsakala, N., Liljedahl, G., Turek, D. (2004) *Greenhouse Gas Emissions Control by Oxygen Firing in Circulating Fluidized Bed Boilers: Phase II – Pilot Scale Testing and Updated Performance and Economics for Oxygen Fired CFB*, PPL Report No. PPL-04-CT-25 under cooperative agreement No. DE-FC26–01NT41146.

Snow, M.J.H., Longwell, J.P., Sarofim, A.F. (1988) Direct Sulfation of Calcium Carbonate, *Industrial and Engineering Chemistry Research* **27**, 268–273.

Stamatelopoulos, G.N., Darling, S. (2008) *Alstom's CFBC Technology*, Proceedings of the 9th International Conference on Circulating Fluidized Beds, in conjunction with the 45th International VGB Workshop, 'Operating Experience with Fluidized Bed Systems', Werther, J., Nowak, W., Wirth, K-E., Hartge, E-U. (eds.), Hamburg, Germany, 13–16 May 2008, pp. 3–9.

Sun, P., Grace, J.R., Lim, C.J., Anthony, E.J. (2007) Removal of CO_2 by Ca-Based Sorbents in the Presence of SO_2, *Energy and Fuels* **21**, 163–170.

Suraniti, S.L., Nsakala, N.Y., Darling, S.L. (2009) Alstom Oxyfuel CFB Boilers: A Promising Option for CO_2 Capture, *Energy Procedia* **1**, 543–548.

Symonds, R., Lu, D., Hughes, R., Anthony, E.J., Macchi, A. (2009) CO_2 Capture from Simulated Syngas *via* Cyclic Carbonation/Calcination for a Naturally Occurring Limestone: Pilot Plant Testing, *Industrial and Engineering Chemistry Research* **48**, 8431–8440.

Tan, Y., Croiset, E., Douglas, M., Thambimuthu, K. (2006) Combustion Characteristics of Coal in a Mixture of Oxygen and Recycled Flue Gas, *Fuel* **85**, 507–512.

Toftegaard, M.B., Brix, J., Jensen, P.A., Glaborg, P., Jensen, A.D. (2010) Oxy-fuel Combustion of Solid Fuels, *Progress in Energy and Combustion Science*, in press.

Wang, C., Jia, L., Tan, Y., Anthony, E.J. (2008) Carbonation of Fly Ash in Oxy-Fuel CFB Combustion, *Fuel* **87**, 1108–1114.

Yaverbaum, L. (1977) *Fluidized Bed Combustion of Coal and Waste Materials*, Noyes Data Corp., Park Ridge, New Jersey, USA.

Part II

Oxy-fuel combustion fundamentals

6

Ignition, flame stability, and char combustion in oxy-fuel combustion

C. SHADDIX, Sandia National Laboratories, USA and
A. MOLINA, National University of Colombia, Colombia

Abstract: This chapter discusses the influence of oxy-fuel combustion conditions on coal ignition, flame stability, char combustion, and carbon burnout. Both experimental results and theoretical analysis are employed to illustrate the predominant influences of elevated concentrations of O_2, CO_2, and H_2O on these important aspects of the combustion process. In some instances substantial differences in the combustion behavior are apparent relative to traditional air-fired combustion practices, whereas in other cases the influence is relatively minor. Both an understanding of the governing physics of the processes and detailed simulations lead to an understanding of which aspects of gas transport properties or reactions involving CO_2 and H_2O are likely to be responsible for the observed trends.

Key words: oxy-fuel combustion, coal, char, ignition, burnout.

6.1 Introduction

Oxy-fuel combustion of coal produces various changes in the combustion process itself. These combustion-related effects are manifest in coal ignition and coal char combustion and burnout and thus play important roles in the practical consideration of flame stability and flame shape, burner turndown, boiler heat transfer distribution, waterwall corrosion, and carbon burnout. For new-build applications, the burners and boiler itself can be designed to account for changes in combustion performance in oxy-fuel combustion, whereas for retrofit applications the available options are more tightly constrained by the existing boiler geometry and heat transfer surfaces.

The changes in the oxy-fuel coal combustion process relative to traditional combustion in air are primarily driven by the differences in the chemical composition of the oxidizer streams that are utilized relative to air. In oxy-fuel combustion, the overall burner inlet oxygen concentration can be controlled by the amount of flue gas that is recycled and mixed with the oxygen produced in the air separation unit. In addition, CO_2 concentrations are very high during oxy-fuel combustion. Whereas the chemical effect of CO_2 is more subtle than that of oxygen concentration, the effect of CO_2 on heat capacity and gas transport properties can be significant, and these properties have a strong influence on the combustion process. Similarly, the common use of wet recycle during oxy-fuel combustion, particularly for the secondary 'air' stream of a burner, results in elevated concentrations of steam in the combustion process, with potential impacts on all aspects of coal combustion.

This chapter summarizes the current understanding of the effects of oxy-fuel combustion conditions on the coal combustion process itself. The chapter begins with a discussion of coal ignition and then progresses to the concept of flame stability, emphasizing the strong roles of oxygen concentration, diluent gas heat capacity, and thermal diffusivity on these processes. This is followed by a discussion of char particle combustion and burnout that highlights the roles of oxygen concentration and oxygen diffusivity on these processes. A current topic of widespread interest and debate concerns the contribution of steam-carbon and CO_2-carbon gasification reactions to char particle consumption during oxy-fuel combustion. The available information regarding this topic is summarized before discussing probable directions of future research.

6.2 Coal ignition

The general topic of coal particle ignition has been reviewed by Essenhigh *et al.* (1989), Wall *et al.* (1991), and Annamalai *et al.* (1994), including considerations of homogeneous (gas-phase) versus heterogeneous (particle-surface) ignition and the effects of particle size, bulk oxygen content, coal volatile content, and particle loading. In this chapter we focus on the key considerations of coal particle ignition during oxy-fuel combustion.

6.2.1 Experimental studies

As early as 1997 Kiga *et al.* (Kiga et al., 1997) aimed to answer the question of why there was poor ignition of coal during oxy-fuel combustion with CO_2 recirculation. In the past five years, several experimental and numerical investigations have been conducted on coal ignition during oxy-fuel combustion to get a better answer to the same question. While some studies have been carried out in laboratory-scale equipment that allows well-controlled conditions (Molina and Shaddix, 2007; Arias *et al.*, 2008; Fan *et al.*, 2008, Shaddix and Molina, 2009; Molina *et al.*, 2009), others have focused on the study of coal flames in burners that more closely resemble the conditions in industrial boilers (Liu *et al.*, 2005a, 2005b; Huang *et al.*, 2008; Khare *et al.*, 2008).

The two studies by Molina and Shaddix (2007; Shaddix and Molina, 2009) focused on ignition of isolated, entrained pulverized coal particles of typical US sub-bituminous and high-volatile bituminous coals, size-classified to 75–106 μm and 106–125 μm. Experiments were performed at furnace temperatures of 1250 K and 1700 K in gas environments consisting of oxygen and 14% steam in N_2 or CO_2 balance gas. Oxygen concentrations were varied between 12% and 36%, including an explicit comparison of ignition in 21% O_2 and 30% O_2. The point of ignition of the particles was determined from either charge-coupled device (CCD) camera images of the CH* chemiluminescence from an accumulated ensemble of particles (Molina and Shaddix, 2007) or from statistical analysis of single-particle intensified CCD camera images of the thermal emission from individual particles (Shaddix and Molina, 2009). In the 1700 K study, the ignition time of the particles varied from 8 ± 1 ms in 12% O_2 down to 5 ± 1 ms in 36% O_2, in the N_2 diluent. In the CO_2 diluent, the ignition times were found to be consistently 1–2 ms longer. Similarly, in the 1250 K study, the ignition times of the coal particles were shorter for the higher oxygen concentration environment and were 2–3 ms longer in the CO_2 environments, as shown in Fig. 6.1. In all cases, the image data suggested that particle ignition occurred from the emitted volatiles mixing and reacting with the hot oxidizing environment (i.e. homogeneous ignition). This result is consistent with the detailed modeling of isolated particle ignition and combustion performed by Lau and Niksa (1992), which only showed heterogeneous ignition occurring for particles less than 7–17 μm in size, for low- to mid-rank coals. Typical thermal images of ignited particles of the bituminous coal are shown in Fig. 6.2. These images highlight the formation of soot within a volatile-fed particle envelope flame under these low-slip, isolated particle conditions.

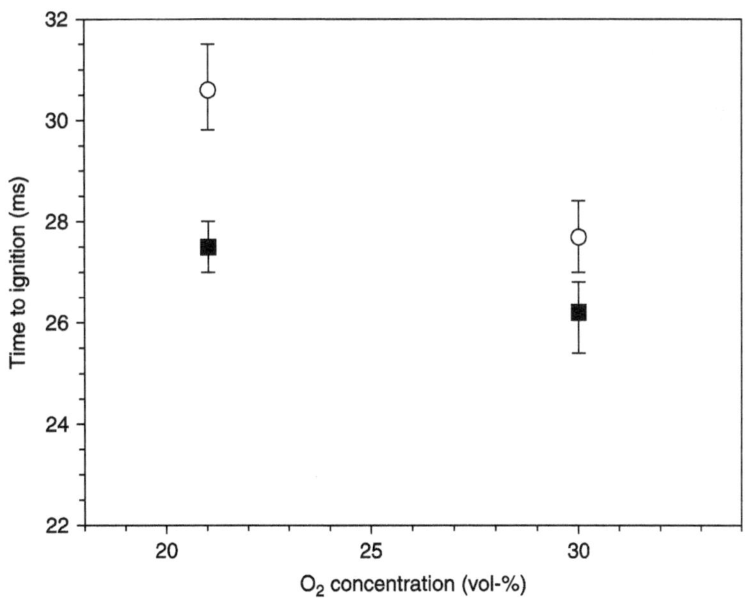

6.1 Mean ignition time of isolated, Pittsburgh high-volatile bituminous coal particles in O_2/N_2 and O_2/CO_2 environments at 1250 K. The particles were sieved to a 106–125 μm size fraction. Error bars correspond to 98% confidence statistical error. (Adapted from Molina and Shaddix, 2007.)

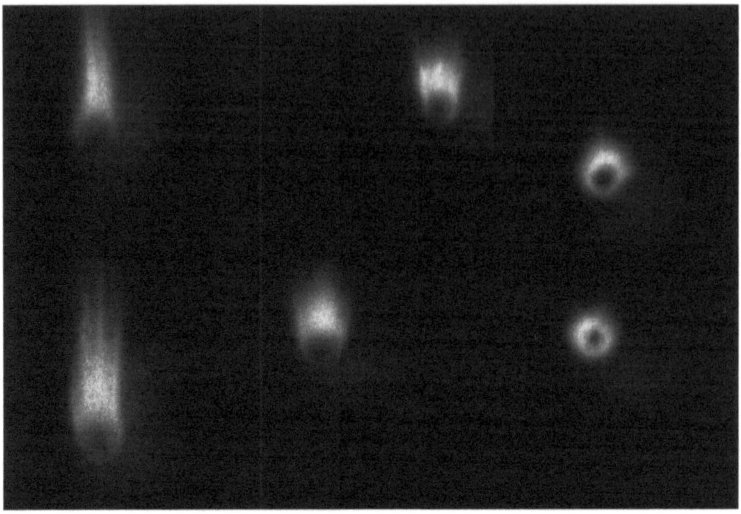

6.2 Thermal emission images captured just after ignition during the combustion of Pittsburgh coal particles in different gaseous environments. The top row shows images for combustion in a N_2 diluent and the bottom row shows images for combustion in a CO_2 diluent. The oxygen content varies from 12 vol-% (left column) to 24 vol-% (middle) to 36 vol-% (right column). Each image is 2.1 mm square. (Adapted from Shaddix and Molina, 2009.)

In a more recent study, Molina *et al.* (2009) injected a dense stream of 75–106 µm high-volatile bituminous coal particles into O_2/N_2 and O_2/CO_2 furnace flows with characteristic temperatures of 1130 K and 1650 K. For the lower temperature condition, the ignition time delay showed a strong sensitivity to both oxygen concentration and the use of N_2 versus CO_2 diluent. At the higher temperature condition, the effect of oxygen concentration was almost non-existent, and the CO_2 diluent only retarded ignition by a few milliseconds.

Huang *et al.* (2008) reported on ignition studies of finely ground bituminous Chinese coal in O_2/CO_2 mixtures at relatively low temperatures, using a thermogravimetric analyzer (TGA). This work emphasized the strong role of oxygen concentration on the ignition temperature when coal samples were heated at ramp rates varying from 10 to 50 K/min. Unfortunately, no experiments were performed comparing the ignition temperature in N_2 and CO_2 environments. Other investigations of minimum ignition temperature, in nitrogen atmospheres, have also shown a significant influence of oxygen concentration (see, for example, Fan *et al.*, 2008). Analysis of the differential thermogravimetric (DTG) weight-loss curves in the Huang study suggests that under the slow heating rate conditions characteristic of the TGA, heterogeneous ignition occurs for low-to-moderate oxygen concentrations, whereas homogeneous ignition occurs for high oxygen concentrations.

Arias *et al.* (2008) used an externally heated entrained flow reactor to determine the ignition temperature of a stream of injected coal particles, size-classified to 75–150 μm, by slowly increasing the reactor temperature and measuring the product gas composition. Two coals were investigated, a semi-anthracite and a high-volatile bituminous coal. Somewhat different results were found for the two coals. For the semi-anthracite, the ignition temperature was substantially higher in 21% O_2 in CO_2 than in 21% O_2 in N_2, and the O_2 content had to be raised to 30% in CO_2 to approximately match the ignition temperature in air. For the bituminous coal, the ignition temperature was practically the same for 21% O_2 in either N_2 or CO_2, and increasing the O_2 content in CO_2 yielded lower ignition temperatures than in air.

In a 20 kW pilot-scale swirl burner investigation of pulverized coal combustion, Liu *et al.* (2005a) found evidence of significantly delayed ignition when burning in a 21% O_2/CO_2 mixture, in comparison with burning in air. Combustion in a 30% O_2/CO_2 mixture yielded a comparable ignition location to burning in air. This result agrees with the earlier findings from the pioneering research on oxy-fuel combustion of pulverized coal conducted by Battelle and Argonne National Laboratory in a 112 kW furnace operating with a non-swirling combustor (Wang *et al.*, 1988).

Experiments at firing rates of 0.5 MW and 0.8 MW in the IHI vertical furnace with a low-swirl, 'type-0' combustor suggested a weak ignition delay when burning an Australian high-volatile bituminous coal in oxy-firing mode (with 27% O_2 at the burner inlet) in comparison with air-firing (Khare *et al.*, 2008). Modeling of the furnace with FLUENT suggested stronger ignition delays for oxy-fuel combustion than measured experimentally and also showed that the delays with oxy-fuel were both a consequence of the different gas properties and the larger ratio of primary to secondary flow momentum (Khare *et al.*, 2008).

6.2.2 Ignition theory

To understand the influence of gas composition on the coal ignition process, one should first consider the initial heating of the injected coal particles until the point at which significant devolatilization occurs. For the case of an isolated, non-reactive, dry particle, this process is described by Eq. 6.1:

$$\frac{dT_p}{dt} = \frac{3}{C_p \rho_p r_p}\left[\varepsilon\sigma\left(T_{surr}^4 - T_p^4\right) + h\left(T_g - T_p\right)\right] \qquad [6.1]$$

where T_p and T_g are the particle and gas temperatures, respectively; C_p, ρ_p, r_p and ε are the particle heat capacity, density, radius and emissivity, respectively; σ is the Stefan-Boltzmann constant; and h is the convective heat transfer coefficient. T_{surr} is the effective temperature of the radiant surroundings. As shown on the right-hand side of Eq. 6.1, heat transfer to the particle occurs from a combination of radiation and convective heat transfer from hot gases (commercial burners typically use recirculation zones to bring hot combustion products into direct contact with freshly injected coal). For near-term or retrofit oxy-fuel applications, the radiation intensity within the furnace is expected to be similar to conventional air-fuel combustion (to produce similar heat transfer distributions to the boiler tubes), so any differences in the particle heat-up process will be driven by differences in convective heat transfer.

Convective heating of the particles is determined by T_g and h. For pulverized coal particles, typically with low particle slip velocity, h can be calculated assuming a Nusselt number ($Nu = hD/\lambda$) of 2, where D is the particle diameter and λ the gas thermal conductivity. Hence, the only gas property that affects the initial particle heat-up is the thermal conductivity. Figure 6.3 shows the thermal conductivity of the most prominent gases in air-fuel and oxy-fuel combustion environments as a function of temperature. CO_2 has a slightly lower thermal conductivity than N_2 at low temperatures, but at 900 K the thermal conductivities are equal, and at higher temperatures CO_2 has a slightly greater thermal conductivity than N_2. For a particle exposed to gas at 1500 K, the rate at which the particle temperature rises up to 1000 K is nearly identical in N_2 and CO_2 atmospheres, as shown in Fig. 6.4. On the other hand, water vapor has the same

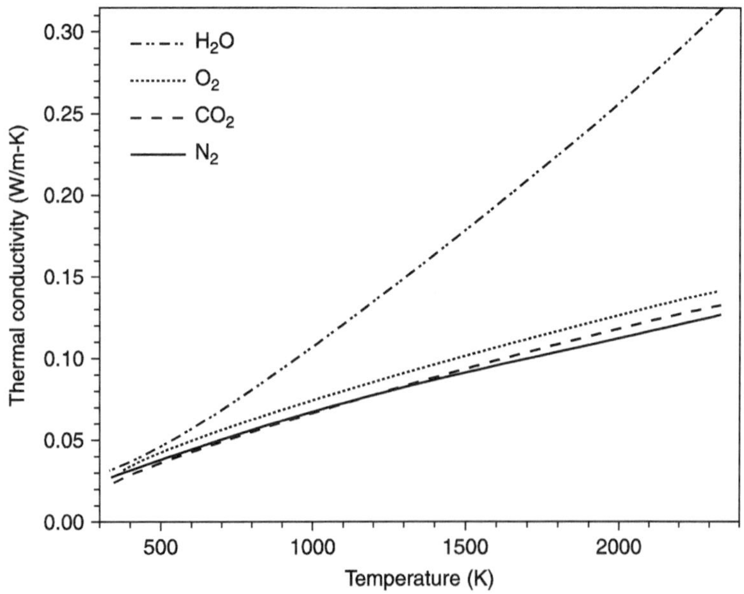

6.3 Thermal conductivity of selected gases as a function of temperature. All gases except water calculated using the modified Eucken technique (Poling *et al.*, 2001). Water thermal conductivity calculated using the recommendation of Paul and Warnatz (1998) for improved accuracy.

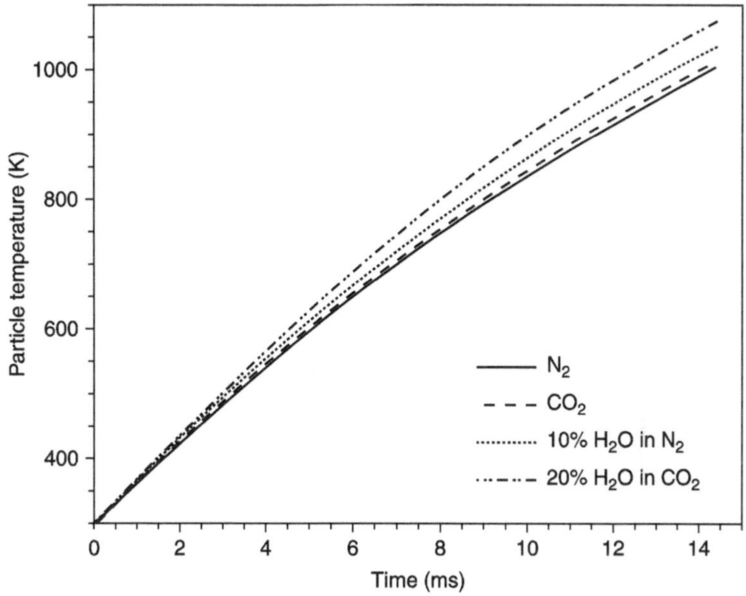

6.4 Calculated particle temperature of a spherical 100 μm coal particle in selected gas environments at 1500 K, with a radiative boundary at 1000 K and a Nusselt number of 2.0. Gas thermal conductivities are those indicated in Fig. 6.3, with gas mixture conductivity estimated according to simple molar mixing ratios.

thermal conductivity as N_2 at room temperature and rapidly attains a much higher thermal conductivity at elevated temperatures. Therefore, implementation of oxy-fuel combustion with a wet recycle loop would be expected to result in a somewhat faster coal particle heat-up rate, all else being equal. In reality, at least for retrofit applications, the furnace will be operated at somewhat lower temperatures in oxy-fuel mode than in air-fired combustion (to keep the radiant heat transfer approximately the same), so that the net effect of using a wet recycle gas will probably be to have a negligible effect on particle heating rates. Rather, the effect of changes in the primary versus secondary gas flow momentum, and resultant impacts on the near-burner fluid mechanics, will likely produce the predominant differences in particle heat-up and devolatilization.

Following initial particle heat-up, either heterogeneous ignition of the coal particle surface or release and ignition of volatiles (homogeneous ignition) will occur. In some cases, initial heterogeneous ignition may be quenched by the eruption of volatiles, an event known as 'heterogeneous–homogeneous ignition' (Essenhigh *et al.*, 1989). Critical variables influencing the mode of ignition are the volatile fraction of the coal, particle size, particle loading, and heating rate. Slow heating, low volatile content, low particle loading, and small particle size all favor heterogeneous ignition, on account of the ease of access of oxygen to the particle surface. Simulations conducted by Zhao *et al.* (2007) suggest that there is an optimal coal particle number density for minimal ignition delay, corresponding to a Group number (G) of ~ 10 (for $G = 2\pi n R_c^2 d_p$, where n is the particle number density, R_c the radius of the particle cloud, and d_p the particle diameter), in rough concurrence with very limited experimental measurements (Du *et al.*, 1995).

For those conditions favoring heterogeneous ignition, the gas-phase oxygen concentration is important, to nearly first-order. Besides that, the only gas-phase property expected to be important is the thermal conductivity, with a low conductivity favoring the heterogeneous ignition process. Based on the data shown in Fig. 6.3, no significant difference in heterogeneous ignition tendency is expected for N_2 and CO_2 atmospheres at the same temperature. In contrast, the high thermal conductivity associated with elevated moisture levels, when employing wet recycle, tends to favor homogeneous ignition of coal particles.

Based on the experimental data on oxy-fuel combustion that has been measured to date, indicating a delay of ignition in CO_2 atmospheres (with the sole exception of one of the coals tested by Arias *et al.* (2008)), it appears that pulverized coal ignition generally occurs via a homogeneous or heterogeneous–homogeneous mechanism. Homogeneous ignition (or the final step of heterogeneous–homogeneous ignition) occurs when the light hydrocarbons and tar material released during devolatilization react as they come into contact with hot, oxygen-containing gases. Coal devolatilization itself is slightly endothermic and the devolatilization rate is dependent on the particle temperature and heating rate (Solomon *et al.*, 1992). Therefore, as with the initial coal heat-up process, the temperature and thermal conductivity of the surrounding gases control the devolatilization rate for a given coal type and particle size.

Once coal volatiles have been ejected from the particles and begin to mix with the surrounding hot gases, the gas-phase ignition process itself may be interpreted via adiabatic thermal explosion theory. For a one-step overall reaction, the autoignition time is given by Eq. 6.2:

$$\tau = \frac{c_v \left(T_0^2 / T_a\right)}{q_c Y_{F,0} A \exp\left(-T_a T_0\right)} \qquad [6.2]$$

where c_v is the specific heat (at constant volume), T_0 is the initial temperature of the local fuel/air mixture, q_c is the combustion heat release per mass of fuel, $Y_{F,0}$ is the initial mass fraction of fuel, and the reactivity of the fuel/air mixture is given by $k = A \times \exp(-T_d/T_0)$ (Law, 2006, p. 309).

According to Eq. 6.2, the ignition delay of volatile gases decreases in proportion to the chemical heat release rate and increases in proportion to the volumetric heat capacity of the gas mixture. The volumetric heat capacity is an important factor in the ignition delay because it is a measure of the local heat sink for energy that is released from a successful combustion reaction and thereby affects the tendency for thermal runaway (to an ignited state). As is well known, the volumetric heat capacity of CO_2 at the elevated temperatures relevant for ignition is 1.7 times that of N_2, so it would be expected that the volatiles released into an N_2 atmosphere will ignite more quickly than those released into a CO_2 environment. Similarly, the heat capacity of H_2O in the temperature range of ignition (1000–1400 K) is 1.3 times that of N_2, so the additional H_2O present in oxyfuel combustion with wet recycle would be expected to induce a slight increase in the ignition delay.

Equation 6.2 also shows that the homogeneous ignition delay decreases in proportion to the reaction rate of the local fuel–oxidizer gas mixture. This reaction rate increases with higher O_2 concentration in the bulk gas (according to the effective reaction order of the governing reactions) and thereby the homogeneous ignition delay decreases for higher O_2 concentrations. Thus, Eq. 6.2 explains the trade-off between higher O_2 levels and CO_2 diluent that can yield similar ignition behavior for oxy-fuel combustion as for combustion in air, as demonstrated in many experimental studies (Wang *et al.*, 1988; Molina and Shaddix, 2007; Arias *et al.*, 2008; Shaddix and Molina, 2009; Molina *et al.*, 2009).

6.2.3 Chemical effects on ignition

Another potential influence of CO_2 on ignition stems from the reduction of the primary CO oxidation reaction rate, $CO + OH \leftrightarrow CO_2 + H$, by promoting the backward reaction, thereby reducing the concentration of the highly reactive and diffusive H radical (Guo and Smallwood, 2008). In addition, both CO_2 and H_2O enhance the rate of key radical recombination reactions such as $H + O_2 + M \leftrightarrow HO_2 + M$ (Maas and Warnatz, 1988; Ashman and Haynes, 1998; Michael *et al.*, 2002). The chaperon efficiency of CO_2 in these recombination reactions is between two and three times that of N_2 and the chaperon efficiency of H_2O is approximately 11 times that of N_2. These chemical kinetic effects are consistent with the observed increase in ignition delay in the CO_2 environments. Therefore, it is unclear whether the primary effect of CO_2 on homogeneous coal particle ignition results from its high specific heat, its suppression of radical concentrations, or its suppression of CO oxidation. For elevated H_2O concentrations, some chemical inhibition of ignition can be expected.

It should be noted that it is possible that elevated concentrations of CO_2 and H_2O could have a chemical effect on coal particles during coal devolatilization. Gale *et al.* (1995b) found some evidence that the surface area (particularly the micropore surface area) increased for chars produced in an atmosphere at 1000–1100 K (over a 0.15–0.5 s dwell time) with a moisture level of 18%, though no net mass conversion from gasification of the char was detectable. Considering the much greater thermal conductivity of water vapor relative to N_2, it is possible that this effect results from higher particle heating rates in wet gas environments (as shown in Fig. 6.4), considering the profound effect of heating rates on devolatilization (Gale *et al.*, 1995a). Rathnam *et al.* (2009) showed that, for pyrolysis at 1673 K with a (long) 0.6 s dwell time, approximately 10% greater apparent volatile yields occur in CO_2 atmospheres (relative to N_2), presumably because of contributions from char gasification.

6.3 Flame stability

Flame stability refers to the ability of an existing flame to remain ignited, despite variations in the fuel and oxidizer flow rates, turbulent shear, fuel quality, etc., and therefore is concerned with flame *extinction* phenomena, rather than ignition. Flame stability is an important concept in burner design and operation, as discussed further in Chapter 7, because burners must stay ignited through a reasonable range of fuel and oxidizer flow characteristics and over a range of heat release rates (i.e. 'turndown'). The potential use during oxy-fuel combustion of a primary gas stream with a low oxygen concentration (to minimize the potential for mill fires or fires at downstream mixing locations) poses a particular challenge to burner designers and boiler operators to maintain flame stability. Consequently, one or more jets of concentrated oxygen are often placed in oxy-fuel burners for the express purpose of improving flame stability (Kimura *et al.*, 1995; Chui *et al.*, 2003).

Toporov and co-workers (Toporov *et al.*, 2008; Heil *et al.*, 2009) have investigated flame stabilization characteristics of a swirl-stabilized burner when burning lignite at a 40 kW load in O_2/N_2 and O_2/CO_2 mixtures. Operation of a conventional air burner in oxy-fuel mode led to poor burnout and a lifted, dark flame for 21 vol-% O_2 in CO_2. Using strong swirl of the secondary gas stream and a redesigned quarl (to enhance the interior recirculation zones), stable combustion could be achieved for both air-firing and oxy-fuel combustion with oxygen concentrations as low as 18 vol-%. In general, low concentrations of O_2 are undesirable in oxy-fuel applications, as they imply higher recycle gas flow rates and therefore higher capital and operating costs associated with the recycle flow. However, Toporov and co-workers have considered the potential for high-temperature, membrane-based oxygen production to provide a low-cost source of oxygen, particularly when integrated into the high-temperature furnace exhaust stream. This particular application motivates oxy-fuel burner operation at relatively low levels of oxygen, to improve the efficiency of the membrane separation process (Toporov *et al.*, 2008).

Unlike coal ignition, for which most characteristics can be adequately revealed through experiments and analysis of single particles or small groups of particles, flame stability is inherently associated with the behavior of large groups of coal particles or, more realistically, a continuous flow of particles. The most fundamental flame property that controls the flame stability of premixed flames is the flame speed. A high flame speed makes a given fuel/oxidizer mixture relatively immune to flame quench, whereas a low flame speed makes a flame susceptible to extinction. For non-premixed flames, the characteristic flame speed is that of the stoichiometric fuel-oxidant mixture formed at the flame sheet.

6.3.1 Flame speed measurements

Suda *et al.* (2007) measured the spherical flame propagation rate for coal dust clouds of 53–63 μm particles suspended in a spherical vessel that was deployed in a drop tower to achieve microgravity conditions. The microgravity conditions allowed the dust to remain suspended and produced spherical flame propagation unperturbed by buoyancy (allowing ready determination of the outward flame propagation rate). Most of the experiments were conducted in atmospheres of 40% oxygen mixed with either N_2 or CO_2 (a few experiments were conducted with Ar diluent and with higher and lower oxygen concentrations). For a high-volatile coal, the flame speed in CO_2 was consistently found to be two times lower than in N_2, over a range of fuel particle loadings. For a medium-volatile coal, the flame speed was substantially lower than for the high-volatile coal, especially in CO_2 atmospheres, wherein the flame speed was nearly five times lower than in N_2.

Kiga *et al.* (1997) also measured the flame propagation speed under microgravity conditions for coal particle combustion in O_2/N_2, O_2/CO_2 and O_2/Ar atmospheres, as a function of oxygen concentration. In all cases the flame speeds increased non-linearly with oxygen content. The flame speed was greatest in Ar atmospheres, intermediate in N_2 atmospheres, and considerably lower in CO_2 atmospheres. These variations in flame speed were suggested to result from differences in the specific heat of the gas mixtures, as the specific heat was the highest for the CO_2/O_2 mixtures and was 38% and 76% lower for the $N_2/O_2$and Ar/O_2 mixtures, respectively.

6.3.2 Extinction theory

For all but the highest rank coals, near-burner flame attachment and stability is controlled by combustion of coal volatiles. The manner in which the volatiles are burned can vary from being primarily non-premixed in character to being partially premixed. Consequently, flame stability and extinction characteristics of both non-premixed and partially premixed volatile flames need to be considered. For both premixed and non-premixed flames, Williams (2000) demonstrated that flame extinction is controlled by a characteristic extinction Damköhler number, which is defined to be the ratio of the characteristic fluid dynamic or mixing timescale to the chemical timescale:

$$Da = \frac{\tau_{fl}}{\tau_{ch}} \qquad [6.3]$$

For a simple strained flame, the fluid dynamic timescale is equal to the inverse of the local strain rate and is primarily controlled by the hydrodynamics of the flow in the burner, with a minor impact from the presence of CO_2 or N_2. The chemical timescale is given by the characteristic gas thermal diffusivity divided by the square of the flame speed:

$$\tau_{ch} = \frac{\alpha}{S_L^2} \qquad [6.4]$$

The flame speed, in turn, is proportional to the square root of the thermal diffusivity, and for a second-order reaction characterized by a single Arrhenius kinetic rate is given by Eq. 6.5 (Glassman and Yetter, 2008):

$$S_L \sim \left[\alpha P_{reac} \left(\frac{T_0^2 T_f}{\left(T_f - T_0\right)^3} \right) \left(\frac{RT_f}{E} \right)^3 A e^{-E/RT_f} \right]^{1/2} \qquad [6.5]$$

where P_{reac} is the partial pressure of the reactants, T_f is the (adiabatic) flame temperature, and A and E are the Arrhenius constants that characterize the kinetic reaction rate. Therefore, for a given flame strain rate, the resistance to extinction (i.e. the flame stability) is characterized by

$$\frac{1}{\tau_{ch}} \sim P_{reac} \left(\frac{T_0^2 T_f}{\left(T_f - T_0\right)^3} \right) \left(\frac{RT_f}{E} \right)^3 A e^{-E/RT_f} \qquad [6.6]$$

Equation 6.6 demonstrates the importance of the reactant concentrations and especially the flame temperature and activation energy to flame stability. For premixed flames, the flame temperature to be used in Eq. 6.6 is that of the bulk reactant gas mixture, whereas for non-premixed flames the characteristic flame temperature is that of a stoichiometric fuel–oxidant mixture.

From Eq. 6.5, flame speeds in CO_2 atmospheres will clearly be lower than those in N_2 atmospheres with the same O_2 content, because the thermal diffusivity is lower in CO_2 (by about 60%) and the adiabatic flame temperature is lower. Figure 6.5 shows the computed adiabatic flame temperature for combustion of coal volatiles in N_2 or CO_2 atmospheres as a function of the bulk oxygen content. Using higher concentrations of O_2 in CO_2 compensates for the reduction in flame speed from CO_2 by increasing the partial pressure of reactants and raising the flame temperature. For experiments in Ar atmospheres, as conducted by Kiga *et al.* (1997) and Suda *et al.* (2007), the flame speed is higher than in N_2 atmospheres because of substantially greater flame temperatures (resulting from the 40% lower specific heat of Ar).

Despite the emphasis in the literature on measurements of flame speed in characterizing trends in flame stability, one should recognize that for flame stability ultimately the important factor is the inverse of the chemical timescale (as shown in Eq. 6.3, 6.4, and 6.6), which is, in fact, equal to the square of the

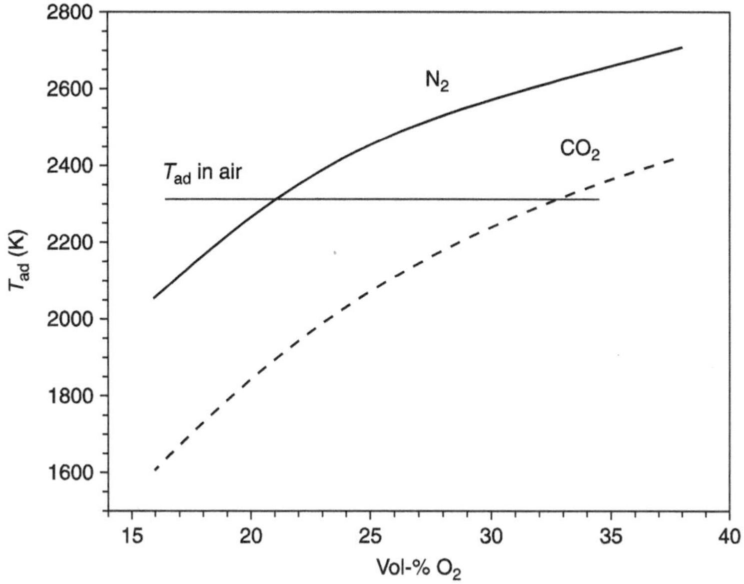

6.5 Computed adiabatic flame temperature T_{ad} at 1 atm of typical coal volatiles burning in O_2/N_2 and O_2/CO_2 environments. Coal volatiles assumed to consist of 20 vol-% C_2H_4, 48 vol-% H_2, 14 vol-% CO, 3 vol-% CO_2, and 15 vol-% H_2O, consistent with the analysis of gas species evolved from the rapid, high-temperature devolatilization of a bituminous coal (Xu and Tomita, 1987). Calculations of adiabatic flame temperature were performed using the NASA CEA chemical equilibrium code (http://www.grc.nasa.gov/WWW/CEAWeb/).

flame speed divided by the thermal diffusivity and which, ironically, is itself independent of thermal diffusivity (as shown in Eq. 6.6). Rather, oxy-fuel conditions affect the flame stability primarily through the strong dependence of the flame speed on the adiabatic flame temperature of the chemical mixture. The inhibitory chemical effects of CO_2 and H_2O discussed in section 6.2.3 also serve to reduce flame stability, through their influence on the overall reaction rate parameters shown in Eq. 6.6.

6.4 Char combustion

The char combustion phase dominates the total burnout time of coal particles and thus plays a significant role in determining the level of unburned carbon (or LOI – loss-on-ignition) and CO emissions from coal boilers. In addition, radiation from burning char particles can play a significant role in boiler heat transfer, so understanding the effects of oxy-fuel combustion conditions on coal char combustion temperatures is important both for retrofit and new-build boiler applications.

6.4.1 Role of oxygen (O_2)

Oxygen partial pressure plays a key role in char combustion, whether the particles are burning with chemical kinetic control (referred to as 'Zone I' combustion), diffusion control ('Zone III' combustion), or, as is most typical, a mixture of the two ('Zone II' combustion). In kinetically controlled combustion, various measurements at low or intermediate temperatures have shown the char combustion rate exhibits a reaction order of 0.6–1.0 towards oxygen (Hurt and Calo, 2001). For diffusion control, the burning rate is proportional to the bulk oxygen partial pressure (i.e. the effective reaction order is 1.0), and in Zone II combustion one expects a dependence on oxygen that falls somewhere between the Zone I and Zone III limiting cases (Thiele, 1939).

A number of investigations have demonstrated an increase in the char particle combustion temperature (which is related to the char burning rate, in a given thermal environment) and a reduction in the char burnout time as the oxygen concentration increases (Timothy *et al.*, 1982; Murphy and Shaddix, 2006; Bejarano and Levendis, 2008; Shaddix and Molina, 2008; Shaddix and Molina, 2009). Moreover, Murphy and Shaddix (2006) showed that the global char kinetic rate expressions that describe char oxidation under conventional, vitiated-air conditions apply equally well to char oxidation in oxygen-enriched environments.

One consideration in oxygen-enriched atmospheres is the suitability of applying a so-called 'single-film' model, with a presumed unreactive boundary layer, to the combustion of the larger size portion of a typical pulverized coal grind. Detailed modeling of the boundary layer chemistry and energy balance (Mitchell *et al.*, 1990; Shaddix and Molina, 2008; Hecht *et al.*, 2010) shows that, for particles as large as 100 μm in diameter, as the particle temperature increases (with increasing oxygen content of the bulk gas) partial conversion of CO in the boundary layer raises the particle temperature and the burning rate relative to what a frozen boundary layer assumption would predict. Therefore, traditionally used single-film models of char combustion may fail for larger particles burning in oxygen-enriched environments.

6.4.2 Effects of enhanced carbon dioxide (CO_2) and water vapour (H_2O) concentrations

There are several different ways in which the presence of a CO_2 bath gas and potentially larger concentrations of water vapor could influence pulverized char combustion:

- By hindering the diffusion of oxygen to the char surface, the presence of CO_2 will reduce the burning rate for all but kinetically controlled burning conditions (Zone I). Conversely, elevated H_2O concentrations tend to promote combustion by augmenting oxygen diffusion (at elevated temperatures, the diffusivity of oxygen is ~ 20% higher in H_2O and 20% lower in CO_2, in comparison with N_2, as shown in Fig. 6.6).

- If there is significant heat release in the char particle boundary layer (from oxidation of CO) that is transferred back to the particle, the higher heat capacities of CO_2 and H_2O will reduce the peak gas temperature and therefore the heat transfer back to the particle, reducing the burning rate. However, the higher thermal conductivity of H_2O-containing gas mixtures will tend to increase the heat transfer back to the particle, making the net effect of H_2O on this phenomenon unclear.
- Dissociative adsorption of CO_2 and H_2O on the char surface (leading to stable surface oxides instead of successful gasification reactions) could result in significant surface coverage and therefore competition for available reaction sites for oxygen, reducing the burning rate.
- Gasification of char carbon by CO_2 and H_2O could augment carbon removal by oxidation under some conditions. However, the endothermicity of the gasification reactions acts to lower the char temperature, thereby lowering the oxidation rate (Hecht *et al.*, 2010).

Experimental measurements of char combustion temperatures in O_2/CO_2 environments have consistently shown lower char combustion temperatures in the

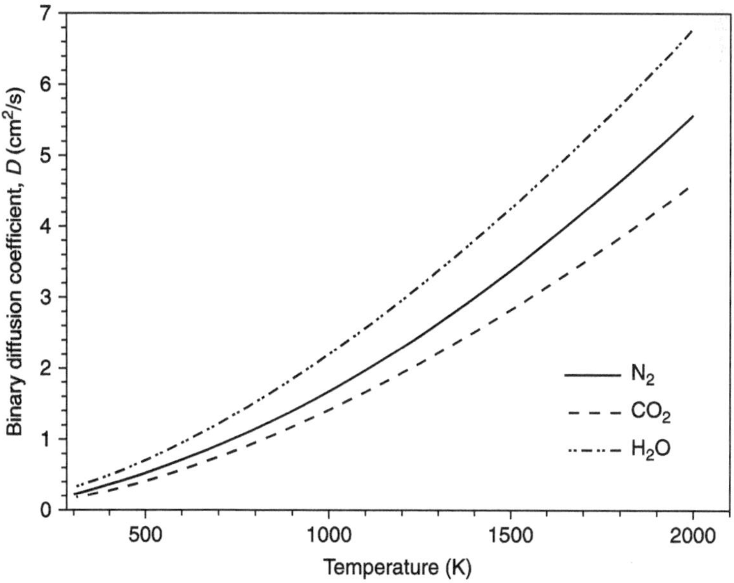

6.6 Binary diffusion coefficient of oxygen in N_2, CO_2, and H_2O as a function of temperature, computed using correlations from Marrero and Mason (1972).

presence of CO_2 (Bejarano and Levendis, 2008; Shaddix and Molina, 2008, 2009). A typical result is shown in Fig. 6.7. For char combustion in mixtures of oxygen and an unreactive diluent, a lower char combustion temperature necessarily implies a lower char burning rate, because of the exothermicity of the char oxidation reaction. However, the char gasification reactions with CO_2 and H_2O are endothermic, as shown in Table 6.1, and this confounds the interpretation of char burning rate based on the char combustion temperature. Unfortunately, the actual kinetic rates of these gasification reactions with coal char are uncertain at the high temperatures relevant for oxy-fuel combustion. At lower temperatures, a number of thermogravimetric measurements have been performed of the gasification rates at 1 atm and have found them to be several orders of magnitude smaller than char oxidation (DeGroot and Richards, 1989; Harris and Smith, 1990). However, the gasification reactions have higher activation energy, such that they may become important at high temperatures, especially considering the high concentrations of these reactant gases during oxy-fuel combustion. An additional consideration is that these gasification reactions may not be amenable to accurate description using a single Arrhenius expression, but rather may require the use of Langmuir–Hinshelwood kinetics to describe the two-step adsorption–desorption process (Everson *et al*., 2006).

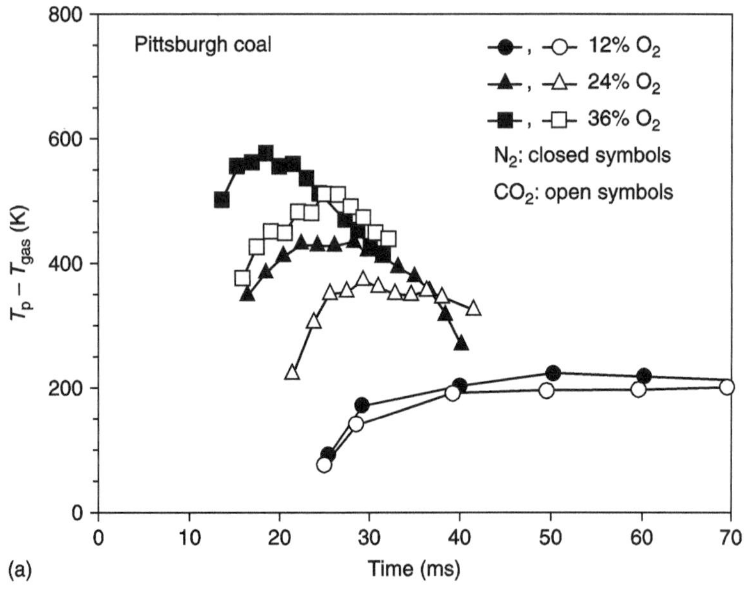

6.7 Measured mean coal char particle temperature as a function of time from injection for different O_2/N_2 and O_2/CO_2 environments at 1700 K. Results shown for Pittsburgh high-volatile bituminous coal (a) and Black Thunder sub-bituminous coal (b), both sieved to a 75–106 μm size fraction. (Adapted from Shaddix and Molina, 2008.)

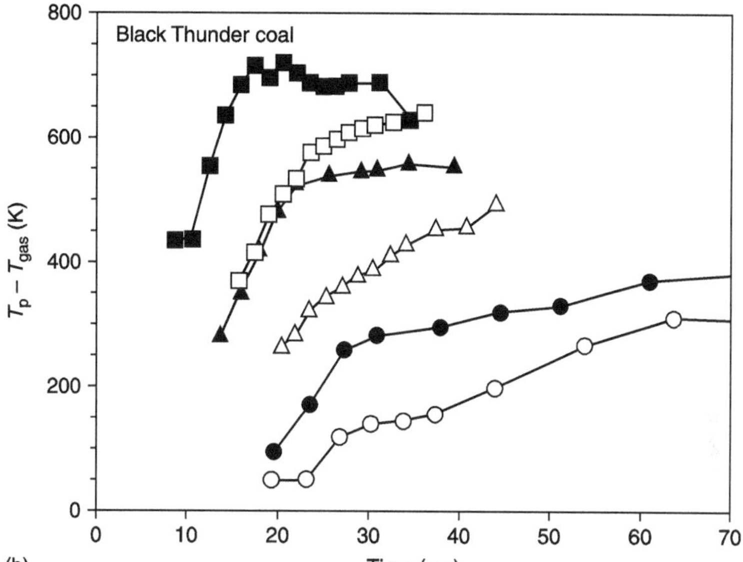

(b)

6.7 Continued.

Table 6.1 Reaction enthalpies for primary char oxidation and gasification reactions

Reaction			ΔH_{rxn} (kJ/mole-C_s)
2C(s) + O_2	→	2CO	−110.5
C(s) + O_2	→	CO_2	−393.5
C(s) + CO_2	→	2CO	172.5
C(s) + H_2O	→	CO + H_2	131.3

Shaddix and Molina (2008) evaluated the impact of reduced O_2 diffusion in CO_2 on single-film, Zone II combustion rates of a bituminous coal char in O_2/CO_2 combustion and showed that it accounted for a 50 K reduction in the char particle combustion temperature and a 10% decrease in the char combustion rate. Using a detailed model for combustion of porous particles that included boundary layer chemistry and heat transfer, they also showed that the diffusivity effect was responsible for all of the observed difference in char particle temperatures for combustion in O_2/N_2 and O_2/CO_2 environments. Conversely, artificially changing the thermal properties of CO_2 to be equal to those of N_2 had a negligible effect on the predicted char particle combustion temperatures. On the basis of these results (shown in Fig. 6.8), it appears that the reduction in oxygen diffusivity in CO_2 is at least the primary cause of the observed lower char burning temperatures in CO_2 environments, though there remains the possibility of a modest influence on char particle temperature from the endothermic gasification reaction with CO_2.

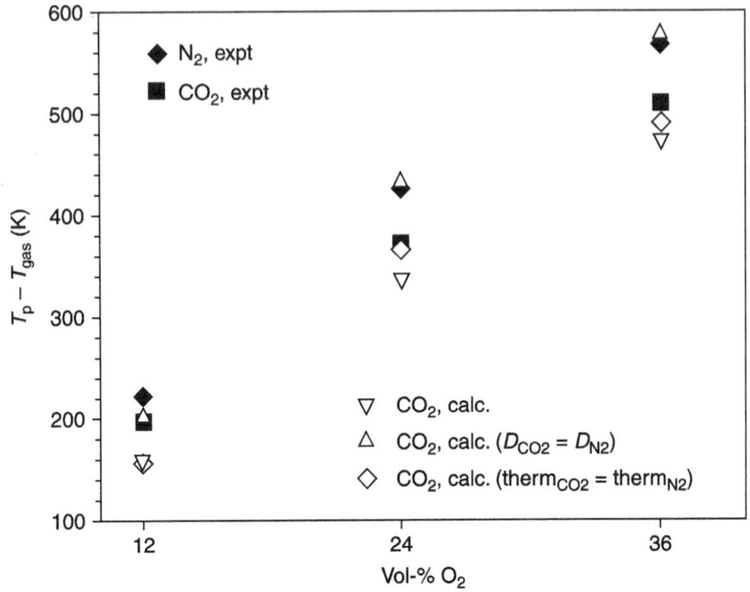

6.8 Experimentally measured and modeled maximum mean Pittsburgh coal char particle temperatures when burning in O_2/N_2 and O_2/CO_2 environments at 1700 K. To determine the primary factors affecting the lower char combustion temperatures in CO_2, the gas transport properties of CO_2 were altered in separate simulations, in one case artificially setting the CO_2 diffusion coefficient equal to that of N_2 and in another artificially setting the thermal properties of CO_2 equal to N_2. (Adapted from Shaddix and Molina, 2008.)

Hecht *et al.* (2010) numerically evaluated the potential range of impact of CO_2 gasification of coal char on the overall char combustion behavior under typical pulverized coal combustion conditions. The endothermicity of the gasification reaction results in lower char combustion temperatures and lower oxidation rates. For sufficiently low oxygen content of the surrounding gas, the total char carbon consumption rate increases because of the gasification reaction, whereas for moderate to high bulk oxygen contents the total char carbon consumption rate *decreases* because of the thermal influence of the gasification reaction. The same general trend is expected from steam gasification of coal char, though the critical concentration of bulk oxygen content (separating enhanced carbon consumption rates from decreased carbon consumption rates) is expected to be somewhat higher, because the steam gasification reaction is not quite as endothermic as the CO_2 gasification reaction.

6.5 Carbon burnout

The overall carbon burnout is important to boiler operation with respect to fly ash utilization, resistivity of fly ash in the electrostatic precipitators, and overall thermal efficiency. The burnout of coal fed into a furnace reflects the integrated influence of coal devolatilization and ignition and char combustion and gasification. In practical boilers, the burnout is often dictated by the large end of the coal particle size distribution, locations of poor fuel–air mixing, and the nature of ash–char interaction during combustion (Cloke *et al.*, 2003; Pallarés *et al.*, 2005; Barranco *et al.*, 2006). With application to oxy-fuel combustion of coal, potential influences on burnout can occur from differences in flame attachment and shape, furnace temperatures, oxygen content, char combustion kinetics, gasification reactions, and overall residence time.

Several laboratory studies have been conducted to assess burnout tendencies in oxy-fuel combustion conditions compared with air-fired combustion. Liu *et al.* (2005a) reported the burnout levels (based on ash-as-tracer) for a high-volatile bituminous pulverized coal burning in a 20 kW insulated furnace, with and without oxidant staging. Combustion in 21% O_2 in CO_2 resulted in lower burnout (96.8% vs. 98.2%) compared with combustion in air (no staging for either case), and combustion in 30% O_2 in CO_2 resulted in substantially greater burnout (99.3%) compared with combustion in air, whether or not staging was performed. Gas temperatures measured in the furnace were nearly identical for combustion in 30% O_2 in CO_2 compared with combustion in air, and were substantially lower for combustion in 21% O_2 in CO_2 (as expected because of the high heat capacity of CO_2). The higher extent of coal burnout in 30% O_2 in CO_2 compared with combustion in air in this furnace was later confirmed for six international medium- to high-volatile bituminous trade coals (Liu *et al.*, 2005b). Wang *et al.* (1988) also found lower burnout for combustion of pulverized high-volatile bituminous coal in their refractory-lined 112 kW furnace when using 21% O_2 in CO_2, in comparison with air, and higher burnout when using 29–31% O_2 in CO_2.

Arias *et al.* (2008) reported on the burnout of the 75–150 μm size fraction of four different coals (varying in rank from high-volatile bituminous coal to anthracite) combusted in a flow reactor operating at 1273 K. The burnout was measured after a residence time of 2.5 s and was reported as a function of the fuel/oxidant equivalence ratio. For all coals, slightly worse burnout was attained for 21% O_2 in CO_2, in comparison with simulated air, and higher burnout was attained when using 30–35% O_2 in CO_2. Note that, in contrast to the insulated furnace results, the data from Arias is for isothermal conditions with no overall temperature effects associated with the use of CO_2.

Rathnam *et al.* (2009) recently reported burnout results from a drop tube furnace that run counter to the predominant trend. The 63–90 μm size fraction of four different coals (varying in rank from sub-bituminous to high-volatile bituminous coal) were burned at various oxygen levels in a furnace nominally at 1673 K. Ash analysis of the particles collected at the bottom of the furnace generally showed stronger burnout for a given oxygen concentration in CO_2 atmospheres compared with N_2. Rathnam *et al.* also performed char combustion experiments in a TGA as a function of oxygen concentration. These experiments showed slightly higher apparent reactivity (i.e. mass loss) in O_2/N_2 environments for oxygen levels of 10% or higher and slightly higher apparent reactivity in O_2/CO_2 environments for low oxygen levels. This result is consistent with the trends computed by Hecht *et al.* (2010), and suggests that CO_2 gasification of the char may enhance mass loss during the final stages of burnout when oxygen levels are low.

6.6 Conclusions and future trends

The different chemical and perhaps thermal environments during oxy-fuel combustion of coal can have significant impacts on coal ignition, flame stability, char combustion, and carbon burnout. In particular, the high concentration of CO_2 during oxy-fuel combustion tends to retard coal ignition relative to air, because of its high molar heat capacity, though this effect can be counteracted with the use of an oxygen content of approximately 30%. These effects can be rationalized by applying autothermal ignition theory to describe the ignition of ejected volatiles in the presence of a hot oxidizing gas. CO_2 and H_2O also tend to inhibit ignition because of their chemical influence in reducing the concentrations of highly reactive radical species. Further experimental and/or computational work evaluating the effects of enhanced CO_2 and H_2O concentrations on the ignition properties of dense streams of coal particles, as present in practical burner flows, would be useful to bridge the gap between several existing studies using dilute particle streams and a number of studies noting ignition properties in burners.

Pulverized coal flame stability may be characterized by the laminar flame speed of the coal, or, more properly, by the ratio of the square of the laminar flame speed to the thermal diffusivity. Experimental measurements of the flame speed have shown it to be lower, sometimes much lower, in the presence of a CO_2 diluent relative to a N_2 diluent (as in air) and to be highly sensitive to the bulk gas oxygen content. Theoretical analysis suggests that one of the reasons for the low flame speeds in CO_2 is the approximately 60% lower thermal diffusivity in N_2, but the thermal diffusivity itself should not be relevant to flame stability. Important factors that influence the flame stability are the concentration of the reactants and the adiabatic flame temperature of the fuel-oxidant mixture, in addition to the presence of strong recirculation zones close to the burner face. Elevated H_2O and, especially, CO_2 concentrations during oxy-fuel combustion therefore tend to reduce flame stability, on account of their high molar heat capacity (which tends to reduce the flame temperature), but the presence of elevated O_2 concentrations can offset this effect. In addition, H_2O and CO_2 may reduce flame stability through their influence on the concentrations of reactive radical species, though an elevated H_2O concentration facilitates heat conduction from hot gases to injected particles. Further clarification of these chemical and thermal effects on flame stability could be achieved through studies of the extinction limits of both partially premixed and non-premixed flames of coal volatiles in different atmospheres containing N_2, O_2 CO_2, and H_2O. Only a minor amount of research concerning flame stability in oxy-fuel combustion has been conducted to-date. Further research, particularly involving coals with a wide range of volatiles content, would be enlightening.

There are several different ways in which the presence of elevated concentrations of H_2O and especially CO_2 may affect oxy-fuel char combustion, by changing the diffusivity of oxygen through the particle boundary layer, through differences in thermal properties, and through participation in char gasification reactions. Experimental measurements show lower char combustion temperatures for a given oxygen level in the presence of CO_2, in comparison with combustion with a N_2 diluent (as in air). Single-particle simulations suggest that this reduction in particle temperature is primarily due to a reduced burning rate, on account of lower oxygen diffusion, but some (endothermic) char gasification may also be occurring, which would also lower the char particle combustion temperature. The use of oxygen-enriched atmospheres during oxy-fuel combustion enhances the char particle temperature and burning rate and also favors partial conversion of CO in the boundary layer surrounding a burning particle. Significant uncertainty currently exists regarding the actual rates of coal char gasification by H_2O and CO_2 at the relevant char combustion temperatures at 1 atm. This information is needed to properly model and assess the influence of these gasification reactions on the char burning rate, particularly as a function of coal type and particle size. Similarly, the influence of boundary layer conversion of CO on char combustion rates during oxy-fuel combustion needs to be better understood, as this influences the validity of the single-film combustion models that are traditionally used to describe pulverized coal combustion.

Carbon burnout in oxy-fuel combustion is generally influenced by the same factors that affect char combustion, with the exception that in practical burners and boilers the overall flame attachment, flame shape, and other mixing parameters often have a strong influence on burnout. Also, the relatively long residence times at low oxygen levels at the back ends of furnaces and boilers means that gasification reactions with enhanced concentrations of CO_2 and H_2O may well influence carbon burnout, even if the gasification rates are too slow to contribute significantly during the main phase of char combustion. Experimentally, carbon burnout is generally seen to be lower when performing oxy-fuel combustion with 21 vol-% O_2 and is higher once the oxygen level is raised to approximately 28 vol-%. Further research is needed to quantify the effects of CO_2 and H_2O gasification reactions over the entire combustion history of a char particle, especially for the larger size char particles that generally control final carbon burnout.

For new-build oxy-fuel combustion plants, there are many options for designing a suitable burner and boiler system. The work of Toporov and co-workers (Toporov *et al.*, 2008; Heil *et al.*, 2009) has demonstrated that reduced oxygen concentrations can be used and still maintain stable flames (at least with lignite), though additional heat transfer surfaces would be required to adequately capture the energy from the resultant lower temperature combustion products. Conversely, high intensity combustion can be employed using significantly elevated levels of oxygen, at least in the secondary 'air' stream, and the furnace can be redesigned with greater radiant energy capture. For retrofit applications, the heat transfer surfaces are fixed, meaning that similar levels of radiant and convective heat transfer need to be provided during oxy-fuel combustion as exist during combustion in air. The overall degree of oxygen enrichment that is required to provide similar heat transfer characteristics during oxy-fuel combustion is somewhat unclear at this point, in lieu of a full-scale oxy-fuel demonstration, and is likely to depend on the moisture level of the recycle gas, the coal that is being used, the size of the boiler, the burner design, etc. (Wall *et al.*, 2009; Toftegaard *et al.*, 2010). Pilot-scale furnace measurements to date suggest similar levels of radiative heat transfer are attained for oxy-fuel combustion with an overall oxygen feed concentration of approximately 25–30% (Toftegaard *et al.*, 2010). As shown in this chapter, the ignition properties, flame stability, char combustion, and char burnout properties are all quite sensitive to the local oxygen concentration. Therefore, even for retrofit applications, oxy-fuel combustion properties are likely to be different from those found for combustion in air.

6.7 References

Annamalai K, Ryan W and Dhanapalan S (1994), 'Interactive processes in gasification and combustion – Part III: Coal/char particle arrays, streams and clouds', *Progr. Energy Combust. Sci.*, 20, 487–618.

Arias B, Pevida C, Rubiera F and Pis J J (2008), 'Effect of biomass blending on coal ignition and burnout during oxy-fuel combustion', *Fuel*, 87, 2753–2759.

Ashman P J and Haynes B S (1998), 'Rate coefficient of $H + O_2 + M \rightarrow HO_2 + M$ (M = H_2O, N_2, Ar, CO_2)', *Proc. Combust. Inst.*, 27, 185–191.

Barranco R, Colechin M, Cloke M, Gibb W and Lester E (2006), 'The effects of pf grind quality on coal burnout in a 1 MW combustion test furnace', *Fuel*, 85, 1111–1116.

Bejarano P A and Levendis Y A (2008), 'Single-coal-particle combustion in O_2/N_2 and O_2/CO_2 environments', *Combust. Flame*, 153, 270–287.

Chui E H, Douglas M A and Tan Y (2003), 'Modeling of oxy-fuel combustion for a western Canadian sub-bituminous coal', *Fuel*, 82, 1201–1210.

Cloke M, Wu T, Barranco R and Lester E (2003), 'Char characterisation and its application in a coal burnout model', *Fuel*, 82, 1989–2000.

DeGroot W F and Richards G N (1989), 'Relative rates of carbon gasification in oxygen, steam and carbon dioxide', *Carbon*, 27, 247–252.

Du X, Gopalakrishnan C and Annamalai K (1995), 'Ignition and combustion of coal particle streams', *Fuel*, 74, 487–494.

Essenhigh R H, Misra M K and Shaw D W (1989), 'Ignition of coal particles: A review', *Combust. Flame*, 77, 3–30.

Everson R C, Neomagus H W J P, Kasaini H and Njapha D (2006), 'Reaction kinetics of pulverized-coal chars derived from inertinite-rich coal discards: Gasification with carbon dioxide and steam', *Fuel*, 85, 1076–1082.

Fan Y-S, Zou Z, Cao Z, Xu Y and Jiang X (2008), 'Ignition characteristics of pulverized coal under high oxygen concentrations', *Energy Fuels*, 22, 892–897.

Gale T K, Bartholomew C H and Fletcher T H (1995a), 'Decreases in the swelling and porosity of bituminous coals during devolatilization at high heating rates', *Combust. Flame*, 100, 94–100.

Gale T K, Fletcher T H and Bartholomew C H (1995b), 'Effects of pyrolysis conditions on internal surface areas and densities of coal chars prepared at high heating rates in reactive and nonreactive atmospheres', *Energy Fuels*, 9, 513–524.

Glassman I and Yetter R A (2008), *Combustion*, 4th Ed., Amsterdam, Elsevier Inc.

Guo H and Smallwood G J (2008), 'A numerical study on the influence of CO_2 addition on soot formation in an ethylene/air diffusion flame', *Combust. Sci. Tech.*, 180, 1695–1708.

Harris D J and Smith I W (1990), 'Intrinsic reactivity of petroleum coke and brown coal char to carbon dioxide, steam and oxygen', *Proc. Combust. Inst.*, 23, 1185–1190.

Hecht E S, Shaddix C R, Molina A, and Haynes B S (2010), 'Effect of CO_2 gasification reaction on oxy-combustion of pulverized coal char', accepted for publication in *Proc. Combust. Inst.*

Heil P, Toporov P, Stadler H., Tschunko S., Förster M. and Kneer, R. (2009), 'Development of an oxycoal swirl burner operating at low O_2 concentrations', *Fuel*, 88, 1269–1274.

Huang X, Jiang X, Han X and Wang H (2008), 'Combustion characteristics of fine- and micro-pulverized coal in the mixture of O_2/CO_2', *Energy Fuels*, 22, 3756–3762.

Hurt R H and Calo J M (2001), 'Semi-global intrinsic kinetics for char combustion modeling', *Combust. Flame*, 125, 1138–1149.

Khare S P, Wall T F, Farida A Z, Liu Y, Moghtaderi B and Gupta R P (2008), 'Factors influencing the ignition of flames from air-fired swirl pf burners retrofitted to oxy-fuel', *Fuel*, 87, 1042–1049.

Kiga T, Takano S, Kimura N, Omata K, Okawa M, Mori T and Kato M. (1997), 'Characteristics of pulverized-coal combustion in the system of oxygen/recycled flue gas combustion', *Energy Convers. Manag.*, 38, 129–134.

Kimura N, Omata K, Kiga T, Takano S and Shikisima S (1995), 'The characteristics of pulverized coal combustion in O_2/CO_2 mixtures for CO_2 recovery', *Energy Convers. Manag.*, 36, 805–808.

Lau C W and Niksa S (1992), 'The combustion of individual particles of various coal types', *Combust. Flame*, 90, 45–70.

Law C K (2006), *Combustion Physics*, Cambridge, UK, Cambridge University Press.

Liu H, Zailani R and Gibbs B M (2005a), 'Comparisons of pulverized coal combustion in air and in mixtures of O_2/CO_2', *Fuel*, 84, 833–840.

Liu H, Zailani R and Gibbs B M (2005b), 'Pulverized coal combustion in air and in O_2/CO_2 mixtures with NOx recycle', *Fuel*, 84, 2109–2115.

Maas U and Warnatz J (1988), 'Ignition processes in carbon-monoxide-hydrogen-oxygen mixtures', *Proc. Combust. Inst.*, 22, 1695–1704.

Marrero T R and E A Mason (1972), 'Gaseous diffusion coefficients', *J. Phys. Chem. Ref. Data*, 1, 3–118.

Michael J V, Su M-C, Sutherland J W, Carroll J J and Wagner A F (2002), 'Rate constants for $H + O_2 + M \rightarrow HO_2 + M$ in seven bath gases', *J. Phys. Chem. A*, 106, 5297–5313.

Mitchell R E, Kee R J, Glarborg P and Coltrin M E (1990), 'The effect of CO conversion in the boundary layers surrounding pulverized-coal char particles', *Proc. Combust. Inst.*, 23, 1169–1176.

Molina A and Shaddix C R (2007), 'Ignition and devolatilization of pulverized bituminous coal particles during oxygen/carbon dioxide coal combustion', *Proc. Combust. Inst.*, 31, 1905–1912.

Molina A, Hecht E S and Shaddix C R (2009), 'Ignition of a group of coal particles in oxyfuel combustion with CO_2 recirculation', *Proc. AIChE Conf.*, 31, 1905–1912.

Murphy J J and Shaddix C R (2006), 'Combustion kinetics of coal chars in oxygen-enriched environments', *Combust. Flame*, 144, 710–729.

Pallarés J, Arauzo I and Díez L I (2005), 'Numerical prediction of unburned carbon levels in large pulverized coal utility boilers', *Fuel*, 84, 2364–2371.

Paul P and Warnatz J (1998), 'A re-evaluation of the means used to calculate transport properties of reacting flows', *Proc. Combust. Inst.*, 27, 495–504.

Poling B E, Prausnitz J M, and O'Connell J P (2001), *The Properties of Gases and Liquids*, 5th Ed., New York, McGraw-Hill.

Rathnam R K, Elliott L K, Wall T F, Liu Y and Moghtaderi B (2009), 'Differences in reactivity of pulverized coal in air (O_2/N_2) and oxy-fuel (O_2/CO_2) conditions', *Fuel Proc. Technol.*, 90, 797–802.

Shaddix C R and Molina A (2008), 'Effect of O_2 and High CO_2 Concentrations on PC Char Burning Rates during Oxy-Fuel Combustion', *Proceedings of the 34th International Technical Conference on Coal Utilization and Fuel Systems*, Clearwater, FL, 1–5 June 2008.

Shaddix C R and Molina A (2009), 'Particle imaging of ignition and devolatilization of pulverized coal during oxy-fuel combustion', *Proc. Combust. Inst.*, 32, 1905–1912.

Solomon P R, Serio M A and Suuberg E U (1992), 'Coal pyrolysis: experiments, kinetic rates and mechanisms', *Prog. Energy Combust. Sci.*, 18, 133–220.

Suda T, Masuko K, Sato J, Yamamoto A and Okazaki K (2007), 'Effect of carbon dioxide on flame propagation of pulverized coal clouds in CO_2/O_2 combustion', *Fuel*, 86, 2008–2015.

Thiele E W (1939), 'Relation between catalytic activity and size of particle', *Ind. Eng. Chem.*, 31, 916–920.

Timothy L D, Sarofim A F and Béer J M (1982), 'Characteristics of single particle coal combustion', *Proc. Combust. Inst.*, 19, 1123–1130.

Toftegaard M B, Brix, J, Jensen P A, Glarborg P and Jensen A D (2010), 'Oxy-fuel combustion of solid fuels', *Prog. Energy Combust. Sci.*, 36, 581–625.

Toporov D, Bocian P, Heil P, Kellermann A, Stadler H, Tschunko S, Förster M and Kneer R (2008), 'Detailed investigation of a pulverized fuel swirl flame in CO_2/O_2 atmosphere', *Combust. Flame*, 155, 605–618.

Wall T F, Gupta R P, Gururajan V S and Zhang D-K (1991), 'The ignition of coal particles', *Fuel*, 70, 1011–1016.

Wall T F, Liu Y, Spero C, Elliott L, Khare S, Rathnam R, Zeenathal F, Moghtaderi B, Buhre B, Sheng C, Gupta R, Yamada T, Makino K and Yu J (2009), 'An overview on oxyfuel coal combustion – State of the art research and technology development', *Chem. Eng. Res. Des.*, 87, 1003–1016.

Wang C S, Berry, G F, Chang, K C and Wolsky A M (1988), 'Combustion of pulverized coal using waste carbon dioxide and oxygen', *Combust. Flame*, 72, 301–310.

Williams F A (2000), 'Progress in knowledge of flamelet structure and extinction', *Prog. Energy Combust. Sci.*, 26, 657–682.

Xu W C and Tomita A (1987), 'Effect of temperature on the flash pyrolysis of various coals', *Fuel*, 66, 632–636.

Zhao Y, Kim H Y and Yoon S S (2007), 'Transient group combustion of the pulverized coal particles in spherical cloud', *Fuel*, 86, 1102–1111.

7
Oxy-coal burner design for utility boilers

J. SHAN, Siemens Energy, USA and A. FRY,
Reaction Engineering International, USA

Abstract: This chapter discusses the design of oxy-coal burners intended for application in utility boilers, with the understanding that this is an emerging technology. Physical and operational constraints on the oxy-fired design are discussed, relative to traditional air-fired burners. These constraints result in an oxy-fired flame with delayed ignition and inhibited flame stability. Additional degrees of freedom are introduced into the burner design and operation with the use of pure oxygen. Leveraging these degrees of freedom allows the design of an air-like oxy-coal burner and firing system that will produce a stable flame with tailored shape and heat transfer profile.

Key words: oxy-fuel combustion, pulverized coal, low-NO_x burner, flame stability, utility boiler.

7.1 Introduction

Oxy-fuel burners are commonplace in high-temperature industrial furnaces today. Applications of oxy-fuel burners include: glass melting furnaces, electric arc furnaces, steel scrap melting, reheat and forging furnaces, soaking pit furnaces, ladle furnaces, aluminum melting furnaces, copper smelting and anode furnaces, hazardous waste incinerators, rotary enamel frit furnaces and lead melting furnaces. Motivating factors for use of oxy-fuel burners include: high flame temperature and heat transfer, fuel savings, reduction in size of gas handling equipment, increase in production, pollutant reduction and reduction in capital cost (Kobayashi and Tsiava, 2004).

Although the use of oxygen in burners is commonplace in industry today, application of oxy-fired burners in coal-fired utility boilers is an emerging technology. Their application in typical steam generation boilers requires design considerations beyond our current experience. Consequently there is no established, or widely accepted, design methodology for burners of this type. This chapter will focus on combustion and burner principles that will influence oxy-coal burner design and how they may differ from air-coal burner design considerations. It is hoped that these principles may be generally applied to oxy-coal burners designed for any application. The information presented here is intended to be an overview of the critical aspects of oxy-coal burner design as it exists today. This is by no means an in-depth survey of all of the details associated with this issue.

7.2 Overview of air-fired burner design methodology

Combustion of pulverized coal in water-walled utility boilers began to emerge in the early 1900s. By the late 1920s coal was being fired as it was directly supplied by online mills. In the 1930s and 1940s the use of pulverized coal as a fuel was widespread. The design of pulverized coal burners was well established in the 1950s and 1960s. At that time burners were typically designed with the coal introduced as an annular jet with a radial coal spreader at the burner tip. Inside the coal jet was an oil burner that was used for ignition at startup. A major portion of the combustion air was introduced in an annulus outside the coal jet. The air in this annulus had both axial and tangential components to its velocity. This burner design was oriented towards maximizing combustion efficiency and heat input per unit volume, and ensuring flame stability, which resulted in rapid mixing burners with high peak temperatures (Stultz and Kitto, 1992). While this design provided good combustion efficiency it had the undesirable byproduct of high NO_x, which today is considered a hazardous pollutant.

Beginning in the 1970s the design methodology for pulverized coal burners was completely revamped in order to address NO_x as a pollutant. It was found that when devolatilization and combustion occurred in a sub-stoichiometric environment, fuel nitrogen was converted to N_2 and thermal NO_x was limited by lower maximum combustion temperatures (Wall, 1987). Today, pulverized coal burners carefully control the mixing of coal with combustion air for the purpose of minimizing NO_x formation while limiting impacts on combustion efficiency and water wall wastage. Often times technologically advanced low-NO_x burners are implemented in concert with operation of the burner region at sub-stoichiometric conditions by introducing a portion of the combustion air into the furnace above the level of the burners through over-fire air (OFA) ports. Many of these firing systems can achieve up to a 70% reduction in NO_x emission compared with the turbulent burners of the 1950s and '60s (Shan *et al.*, 2009). NO_x emissions from a well-designed low-NO_x burner will on the order of 0.20 lb/MBtu for sub-bituminous coals and 0.25 to 0.30 lb/MBtu for bituminous coals without over-fire air (OFA), which can achieve an additional 30% reduction in NO_x. A good burner design should also provide a stable flame originating at the exit of the fuel injector without an oil or gas pilot flame for support. The combustion efficiency should be high with burnout of organic material in excess of 99%. More information concerning the evolution of coal-fired burners for utility boilers may be found elsewhere (Wall, 1987; Stultz and Kitto, 1992).

7.2.1 Physical design characteristics of low-nitrogen-oxide (NO_x) burners

There are three key physical processes that are important to the design of low-NO_x burners. These processes have been described in detail in Chapter 6 and in other works and will only be briefly summarized here as they pertain to low-NO_x burners.

Heat transfer

The first step towards the ignition of pulverized coal particles is the heat-up of the particles. This process begins to occur as the coal jet enters the furnace and entrains recirculated flue gases. The primary mode of heat transfer to the coal particles is convection, as discussed in Chapter 6, and is dependent on the temperature of the combustion gases that are recirculated within the coal jet and on the thermal conductivity of those gases. Heating rates in the range of 10^4–10^5°F/s are common in pulverized coal burners (Wall, 1987).

Chemistry

When the coal particles reach a temperature of about 750°F, volatile organic components rich in hydrogen begin to vaporize and leave the solid, which becomes more carbon rich. This process continues until only carbon and mineral matter remain in solid form, which is referred to as char. The devolatilized gases mix with the oxygen available in the coal carrier gas and gas from the secondary register of the burner. Ignition of this combustible mixture occurs when the flammability limit and ignition temperature is reached. The ignition of the coal particle is strongly dependent on local oxygen concentration and gas temperature.

The carbon char that remains after devolatilization reacts heterogeneously with species in the gas phase, mainly O_2 and CO_2, to form CO and CO_2, The CO produced further oxidizes in the gas phase by reacting with O_2. The rate of char oxidation depends on the concentration of O_2 in the local gases and on the matrix of gas through which O_2 must diffuse to arrive at the surface.

NO_x formation occurs in pulverized coal combustion systems primarily through two mechanisms. Approximately 75% of the NO_x is produced as fuel bound nitrogen is oxidized in the fuel lean portion of the combustion zone. This type of NO_x is often referred to as fuel NO_x. The remainder of the NO_x is formed as N_2 is oxidized by at high temperatures and is referred to as thermal NO_x. NO_x formation can be reduced drastically when the bulk of the combustion occurs within a zone where there is limited oxygen. Under these conditions NO is effectively reduced to N_2. To produce these conditions the oxygen must be only gradually mixed with the reacting coal particles. This mode of combustion also results in lower peak combustion temperatures limiting thermal NO_x. When the rate of particle heating is very high, the amount of volatiles released may exceed the measured value in the proximate analysis. This phenomenon further reduces the stoichiometry in the combustion zone, further reducing NO_x.

Aerodynamics

Aerodynamics of the burner is the means by which air, coal particles and flue gas from the furnace are mixed together. In order to control the NO_x chemistry, the mixing of air and coal must be performed gradually. In order to produce a flame that is strongly attached to the burner, hot flue gas must be recirculated back to the burner throat to heat the coal particles. Aerodynamics are used to enhance the heat transfer, stage the combustion of coal and control the reaction and flame shape. This is typically done by introducing swirled air through different annular regions within the burner and adding stabilizers. The coal transport gases are typically in the center and are referred to as the primary gas stream or coal jet. These gases are introduced through the burner in the primary register or coal injector. The amount of air used to transport the pulverized fuel into the furnace is far less than what is required for full combustion. Additional air is introduced through annuli surrounding the primary register. These air streams are referred to as the secondary air and they originate from the burner in the secondary register.

The mixing of these streams is controlled by two methods. First the velocities of each of the gas streams are significantly different causing shear at the boundary layer and therefore mixing. Second, swirling flows can be used to enhance mixing. Swirl is created when a tangential velocity component is introduced to the secondary air. Swirl causes internal recirculation zones which bring hot combustion gas into the primary to heat the particles and case mixing between the secondary and primary gases. Significant research has been invested in the study of burner aerodynamics. Details can be found elsewhere (Beer and Chigier, 1972).

7.2.2 Low-nitrogen-oxide (NO_x) burner design

Low-NO_x burners from various manufacturers today have a different design philosophy and characteristics than the first pulverized coal burners. However, they share some design aspects. Mixing of air with coal is controlled in stages through the use of discrete registers. Coal is transported to the burner entrained in air and is introduced through the primary register or fuel injector, which is collocated with the centerline of the burner. Surrounding the primary register are two annuli which together comprise the dual stage secondary register. The inner and outer secondary registers each have means to induce tangential velocity, or swirl, which can be independently adjusted. For modern designs the flow rate of air to each of the secondary registers can also be independently controlled. The angled or cone shaped refractory area at the burner exit is referred to as the quarl. A cross sectional view of a Siemens low-NO_x burner is presented in Fig. 7.1. Each of the previously mentioned components is depicted in the figure.

Fuel injector

The fuel injector is designed to introduce the dispersed coal particles in a medium of air into the furnace. The mass ratio of air to coal is dependent on the coal mill manufacturer and usually ranges from 1.75 to 2.2 with a typical value of 2.0. An air to fuel mass ratio of 1.8 produces a primary stoichiometric ratio of approximately

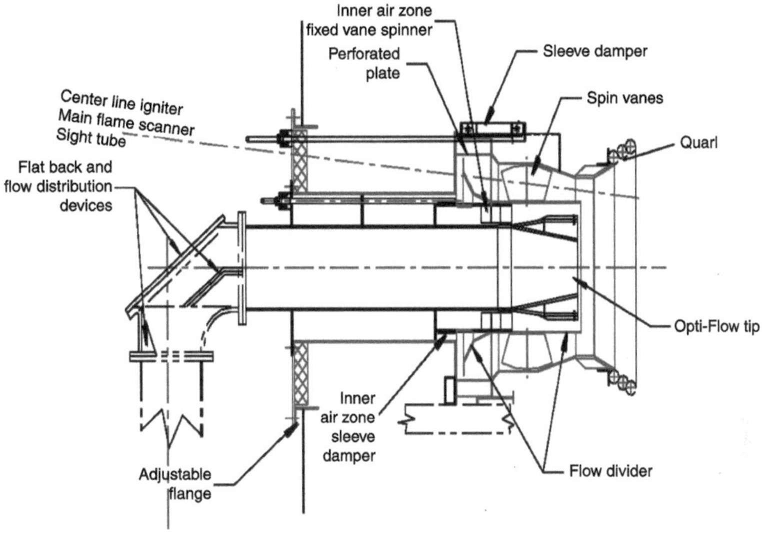

7.1 Cross sectional view of the Siemens Opti-Flow™ low-NO_x burner, showing discrete registers for fuel and air injection.

0.16, or 16% of the air necessary for complete combustion of the coal. According to the previous discussion of NO_x formation chemistry it is expected that lower NO_x concentrations are achievable with lower primary gas/fuel ratios. The diameter of the coal transport line is constrained by the minimum velocity at which coal particles remain entrained in the carrier gas, or the coal layout velocity. This velocity is generally accepted to be 50 ft/s (Wall, 1987). The dimension of the fuel injector itself is selected by the burner manufacturer to provide the desired gas and particle velocity at the exit of the burner. The velocity here is anywhere from 50 to 115 ft/s and is chosen to provide the desired near flame aerodynamics impacting the mixing between the primary and secondary air. In many applications, there is an elbow, scroll or turning head in the coal pipe at the burner inlet. Such inlet devices result in 'roping', or an uneven distribution of coal within the fuel injector. Many manufacturers use components to redistribute the coal particles with an even density around the circumference of the fuel injector at its exit. A uniform distribution is typically desired to minimize NO_x while maximizing combustion efficiency. The material of the fuel injector is chosen to be reliable under high temperatures and erosive conditions and is often a high grade of stainless steel. Another component of the fuel injector that is found on many commercial low NO_x burners is a flame stabilizer. The function of this feature is to provide a stagnation zone at the fuel injector exit on the boundary between the primary and secondary air where small-scale mixing of coal and air occurs, providing ideal conditions for ignition and flame attachment.

Dual stage secondary air register

The purpose of the secondary air register is to introduce the bulk of the combustion gas into the furnace in a manner that delays mixing with the primary gas stream. Approximately 74% of the stoichiometric combustion air is introduced through this register when the burner zone is staged to a stoichiometric ratio of 0.9. Typical low-NO_x operating conditions put 60–90% of the secondary air through the outer register, resulting in delayed mixing. The size of each of these registers is designed to target velocities that provide ideal aerodynamics for gradual mixing. The inner register will have an axial velocity that is significantly less than the primary with a typical value of about 65 ft/s. The outer register will typically have an axial velocity that is significantly more than the primary with a typical value in the range of 100 to 200 ft/s. The swirl of the secondary registers is used to induce an internal recirculation zone within the burner, heating the coal particles with hot combustion gases, causing rapid devolatilization and ignition near the fuel injector tip. Coals with low volatile matter are typically difficult to ignite and a stronger swirl can be used to tailor the ignition location and the shape of the flame.

Quarl

The burner quarl is the surface of the conical expansion occurring downstream of the dual stage air register. This component of the burner is another important element in tailoring flame aerodynamics. The rapid expansion of the cross section induces a recirculation zone external to the flame, which also brings hot combustion gases into the near burner region.

Over-fire air (OFA)

OFA is an essential component in a low-NO_x firing system in order to achieve reductions in NO_x emission on the order of 50% or greater. In a typical OFA configuration less than 100% of the stoichiometric combustion air is introduced through the burners. The remainder of the stoichiometric air and excess air is introduced into the furnace at an elevation significantly above the top row of burners, creating a lower furnace that is fuel rich with enough residence time to significantly reduce NO_x to N_2. The OFA is introduced through strategically placed ports that are often designed with two registers. Air introduced through the inner register will have an axial velocity of greater than 100 ft/s while the air introduced through the outer register will have an axial velocity in the range of 130–200 ft/s. These values vary widely depending on installation and manufacturer. Swirl is typically added to the outer register of the OFA. The design criterion for placement and velocity of OFA ports is to provide enough residence time for NO_x reduction and rapidly mix the O_2 into the rich gas from the lower furnace in order to complete combustion, minimizing CO and unburned carbon in fly ash.

Design and configuration of a combination of low-NO_x burners and OFA ports, a 'firing system', is boiler and fuel specific. An improperly designed system may achieve significant reduction in NO_x but may also cause problems with: slagging (deposition of mineral matter in the radiant section of the boiler, usually in a molten form), fouling (deposition of mineral matter in the convective section of the boiler, initiated by sticky alkali sulfate condensation), water wall corrosion, reduction in combustion efficiency (increase in unburned carbon and CO) and heat maldistribution. Proper design of a firing system can minimize these impacts while maintaining low NO_x emission.

The difference between a properly and improperly designed firing system is usually subtle, consisting of all of the same components with differences only in the operating conditions or placement. A tool commonly used for design and optimization of firing systems is computational fluid dynamics (CFD). CFD modeling can provide valuable insight into burner aerodynamics and mixing strategies. When coupled with mechanisms to describe all of the other physical and chemical processes occurring in solid fuel firing systems CFD modeling can be used to predict and understand details concerning the combustion characteristics, heat distribution, deposition and corrosion. CFD provides a method to evaluate alternative design configurations and operating scenarios in order to optimize performance and reduce the risk of modifying the system after installation. Specifically it can be used to: optimize air flow distribution and minimize pressure loss within the existing duct or windbox, evaluate combustion performance and OFA port locations for optimizing NO_x, CO, O_2 and unburned carbon, and evaluate the potential for increased slagging and/or water wall corrosion. During the process of developing a new burner, a detailed CFD model of the burner should be produced specific to the target furnace to evaluate the flame shape and ignition location and burner-to-burner interactions.

Figures 7.2 and 7.3 summarize CFD modeling results of a low-NO_x burner and OFA firing system in a 700 MW_e utility boiler. The boiler depicted in these figures is a wall-fired unit with four rows of burners on the front wall and three on the rear wall. One row of OFA ports is located on the front and rear walls at an elevation significantly above the upper row of burners. The burners are staged to a stoichiometric ratio of 0.9, with the remainder of the air introduced through the OFA ports. From the O_2 concentration plots in Fig. 7.3, it can be seen that the lower furnace is sub-stoichiometric, with an O_2 concentration of zero everywhere except directly in front of the burners. The mixed region above the OFA has an average O_2 concentration of about 3%.

Commercial burners and firing systems have been developed and installed worldwide by boiler manufacturers and engineering companies such as B&W, Foster Wheeler, Alstom, Siemens and others. As an example Siemens has produced the Opti-Flow™ low-NO_x burner. The specific piece of equipment shown in Fig. 7.4 has a nominal heat input of 222 MBtu/h with an approximate coal flow rate 18,950 lb/h and total secondary air flow rate 121,000 lb/h. Figure 7.5 shows a flame

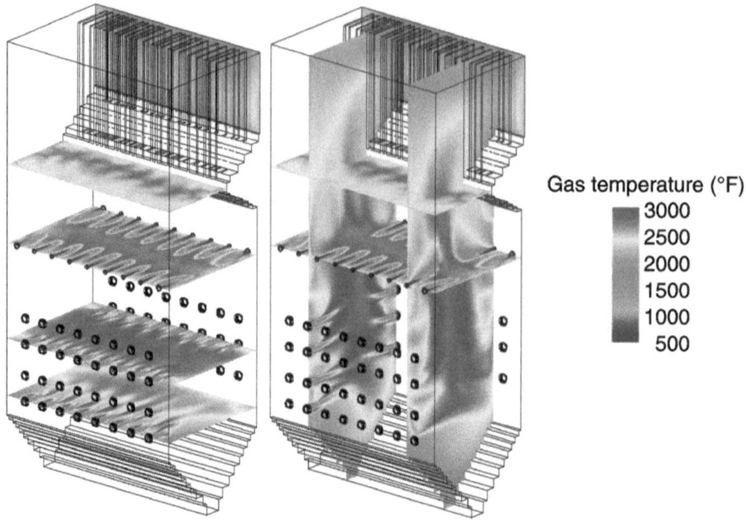

7.2 Predicted gas temperature distribution of a low-NO_x burner and OFA firing system in a 700 MW_e utility boiler using CFD. (Courtesy Reaction Engineering International, visualization performed with FIELDVIEW by Intelligent Light.)

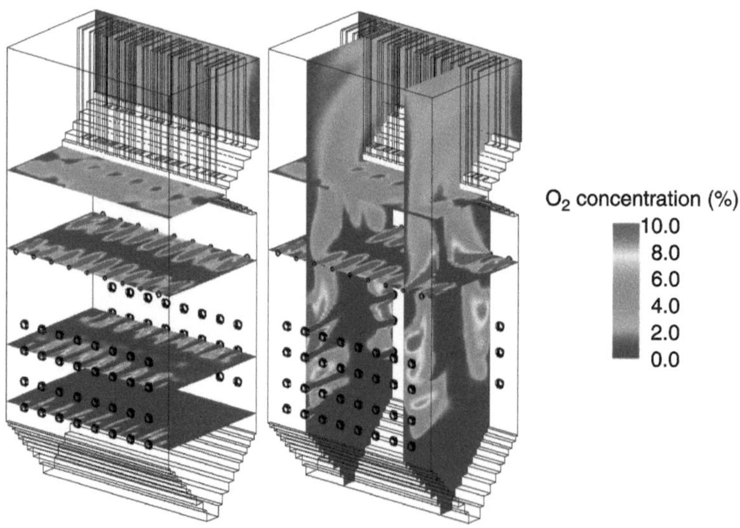

7.3 Predicted O_2 concentration distribution of a low-NO_x burner and OFA firing system in a 700 MW_e utility boiler using CFD. (Courtesy Reaction Engineering International, visualization performed with FIELDVIEW by Intelligent Light.)

7.4 Siemens' low-NO_x Opti-Flow™ burner.

7.5 Flame produced by Siemens' Opti-Flow™ low-NO_x burner operating in a 950 MW_e wall-fired utility boiler.

from Siemens Opti-Flow™ low-NO_x burner installed in a 950 MW wall-fired utility boiler. The clear flame definition from the background gas indicates the intensity of the flame. The bright flame and excellent flame stability are key indicators that the majority of volatile matter in the coal is released and ignited in the early phase of combustion.

7.3 Changes to burner design criteria and constraints

In order to discuss the impacts of oxy-coal combustion on burner design, it is first necessary to understand the design constraints and environment for which the burner will be applied. The greenfield design of a utility boiler for oxy-coal combustion has attractive possibilities. By designing for high temperatures, or mitigating temperatures by invoking internal gas recirculation into the flame, the size and capital cost of many of the fireside components of a steam generator may be reduced drastically. However, application of oxy-coal combustion in utility boilers will occur initially only in boilers of traditional air-fired design. This scenario is discussed extensively in other chapters and will only be summarized here. In an effort to match the heat transfer profile with an air-fired condition and to temper the flame temperature, approximately two-thirds of the flue gas in the boiler will be recycled back to the burner. The specific amount of flue gas recycle (FGR) will likely be chosen in order to match overall heat transfer with the air condition as opposed to adiabatic flame temperature. This will occur at approximately 27% O_2 in the overall O_2/FGR mixture.

7.3.1 Changes to the physical characteristics impacting burner design

The changes to the physical processes between air- and oxy-combustion have been discussed in detail in Chapter 6. This will be briefly reviewed here with a perspective on the design and operation of an oxy-coal burner.

Heat transfer

At 27% O_2 in the overall O_2/FGR mixture, the radiative heat transfer is expected to match with the air-fired condition. At this condition, the adiabatic flame temperature and the resulting temperature of the combustion flue gas is expected to be lower than the air-fired case. This has been shown in experimental studies and predicted in theoretical studies (Liu *et al.*, 2005a, 2005b; Khare *et al.*, 2008). In Chapter 6 it was shown that heat transfer to the particles is dominated by convection. With a decrease in gas temperature there would also be a decrease in convective heat transfer to the particles, resulting in a longer particle heat-up time. The emissivity of the flue gases has also increased due to the large partial pressures of CO_2 and H_2O. This is expected to impact the radiative heat transfer from the combustion gases to the water walls. However, the relative contribution of radiative heat transfer from the hot gas relative to that from the combusting particles and soot is still unclear.

Chemistry

The flame propagation speed in a coal-dust cloud using a microgravity facility has been investigated (Kiga *et al.*, 1997). It was determined that the flame propagation speed in an O_2/CO_2 environment is lower than in an air environment, which was attributed to the higher heat capacity of CO_2 compared with that of N_2. However, flame speed is strongly dependent on O_2 concentration. In the expected oxy-combustion environment these two differences would have an offsetting impact. Chapter 6 noted that the time delay for ignition is extended by the increased heat capacity of the gases and decreased by the faster reaction rate due to higher O_2 concentrations. Likely this is the physical reason for the reduction in flame speed. It was shown in Chapter 6 that devolatilization of the coal is slightly endothermic and is dependent on the particle temperature and heat-up rate. It follows that the rate of devolatilization will be reduced. However, the volatile yields of various coals in a medium of CO_2 and under a high heating rate are higher than those in air. The yields are also higher than the coal proximate analysis volatile value. The higher volatile yield in CO_2 is likely due to char-CO_2 gasification (Wall *et al.*, 2009). Char gasification reactions are endothermic and can reduce peak coal particle temperatures, thereby reducing the overall rate of char conversion at certain conditions (Shaddix *et al.*, 2010). The diffusivity of O_2 in CO_2 is lower than in N_2. However, the driving force for diffusion is potentially higher due to higher O_2 concentrations in some regions of the burner.

Apart from the chemistry which affects the flame directly, there are expected changes in the way that NO_x is formed and destroyed. NO_x is recycled to the burner with the FGR. If the FGR is introduced in a hot and fuel rich environment, the recycled NO_x is effectively reburned (Fry *et al.*, 2010). It has been shown that NO_x emissions are significantly lower for oxy-coal combustion with recycle (Kiga *et al.*, 1997; Fry *et al.* 2010). This is primarily due to the reduced mass of flue gas (including NO_x) exiting the boiler stack under oxy-fired conditions. It has also been shown that the magnitude of NO_x emission reduction is strongly dependent on the burner swirl (Chui *et al.*, 2003; 2004). NO_x emission is also affected by the flame attachment for both air and oxy-combustion cases (Shan *et al.*, 2010). Emission of other pollutants will not be addressed here as they are less dependent on burner design and operation.

Aerodynamics

The volume of gases introduced through the burner is lower in the oxy-fired case than in the air-fired case. This is primarily due to the higher overall O_2 concentration or smaller volume of dilution gas (CO_2 vs. N_2). This difference will have impacts on velocities in the burner, unless the burner is sized specifically for the oxy-combustion condition. The difference in velocities both in the burner primary and secondary registers may have impacts on swirl and the formation of internal recirculation zones (Khare *et al.*, 2008).

The overall impact of these changes is that the ignition of pulverized coal particles is delayed and the flame stability is inhibited in the oxy-fired case when compared with the air-fired condition (Khare *et al.*, 2008; Fry *et al.*, 2010).

7.3.2 Changes to the design constraints impacting burner design

Potentially more important than changes to the heat transfer, chemistry and aerodynamics are changes to burner design constraints. These constraints are significantly impacted by the use of pure oxygen and by the operating conditions. In air-fired combustion, air and coal are the only two components available to mix in the burner. For oxy-combustion the number of components increases to three with coal, oxygen and FGR. This provides many new options, or degrees of freedom, with how and where these components are introduced. Air-fired combustion is limited to a mixture of oxygen and nitrogen containing 21% O_2 by volume. Oxy-fired combustion with FGR can achieve a range of O_2 concentrations in the mixture of O_2 and FGR that can vary depending on the location within the burner. In addition, the composition of the FGR can be varied depending on where the recycle stream originates and what cleanup equipment is installed. These additional degrees of freedom can add challenges and complexity to the hardware and to the operating conditions, but can also be advantageous for overcoming delayed ignition discussed in the previous section.

The Occupational Safety and Health Administration (OSHA) classifies a mixture of gases containing more than 23.5% O_2 by volume an oxidizer (OSHA, 1992). In addition, utilizing gas mixtures with O_2 concentrations greater than 23.5% in non-fuel laden gas streams requires consideration of materials of construction for the handling equipment and necessitates special procedures for preparing/cleaning that equipment for service (Compressed Gas Association, Inc., 2004). These regulations have profound impacts on burner design and operation. According to these safety regulations, gas mixtures containing more than 23.5% O_2 should not be contacted with combustible materials, providing a clear indication of the maximum O_2 concentration that can be used in the coal transport gases and the burner primary. Upon more detailed consideration, the limitation may be much more constrictive than 23.5% for the burner primary. In order to avoid contacting pulverized coal particles with local regions of high O_2 concentration in the coal transport gas, the blending of O_2 and FGR must occur upstream of the coal mills. Under such a configuration, there exists the possibility that, despite fail safe controls and protocols, gases enriched in O_2, or pure O_2, could be introduced into the coal mills. This inherently dangerous scenario will likely be avoided for oxy-coal combustion in utility boilers (Zanganeh and Shafeen, 2007; Toftegaard et al., 2010). Therefore oxygen concentrations in the coal transport gas will be limited to that in the FGR, typically about 3% wet O_2.

It is likely that the constraints on the range of mill operation will limit the gas to fuel ratios attainable for oxy-fired conditions. It is not clear how the behavior of the mills will change when operated with FGR as the carrier. This is a topic that requires further research. This constraint will impact the operating conditions of the burner fuel injector.

7.4 Oxy-coal burner principles

Constraining an oxy-coal burner to operate with sufficient FGR to achieve 27% O_2 in the O_2/FGR mixture affords the designer the opportunity to create a burner that is strikingly similar to air-coal. As previously discussed this is the condition where radiative heat transfer is expected to be matched and is likely to be used when applying oxy-coal combustion in a boiler designed for air combustion. Indeed this pathway may not fully leverage the new degrees of freedom and completely optimize oxy-coal combustion with FGR. However, it provides the simplest and most familiar path forward and is a natural evolution from the low-NO_x burner designs as they exist today. In addition, it may be advantageous or even necessary to construct a utility boiler that is dual-fired, or can handle combustion of coal with either air or oxygen (McDonald and Tranier, 2010).

The remainder of this chapter will focus on principles of burner operation and design that include strategies of mixing coal, O_2 and FGR that can produce a stable oxy-coal flame in a utility boiler environment.

7.4.1 Delayed ignition

It has been previously noted that ignition of an oxy-coal flame is delayed when compared with an air-fired flame. This was due primarily to the impacts on aerodynamics, chemistry and heat transfer when replacing N_2 with CO_2. Delayed ignition may easily be overcome by leveraging these differences even within a traditional air-fired burner.

Aerodynamics

It is not necessary for the velocities in the oxy-fuel burner to be the same as the air-combustion case. In fact, the smaller volume of gas at 27% overall oxygen concentration makes this an inherent condition in a dual-fired capable burner. It has been demonstrated through pilot-scale experimentation that matching velocity in the burner fuel injector, or primary register, with the air-fired condition may result in a detached flame. Conversely, operating with a matched momentum or mass to fuel ratio can provide an attached flame (Fry *et al.*, 2010). Optimum operation of the coal mills with FGR will dictate the primary gas to fuel ratio achievable for the oxy-coal condition. The size of the burner fuel injector may then be assigned in order to achieve an acceptable primary velocity for the oxy-fired case. The magnitude of the acceptable primary velocity will vary by burner manufacturer and specific burner components. Pilot-scale experiments suggest that this velocity may be estimated by matching the gas to fuel ratio or momentum with the air-fired conditions for a given fuel injector cross sectional area (Fry *et al.*, 2010). In the pilot-scale experiments a reduction in velocity of approximately 13% was necessary to strongly attach the flame under oxy-coal conditions.

Adjustment of the swirl vanes may be necessary in order to reestablish the internal recirculation zone in the burner and attach the flame when applying the lower velocities for the oxy-coal condition. Increasing swirl is also expected to increase the circulation of hot flue gases back to the burner face thereby increasing the heat transfer to the particles and the particle heat-up rate. This strategy of overcoming the delayed ignition may be limited by pressure drop or hardware configuration.

Chemistry and heat transfer

Leveraging the new degrees of freedom may provide a more effective method of overcoming delayed ignition when oxy-firing a dual-fired burner. Previous discussions of chemistry and heat transfer impacts showed that delay in ignition was influenced by many factors including: delayed particle heat-up due to reduced heat transfer – which also reduces the rate of devolatilization – lower flame propagation speed and lower diffusivity of O_2 in CO_2. The negative impacts of these factors may be overcome by operating with locally high O_2 concentrations. With higher O_2 concentrations, the adiabatic flame temperature increases, the flame propagation speed increases and the driving force for diffusion increases. Further enrichment of the primary with O_2 is unlikely due to the cited safety concerns. However, enrichment of the inner secondary register with O_2 is easily accomplished and provides the necessary oxygen on the boundary of the primary and secondary gas streams where the flame is initiated. In pilot-scale testing, this has also been proven to strongly attach an oxy-coal flame (Fry *et al.*, 2010).

7.4.2 Oxygen (O_2) injection

Pure oxygen injection at the burner face may be an attractive alternative to blending O_2 in with the FGR prior to the burner under some conditions. This certainly may be the case with the coal transport gases. Such injection may also be used to stabilize or strongly attach a flame. Pilot-scale testing has demonstrated this in several different cases (Kimura *et al.*, 1995; Croiset *et al.*, 2000; Bool *et al.*,

2009; Fry *et al.*, 2010). A key consideration of oxygen injection at the burner face is where to put the oxygen. According to the experiments performed, O_2 injection is most effective when targeting the boundary layer between the primary and secondary where ignition is likely to occur. Effective utilization of this method may require appropriate mixing conditions along this boundary initiated by velocity differential and shear.

7.4.3 Primary with no oxygen (O_2) enrichment

It has been previously discussed that enrichment of the coal transport FGR with oxygen is unlikely. This will result in an O_2 concentration in the primary register of approximately 3%. This condition has been used in pilot-scale experiments and resulted in a stable and attached flame (Fry *et al.*, 2010). The principles described elsewhere in this section may be used to stabilize and shape the flame without O_2 enrichment in the primary.

7.4.4 Integration of burner and over-fire air (OFA) firing system

The influence of burner staging on NO_x emission as been previously discussed. Under air- and oxy-fired conditions, burner staging is necessary to achieve minimum NO_x emission. While this objective may remain important in an oxy-coal utility boiler, there is another attractive incentive to burner staging. Using an FGR rate sufficient to produce 27% O_2 in the O_2/FGR mixture is necessary in order to approximately match the heat transfer in the radiative section of the boiler. This same condition may not be optimal to match heat transfer in the convective section, due to the lower volume of flue gas. Under oxy-coal combustion conditions, integrating the burners and OFA as a firing system will allow the flexibility to tune heat transfer independently for both the radiative and convective sections of the furnace. Additional O_2 and FGR may be introduced through the OFA ports with selective ratio designed to maintain a desired boiler outlet O_2 concentration while boosting the volume of gases passing through the convective section of the boiler.

7.5 Commercial oxy-coal burners

There have been several demonstrations of commercial oxy-coal burners in boilers and furnaces of varying size (Strömberg *et al.*, 2008; Cameron *et al.*, 2010; Fry *et al.*, 2010; Weirich *et al.*, 2010). Examples of the burners used in two of these demonstrations are shown in Fig. 7.6 and 7.7 respectively. Figure 7.6 shows a picture of the oxy-coal research burner developed by Siemens and Reaction Engineering International for the 1.2 MW pulverized coal furnace operating at the University of Utah. This research burner was developed based on existing Siemens

7.6 Oxy-coal burner (left) developed by Siemens and Reaction Engineering International for the 1.2 MW furnace at the University of Utah (right).

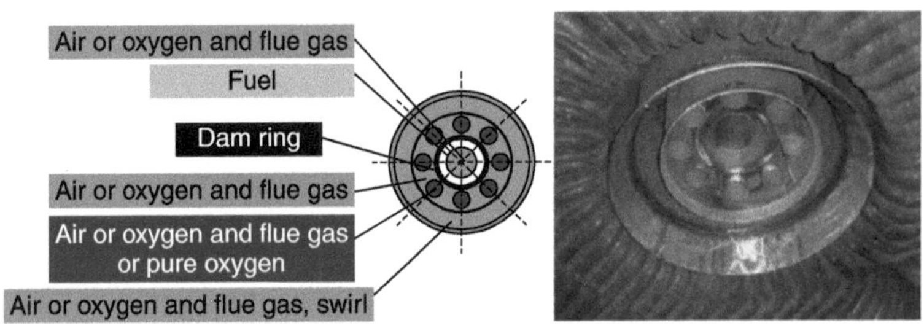

7.7 30 MW_{TH} oxy-coal burner developed by Alstom for Vattenfall's Schwarze Pumpe demonstration facility.

patented oxy-fuel burner technology and contains all of the design characteristics of a traditional low-NO_x burner. CFD modeling provided valuable insight regarding flame shape, particle heating rate and ignition location while designing this pilot-scale burner. This burner is capable of operating over a range of substoichiometric and superstoichiometric conditions and allows introduction of oxygen mixed with FGR in the primary and secondary registers. Pure oxygen can also be added on the burner centerline and at the primary/secondary boundary. Alstom has developed a 30 MW_{TH} oxy-coal burner for Vattenfall's Schwarze Pumpe oxy-coal combustion and CO_2 capture demonstration plant in Germany. Alstom's oxy-coal burner is depicted in Fig. 7.7.

This burner has a primary register for coal injection with an oxygen or air injector down the center. Around the outside of the exit to the primary are dam rings to stabilize the flame. Outside the primary are two secondary registers, The inner register has injection pipes to adjust the oxygen concentration as needed. The primary transport gas is recycled gas with low O_2. The outer zone with swirl and dam ring can create the gas recirculation to help the flame stability. The injection pipes with higher oxygen concentration or even pure oxygen can enhance the local combustion and generate a bright and stable flame even with low O_2 in the primary zone. The preliminary testing results verified that higher O_2 concentration can help the flame stability. Although this burner seems to have deviated significantly from traditional low-NO_x burner design, there is a fuel injector, or primary register, at the center of the burner and two secondary registers, an inner and outer annulus around the primary. This burner deviates from traditional design in that there are injectors for air or oxygen in the center of the primary and distributed through the inner secondary registers (Strömberg *et al.*, 2008).

Both the Siemens/REI burner and the Alstom burner successfully utilized design principles common to traditional air-fired burners (primary, dual-stage secondary, etc.) combined with the flexibility of oxygen injection at multiple locations to investigate burner performance under different oxy-firing scenarios. The Alstom burner, while fired in Vattenfall's Schwarze Pumpe demonstration plant, has been used to produce process steam under both air- and oxy-fired conditions over an extended period of time. These programs along with other commercial burner programs, cited previously, represent state-of-the-art in oxy-coal burners today. The next step in the development and testing of oxy-coal burners will be to implement them in the demonstration of full-scale boilers with multiple burners. At this scale and configuration further research may be performed concerning: burner–burner interaction and flame aerodynamics, integration of OFA for heat transfer tuning and water wall tube corrosion.

7.6 Conclusions

Oxy-fuel burners are currently in widespread use in industrial applications. However, oxy-coal burners for utility boilers are an emerging technology. Currently there is no universally accepted design methodology for oxy-coal burner design.

Air-coal burners have evolved over the century that they have been in use. Initially the design philosophy was to rapidly mix the coal and air, producing a short, stable flame with high heat density and temperatures. Awareness of NO_x as a pollutant and the involvement of burners in its formation prescribed the design methodology for modern air-coal burners. Air-coal burners today internally stage combustion, which lowers heat release density, temperatures and NO_x.

Initially, oxy-coal firing will occur only in utility boilers of traditional air-fired design. To maintain a typical heat transfer profile approximately two-thirds of the flue gas will be recycled and introduced through the burners to produce an O_2 concentration of approximately 27% in the overall mixture of O_2 and FGR. This is expected to provide similar heat transfer in the radiative section of the boiler with air-fired operation, but will also result in a lower adiabatic flame temperature than the air-fired condition. In addition, the first oxy-coal utility boilers will likely be dual-fired, or capable of air- or oxy-coal combustion. These constraints will result in a burner design that is strikingly air-like.

There are three general physical and chemical characteristics that will be impacted by a conversion from air- to oxy-coal combustion that have strong influence over burner design and operation. These characteristics and the expected changes are:

1 Heat transfer
   (a) Heat transfer to pulverized coal particles is dominated by convection and is therefore a function of gas temperature and thermal conductivity.
   (b) Gas temperature will be reduced; therefore particle heat-up time will be increased.
2 Chemistry
   (a) Flame propagation speed is reduced, but is strongly dependent on O_2 concentration.
   (b) Time delay for ignition is controlled by heat capacity of the gases and is increased.
   (c) Devolatilization is a function of particle heating; therefore it will decrease.
   (d) Diffusivity of O_2 is lower in CO_2 than N_2.
3 Aerodynamics
   (a) The volume of gases through the burner is reduced; therefore the velocity of the gases will be reduced.
   (b) Lower velocities can reduce swirl and upset the internal recirculation zones in the burner.

The overall impact of the changes in physical and chemical characteristics of combustion is to delay ignition and inhibit flame stability.

There are new operational constraints that will impact burner design and operation. The number of independently controlled streams entering the burner has increased by one for each register of the burner with air being replaced by oxygen and FGR. Therefore there are now more degrees of freedom in the operation of the burner. Oxygen concentration within discrete zones of the near burner region can now be varied across a broad range. Enriching the coal transport gas with oxygen presents an acute safety hazard and is unlikely to occur. Therefore the primary zone of the burner will be limited to the oxygen concentration of the exit flue gas.

Principles have been suggested for the design and operation of oxy-coal burners. These principles are as follows:

1 Delayed ignition and inhibited flame stability can be easily overcome by leveraging new degrees of freedom and by slightly modifying air-fired design or operation.
2 The velocity of the primary gas stream should be reduced from the air-fired condition by approximately 13% to stabilize the flame.
3 Swirl should be adjusted to reestablish internal recirculation zones that were disrupted due to drop in burner velocities.

4 Oxygen enrichment and pure oxygen injection can be used to:
   (a) create regions of high temperature to increase the rate of particle heat-up and devolatilization,
   (b) increase flame propagation speed,
   (c) overcome time delay for ignition.
5 Oxygen enrichment and oxygen injection are most effective when introduced on the boundary layer between the primary and secondary.
6 The coal transport gases and burner primary should not be enriched with oxygen:
   (a) it is not necessary for flame stability and flame shape and burnout can be tailored using other burner principles.
7 Burners and OFA should be integrated into a firing system used to tailor heat transfer profile through the radiative and convective heat transfer sections of the boiler.

Commercial oxy-coal burners have been designed and tested at various scales. These burners share design principles with modern low-NO_x burners and have been modified to produce stable oxy-coal flames.

7.7 References

Beer J M and Chigier N A (1972), *Combustion Aerodynamics*, London: Applied Science Publishers.

Bool L, Rosen L, Laux S and Kobayashi S (2009), IFRF Members Conference, *Direct Injection of Oxygen in Pulverized Coal Burners for Flame Optimization* (Technical Session), 8 June 2009, Boston, MA.

Cameron E D, Sturgeon D W, McGhie C and Fitzgerald F D (2010), 'Demonstration of the Doosan Babcock 40MW_T OxyCoal™ Combustion System', Proceedings of the 35th International Technical Conference on Coal Utilization and Fuel Systems, Clearwater, FL, 7–10 June 2010.

Chui E H, Douglas M A and Tan Y (2003), 'Modeling of oxy-fuel combustion for a western Canadian sub-bituminous coal', *Fuel*, 82, 1201–1210.

Chui E H, Majeski A J, Douglas M A, Tan Y and Thambimuthu K V (2004), 'Numerical investigation of oxy-coal combustion to evaluate burner and combustor design concepts', *Energy*, 29, 1285–1296.

Compressed Gas Association, Inc. (2004), *CGA G-4.1–2004 Cleaning Equipment for Oxygen Service*, 5th edition, Chantilly, VA.

Croiset E, Thambimuthu K V and Palmer A (2000), 'Coal combustion in O_2/CO_2 mixtures compared with air', *Can J Chem Eng*, 78, 402–407.

Fry A, Adams B and Shan J (2010), 'Oxy-burner retrofit principles for existing coal-fired utility boilers', Proceedings of the 35th International Technical Conference on Coal Utilization and Fuel Systems, Clearwater, FL, 7–10 June 2010.

Khare S P, Wall T F, Farida A Z, Liu Y, Moghtaderi B and Gupta R P (2008), 'Factors influencing the ignition of flames from air-fired swirl pf burners retrofitted to oxy-fuel', *Fuel*, 87, 1042–1049.

Kiga T, Takano S, Kimura N, Omata K, Okawa M, Mori T and Kato M (1997), 'Characteristics of pulverized-coal combustion in the system of oxygen/recycled flue gas combustion', *Energy Convers Manage*, 38, 129–134.

Kimura N, Omata K, Kiga T, Takano S and Shikisima S (1995), 'The characteristics of pulverized coal combustion in O_2/CO_2 mixtures for CO_2 recovery', *Energy Convers Manage*, 36, 805–808.

Kobayashi H and Tsiava R (2004), 'Oxy-Fuel Burners', in Baukal C E (ed.), *Industrial Burners Handbook*, New York: CRC Press, 693–723.

Liu H, Zailani R and Gibbs B M (2005a), 'Comparisons of pulverized coal combustion in air and in mixtures of O2/CO2', *Fuel*, 84, 833–840.

Liu H, Zailani R and Gibbs B M (2005b), 'Pulverized coal combustion in air and in O2/CO2 mixtures with NOx recycle', *Fuel*, 84, 2109–2115.

McDonald D and Tranier J P (2010), 'Oxy-coal is ready for demonstration', Proceedings of the 35th International Technical Conference on Coal Utilization and Fuel Systems, Clearwater, FL, 7–10 June 2010.

OSHA (1992) (ca.), GHS – OSHA HCS Comparison of Hazard Communication Requirements. Available from: http://www.osha.gov/dsg/hazcom/docs/ghsoshacomparison.pdf (Accessed 17 June 2010).

Shaddix C R, Hecht E S, Geier M, Molina A and Haynes B S (2010), 'Effect of gasification reactions on oxy-fuel combustion of pulverized coal char', Proceedings of the 35th International Technical Conference on Coal Utilization and Fuel Systems, Clearwater, FL, 7–10 June 2010.

Shan J, Fry A and Adams B (2010), 'Oxy-burner retrofit principles for existing coal-fired utility boilers', Proceedings of the 35th International Technical Conference on Coal Utilization and Fuel Systems, Clearwater, FL, 7–10 June 2010.

Shan J, Hart N, Schiazza G, Lindeman M and Vatsky J (2009), 'Recent field experiences with the Mark II Opti-Flow™ burner in minimizing unburned carbon and NO_x', Proceedings of the 34th International Technical Conference on Coal Utilization and Fuel Systems, Clearwater, FL, 31 May–4 June 2009.

Strömberg L, Lindgren G, Jacoby J, Giering R, Anheden M, Burchhardt U, Altmann H, Kluger F and Stamatelopoulos G (2008), 'First test results from Vattenfall's 30 MW_{th} oxyfuel pilot plant in Schwarze Pumpe', Proceedings of the 34th International Technical Conference on Coal Utilization and Fuel Systems, Clearwater, FL, 31 May–4 June 2009.

Stultz S C and Kitto J B (eds.) (1992), *Steam: Its Generation and Use*, 40th edition, Ohio: Babcock and Wilcox, 13.

Toftegaard M B, Brix J, Jensen P A, Glarborg P and Jensen A D (2010), 'Oxy-fuel combustion of solid fuels', *Prog Energy Combust Sci*, 36, 581–625.

Wall T F (1987), 'The Combustion of Coal as Pulverized Fuel through Swirl Burners', in Lawn C J (ed.), *Principles of Combustion Engineering for Boilers*, London: Academic Press, 197–335.

Wall T F, Liu Y, Spero C, Elliott L, Khare S, Rathnam R, Zeenathal F, Moghtaderi B, Buhre B, Sheng C, Gupta R, Yamada T, Makino K and Yu J (2009), 'An overview on oxyfuel coal combustion – state of the art research and technology development', *Chem Eng Res Des*, 87, 1003–1016.

Weirich T, Leisse A, Niesbach J, Rehfeldt S, Kuhr C and Koczorowski H J (2010), 'Flame investigations of coal and biomass combustion with a 35 MW DS – burner and modifications for indirect firing', Proceedings of the 35th International Technical Conference on Coal Utilization and Fuel Systems, Clearwater, FL, 7–10 June 2010.

Zanganeh K E and Shafeen A (2007), 'A novel process integration, optimization and design approach for large-scale implementation of oxy-fired coal power plants with CO_2 capture', *Int J Greenhouse Gas Control*, 1, 47–54.

8

Pollutant formation and emissions from oxy-coal power plants

Y. TAN, CanmetENERGY, Natural Resources Canada, Canada

Abstract: Oxy-fuel combustion presents some unique features that can impact the emissions of major air pollutants, such as NO_x, SO_x, trace metals and fine particulate. Some pollutants (NO_x) are positively influenced in that their emissions can significantly decrease as compared with the conventional air-blown combustion mode. Other pollutants (SO_x) can accumulate in the flue gas recycle loop in such a way that their concentrations can double or even quadruple. This chapter will look at the impact of oxy-fuel combustion on these pollutants and their subsequent impact on oxy-fuel plant design and operation, including approaches for flue gas recirculation and some innovative integrated emissions control concepts.

Key words: pollutant emissions, NO_x and SO_x, mercury, ash deposits, integrated emission control.

8.1 Introduction

Oxy-fuel combustion presents some interesting issues when dealing with emissions. In a conventional power plant, the flue gas leaves the stack along with all the air pollutants that it contains. In an oxy-fuel fired power plant, the purpose is to capture the flue gas that is highly enriched in CO_2 and then compress, transport and dispose of the CO_2. As such, under ideal conditions, oxy-fuel combustion can be considered as almost zero-emissions technology with no pollutants emitted to the atmosphere. Considering this, one might think that emissions and their controls would not constitute a significant issue. This, however, is not the case.

At this time, the oxy-fuel-fired power plant requires recycling a significant portion of the flue gas back into the feed gas stream to control the combustion temperature, which would otherwise be too high, and furnace heat transfer. Depending on where this recycled flue gas stream is drawn, it will contain various amounts of pollutants, contributing to accumulations of certain species, notably sulphur compounds. This accumulation may have deleterious effects on a number of equipments as well as plant operation and, in many cases, these pollutants need to be minimized or removed before the flue gas can be recycled.

Another issue that needs to be considered here is the impact of pollutants on the downstream flue gas compression train. Most compression trains are based on existing technologies that were previously used for relatively pure CO_2 compression, and although current designs are beginning to take the existence of pollutants into account, at this time it seems that it is desirable to reduce flue gas impurities to as low a level as possible.

Oxy-fuel combustion produces the same major pollutants as conventional air-fired combustion, namely SO_x, NO_x, CO, trace metals including mercury and particulate matter. The fundamental chemical mechanisms for the formation and destruction of these pollutants do not change very much in oxy-fuel combustion, although the enhanced flue gas recirculation and locally high oxygen concentration may have significant effects on certain aspects of these fundamental mechanisms for specific species, such as NO_x.

One of the major differences between oxy-fuel combustion and air-fired combustion in terms of emissions is, of course, the high concentrations of carbon dioxide in the flue gas. However, since CO_2 is a relatively stable species under normal conditions, this does not alter the overall mechanisms for major pollutants in significant ways, except for CO in the flame zone.

One of the many advantages of an oxy-fuel power plant is that it requires very little modification relative to an air-fired power plant, and this is especially true when it comes to pollution control. All the well-established pollution control equipment such as electrostatic precipitator (ESP), fabric filter and scrubbers can be used either directly or in slightly adapted forms for oxy-fuel-fired power plants. One of the main exceptions to this is the low-NO_x burner. Due to the higher oxygen concentration required (~30% by volume dry depending on the characteristics of coal and mode of flue gas recirculation), there can be a number of ways to introduce oxygen, and this will certainly require modifications to the burner design. Since NO_x formation is highly sensitive to oxygen, any burner design modifications can have consequences on the NO_x being produced.

In the following sections of this chapter, we will focus our discussions on the major air pollutants, i.e., NO_x, SO_x, trace metals including mercury and particulate, in the context of oxy-fuel combustion.

8.2 Nitrogen oxide (NO_x) emissions

Emissions of NO_x and its mechanisms have been the subject of extensive studies, resulting in a large amount of literature. Most of these findings are still applicable to oxy-fuel combustion. However, with the increased oxygen concentration and flue gas recirculation, some specific aspects of its mechanism could be enhanced or reduced depending on circumstances. Before we proceed to explain the differences, it is useful to briefly review NO_x mechanisms in conventional air-fired combustion. For a more comprehensive understanding of the NO_x mechanism, readers are encouraged to refer to Glarborg *et al.* (2003).[1]

It is well accepted that NO_x resulting from coal combustion arises from three sources: thermal NO_x, fuel NO_x and prompt NO_x. Thermal NO_x refers to NO_x formed with reactions between molecular nitrogen and oxygen. Thermal NO_x formation is a high-temperature process, effective at temperatures above 1400°C.

Its mechanism, also known as the Zeldovich mechanism, can be described with these reactions:

$$N + O_2 = NO + O$$
$$N + NO = N_2 + O \qquad [8.1]$$
$$N + OH = NO + H$$

It should be noted that the Zeldovich mechanism can contribute to the formation as well as the destruction of NO, depending on specific circumstances.

Prompt NO refers to the reaction between N_2 and hydrocarbon radicals such as CH and CH_2. Prompt NO usually accounts for a very small part of overall NO_x emissions from coal combustion under typical operating conditions.

Fuel NO comes from N_2 species bound in fuel, which, upon devolatilization, are divided into two routes – volatile-N, such as HCN and NH_3, and char-N. Volatile-N reacts with other species during the combustion processes to produce either NO_x or N_2, depending on various factors, such as the availability of oxygen. Char-N goes through heterogeneous reactions along with char oxidation. At typical coal combustion temperatures (below 1400°C), fuel NO is by far the dominant source for NO production with thermal NO_x as a minor contributor. However, as temperature rises above 1400°C, the contribution from thermal NO_x dramatically increases and can eventually surpass that of fuel NO_x when temperature rises above 1500°C.

The literature has pointed out that NO_x emissions in coal combustion are heavily influenced by combustion temperature and atmospheric environment; as a result, some NO_x reduction technologies are based on these interactions. Air staging, for example, involves diverting a portion of combustion air further downstream of combustion to create an environment that drives nitrogenous species to form preferentially N_2 instead of NO_x. The drawback of some of these approaches is slight loss of combustion efficiency, since fuel burnout will be affected; however, improved applications of these technologies, such as fuel staging, have greatly reduced this issue.

As far as oxy-fuel combustion is concerned, all conventional NO_x reduction strategies should, in principle, be applicable. However, with pure oxygen and recycled flue gas being introduced into the furnace, oxy-fuel combustion offers some unique opportunities for aggressive NO_x reduction. One of the widely reported observations concerning oxy-fuel coal combustion is the considerable reduction in NO_x emissions compared with air-fired combustion under similar conditions. These reductions varied from 30% to 60%, depending on the type of coal, burner configurations and flue gas recirculation ratio.

The most readily available reason for this reduction is obviously the suppression of the thermal NO_x mechanism due to the lack of nitrogen in the feed gas. However, since thermal NO_x is not a major source of NO_x emissions, the extent of the reduction of NO_x emissions is such that this cannot be the only reason.

Okazaki and Ando[2] investigated the NO_x reduction mechanism in oxy-fuel mode using a one-dimensional, premixed reactor. Their results showed that the effect of increased CO_2 concentration on NO_x emissions is, as expected, insignificant. On the other hand, most of the observed NO_x reduction is due to the fact that a large portion of the flue gas is being recycled and, through chemical reactions, more than 50% of NO in the recycled flue gas is reduced to N_2 in the zone of combustion of coal volatile matter.

While Okazaki and Ando did not specifically mention the chemical reactions involved in NO_x reductions, it has been reported in the literature that NO_x emissions could be lowered through what is commonly termed the NO_x reburn mechanism. This involves reactions between recycled NO and devolatilized hydrocarbons and nitrogenous species in the flame leading to the destruction of NO and the formation of N_2:[3]

$$\begin{aligned} &NO + CH_i \rightarrow HCN + O \\ &NO + NH_i \rightarrow N_2 + H_2O \end{aligned} \qquad [8.2]$$

HCN is then converted to N_2 in the flame through:

$$\begin{aligned} &HCN + O = NCO + H \\ &NCO + H = NH + CO \\ &NH + H = N + H_2 \\ &N + NO = N_2 + O \end{aligned} \qquad [8.3]$$

Another possible contributor to the NO_x reduction in coal combustion is the reactions between char and NO:[4]

$$\begin{aligned} &2(\text{char})\text{-C} + NO \rightarrow (\text{char})\text{-C-O} + (\text{char})\text{-C-N} \\ &(\text{char})\text{-C-O} \rightarrow (\text{char}) + CO \\ &(\text{char})\text{-C-N} + NO \rightarrow (\text{char})\text{-O} + N_2 \end{aligned} \qquad [8.4]$$

Of importance in the context of NO–carbon reaction kinetics is how other species, such as CO, CO_2, O_2 and H_2O, might affect this kinetic mechanism. For a detailed description of the char–NO reaction mechanism, readers should refer to Glaborg *et al.*[1] and Aarna and Suuberg.[5]

One observation which has relevance to oxy-fuel combustion conditions with its tendency to produce higher CO concentration (results from CanmetENERGY indicate that concentrations of CO could exceed 10,000 ppm over a large region in the flame zone) is that the reaction between CO and NO has been noted as significant. In the high-temperature regime, added CO can enhance the NO–carbon reaction rate. One explanation is the following reaction:

$$CO + C(O) \rightarrow CO_2 + C^* \qquad [8.5]$$

where C(O) represents a surface carbon oxide and C* a 'free' site, active towards NO reduction through the above NO–char reaction pathway. There is also the possibility that CO may play an additional mechanistic role, since it is well established that surface-catalyzed routes exist that involve no surface carbon oxides:

$$CO + NO \rightarrow CO_2 + 0.5N_2 \quad [8.6]$$

A recent paper by Aarna and Suuberg[5] described an extensive investigation into this reaction. Their results confirmed previous reports of surface-catalyzed reactions between NO and CO, thought to be catalyzed by char, iron oxide and transition metal oxides, carbon-supported alkali, limestones, alumina and quartz. The reaction appears to happen in parallel with the NO–char reactions with an activation energy of 116 kJ/mol.

Like other reactions in coal combustion, reactions between char and NO are heavily influenced by the mineral matter in coal. One study[6] shows a distinct influence of Na and K on the reaction rate. However, the effect seems to disappear in the high-temperature regime. The results of Guo and Hecker[7] strongly suggest that calcium is a very active catalyst, based on its ability to partly restore the activity of the acid-washed char by reloading with calcium. The main effect of added Ca was seen in the high-temperature regime.

It is important to note that some of the kinetic findings above[5–7] were obtained in fluidized bed environments where temperatures are relatively low; thus their applicability to the pulverized coal combustion scenario remains to be seen.

Reductions of NO_x emissions under oxy-fuel conditions have been reported by a number of research groups. Andersson *et al.* published a paper in which they studied NO emissions in oxy-fuel combustion with a lignite coal.[8] They reported NO emissions reduction of 70–75% per unit of energy supplied when burning lignite coal under oxy-fuel conditions (oxygen concentration between 25 and 29%) compared with air-blown conditions. In their tests, they were able to observe NO emissions rates just below 50 ng/J under oxy-fuel conditions compared with more than 150 ng/J under air-blown conditions. They also conducted experiments with simulated air ingress and showed, along with a modelling study, that NO reduction was mainly due to the increased destruction of both recycled and freshly formed NO, even though fuel N to NO conversion was similar or slightly higher in the oxy-fuel case.

Hao *et al.* conducted a series of experiments in a 20 kW vertical combustor with pulverized coal combustion and compared NO_x emissions in both air-fired and oxygen-fired conditions with 30% O_2 and 70% CO_2. Even though the authors used an artificial mixture of O_2/CO_2 instead of recycled flue gas, they did add NO and SO_2 in the feed gas stream to simulate the effect of flue gas recirculation on NO_x and SO_x emissions. They also found[9] that the conversion of coal N to NO_x, for coal combustion in the mixture of O_2/CO_2, increases with oxygen concentration in the mixture, consistent with the findings of Andersson *et al.*, cited above.[8] Hao *et al.* also commented that more NO_x was reduced with oxidant staging than without oxidant staging and concluded that NO_x reduction depends on combustion media, combustion mode (staging or non-staging) and recycling location. They concluded in another study that, under optimal conditions, NO_x could be reduced by up to 92%.[10]

Farzan *et al.*[11] conducted extensive investigations on a boiler test bed with 1.46 MW heat input to produce NO_x emissions data that are more representative of industrial facilities. Oxy-fuel firing was conducted with flue gas recirculation and a low-sulphur sub-bituminous coal. Burner optimization was performed to obtain NO_x reductions while maintaining heat transfer in the furnace and the convection pass. Their results showed that NO_x emissions were 65% lower in oxy-fuel mode than in air mode, varying from 28 ng/J to 45 ng/J, much below the 64 ng/J limit suggested by the United States Environmental Protection Agency (US EPA). In addition, they observed that NO_x emissions could be optimized by adjusting burner stoichiometry, oxygen flow rate into the primary air zone and overall flue gas recirculation. For instance, they observed that NO_x emissions were reduced with reduced flow rate of recycled flue gas due to the ensuing higher temperature, which promoted NO_x reduction to N_2 in the reducing zone of the burner. They also observed that lower burner stoichiometry led to lower NO_x emissions. These findings can be instructive in designing burners specific to oxy-fuel combustion.

One important observation that can be drawn from results of both Hao *et al.*[10] and Farzan *et al.*[11] is that, for oxy-fuel combustion, despite the lack of N_2 in the combustion feed gas and the enhanced NO_x reburn mechanism, optimal burner design and operation are still important to minimize NO_x emissions.

CanmetENERGY has been one of the first research institutions to investigate pollutant emissions under oxy-fuel conditions. They compared[12] NO_x emissions for both air-fired and oxy-fired conditions using the Vertical Combustor Research Facility (VCRF, Fig. 8.1) with three different coals – bituminous, sub-bituminous and lignite. They compared the measured concentrations (dry volume basis) of NO (concentrations of NO_2 being negligible under these test conditions) for combustion of all three fuels in oxy-fuel (with 35% (vol. dry) O_2 in the feed gas) and air, respectively. The NO emission rates to the stack on a per unit heat input basis can be seen in Table 8.1.

Table 8.1 shows that, for bituminous coal, switching from air firing to oxy-fuel firing did not bring any reduction to NO_x emissions on a per unit heat input basis with the burner configuration used in this work. In fact, NO_x emissions actually *increased* by about 10%. On the other hand, tests with the same burner configuration for the sub-bituminous coal showed a 37% reduction in NO_x emissions. However, the most dramatic reduction was observed for lignite coal using an *improved* burner design; the NO_x reduction in that case was close to 75%.

These data convey an important message: even with strong mechanisms that can potentially reduce NO_x emissions in the oxy-fuel case, it will only happen with sound burner design to maximize the NO_x reducing potential of oxy-fuel

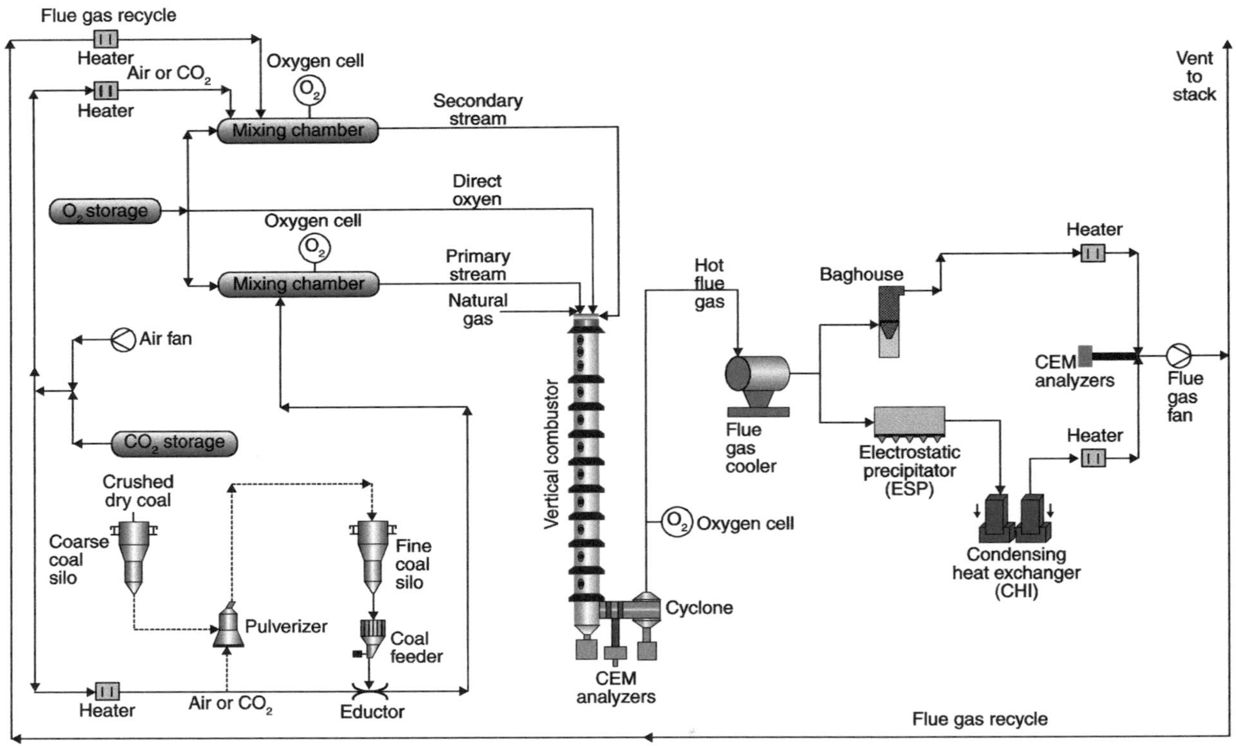

8.1 Schematic of CanmetENERGY's Vertical Combustor Research Facility (VCRF).

Table 8.1 Measured NO_x emissions for selected tests

Fuel and combustion mode	O_2, %	CO_2, %	CO, ppm	NO, ng/J
Bituminous–air	2.0	17	51	211
Bituminous–oxy-fuel	2.1	97	85	233
Sub-bituminous–air	2.0	17	9	236
Sub-bituminous–oxy-fuel	2.2	97	75	148
Lignite–air	3.3	17	1	269
Lignite–oxy-fuel	2.7	92	14	68

combustion. For example, it is well known that NO_x can be effectively reduced by creating a high-temperature, fuel-rich region generated in the central core of the flame. Thus, by strategically locating the oxygen injection points, it is possible to further enhance this effect, leading to significant NO_x reduction that could be more difficult to achieve with just air staging. This was the principle adopted in designing the new burner that was used for the lignite test.

Based on this notion that judicial use of pure oxygen injection in oxy-fuel combustion can provide a very effective means to reduce NO_x emissions, CanmetENERGY has shown[13] that, by controlling the pure oxygen jet so that a fuel-rich core is formed and is surrounded by an envelope of fuel-lean atmosphere, NO_x emissions can be significantly reduced. In this kind of design, coal is devolatilized in a reducing atmosphere, which prevents the formation of NO_x species. Unburned fuel and carbon monoxide are combusted further downstream as flue gas and oxygen react in a lower temperature range.

Other conventional technologies employed in low-NO_x burner designs, such as the oxidant overfire approach, where part of the oxygen-rich stream is diverted through an overfire port downstream of the main burner, can also be successfully used to reduce NO_x emissions. It also should be noted that the extent of NO_x reduction partially depends on coal characteristics: coals with higher volatiles respond better to oxygen staging as can be seen from Table 8.1 when one compares NO_x emissions of bituminous and sub-bituminous coals.

Overall, numerous studies have shown convincingly that NO_x emissions under oxy-fuel conditions can be significantly reduced through proper burner design and there could be opportunities for elimination of dedicated NO_x control devices such as selective catalytic or non-catalytic reduction (SCR or SNCR) under certain circumstances to achieve various government emissions regulations. However, with increasingly stringent air quality rules, it seems that some form of SCR or SNCR may be needed. In these cases, significant cost savings can still be realized by installing them outside of the flue gas recycle loop so that it only needs to process a small flue gas stream going to the compression train. Additionally, as will be pointed out later in this chapter, some compression train designs will be able to eliminate NO_x.

8.3 Sulphur oxide (SO_x) emissions

All coals contain sulphur, which can exist either in pyritic (bound with iron) or organic forms. Upon combustion, the two forms of sulphur may not release at the same time, although the effect of this asynchronous release has no practical impact on the eventual SO_x emissions. The sulphur content varies greatly depending on the coal seam, geological age and geographical location. Typically, but not always, young coals such as lignite and sub-bituminous coals contain less sulphur (<1% by weight) than bituminous coals, some of which may contain more than 6% sulphur. When coal is burned in the furnace, sulphur is released and reacts with oxygen to form sulphur dioxide, SO_2:

$$S + O_2 = SO_2 \tag{8.7}$$

Under high temperature and with oxygen radicals present, SO_2 can further react with oxygen radicals to form SO_3:

$$SO_2 + O + M \leftrightarrow SO_3 + M \tag{8.8}$$

However, the equilibrium conversion of SO_2 to SO_3 is low under high temperatures and the reaction rate slows as gas temperature cools; as a result, the conversion of SO_2 to SO_3 is no more than 2–5% under typical coal combustion conditions. Nevertheless, the presence of SO_3 is significant because SO_3 can react with water vapour to produce sulphuric acid aerosol, which is strongly corrosive to various materials and boiler components, such as the air heater. In addition, SO_3 leads to increased formation of sulphate, a known precursor for PM2.5 particulate.

Similar to NO_x, emissions of SO_x constitute another major pollution concern that requires expensive means to control in any conventional power plant. On a pulverized coal (PC) unit, wet flue gas desulphurization (FGD) units are often used to remove sulphur species, with limestone as the most widely used agent, through the following global steps:

$$\begin{aligned} &SO_2(g) + CaCO_3(s) + H_2O(l) = CaSO_3(s) + CO_2(g) + H_2O(l) \\ &CaSO_3(s) + H_2O(l) + \tfrac{1}{2}O_2(g) = CaSO_4(g) + H_2O \end{aligned} \tag{8.9}$$

Sulphur removal rates greater than 90% can be achieved with modern wet lime-based FGD. The sulphur removal rate with dry FGD is typically lower, at about 80%, though more modern designs can achieve close to 90%. Other advanced and more expensive technologies such as amine scrubbing can remove close to 100% sulphur.

Unlike NO_x, there are no inherent chemical reaction pathways to promote the destruction of SO_x under oxy-fuel combustion conditions. However, a number of studies did observe enhanced sulphur removal under such conditions. For example, Kiga *et al.* observed a 50% drop in SO_2 mass emissions at the stack during oxy-fuel mode as compared with air-fired mode.[14] CanmetENERGY reported[15] similar results for oxy-fuel tests done using its VCRF.

These observed reductions in SO_2 emissions under oxy-fuel combustion conditions are currently attributed to additional sulphur retention on the fly ash. Maier *et al.* tested this hypothesis using a 20 kW once-through furnace.[16] SO_2 was measured both at the end of the radiative section of the furnace (~1150°C) and in the flue gas path before the filter (~450°C) under air-blown as well as oxy-fuel conditions. Their results showed that SO_2 was not reduced in the radiative section of the furnace because high temperature prevents most sulphate-forming elements (Ca, Mg, Na, etc.) from being stable. However, some SO_2 can be reduced in the lower-temperature region of the furnace if these sulphate-forming elements are available. The authors also noted that higher SO_2 concentrations in the flue gas under oxy-fuel conditions may also improve the capture of SO_2 on ash deposits. However, this effect of additional ash retention depends on a number of factors, especially the characteristics of ash deposits – higher alkaline ash deposits will lead to enhanced sulphur retention. As a result, it is not as consistent as the reburn mechanism for NO_x reduction and the emission rate of SO_x will not be reduced to the same extent as that of NO_x. It is noteworthy to point out that the increased fly ash retention of sulphur may cause additional problems such as fly ash utilization in cement and concrete production, causing utilities to lose an important stream of revenue.

Typically, under oxy-fuel conditions, concentrations of SO_2 and SO_3 will considerably build up due to accumulation through flue gas recirculation if they are not removed before being recycled, which has been demonstrated by Wall *et al.*[17] Studies done by CanmetENERGY have also confirmed that SO_2 concentrations can more than double under oxy-fuel conditions, as shown in Table 8.2.

It should be noted that coal analyses showed that the bituminous coal used in this work had a sulphur content of 0.8% and the sub-bituminous coal had a sulphur content of 0.2%. In case of higher sulphur coal or petroleum coke (petcoke) with sulphur content as high as 6%, observations made at CanmetENERGY showed that SO_2 concentration could easily be in the range of 1–2% in the flue gas recycle loop if no sulphur mitigation steps were taken.

Table 8.2 Measured flue gas SO_2 emissions for selected tests

Test ID	SO_2, ppm, vol. dry	Mass flow, mg/min		Mass flow, ng/J	
		SO_2	SO_3	SO_2	SO_3
Bituminous–air	615	7115	325	590	27
Bituminous–oxy-fuel	1431	16828	1108	1350	75
Sub-bituminous–air	175				
Sub-bituminous–oxy-fuel	372	2429	1.7	193	0.14
Lignite–air	277				
Lignite–oxy-fuel	785				

One positive effect of increased concentrations of sulphur is its enhanced reduction effect on NO_x, mainly owing to the reaction HS + NO = NS + OH, and this effect has been demonstrated.[18] However, this effect is so small that it does not constitute a major pathway for NO_x reduction.

The increase in SO_2 concentrations can have significant impact on the operation of the power plant, especially due to the proportionally higher SO_3 concentrations, as is shown in Table 8.2. According to Table 8.2, for the bituminous coal, SO_3 accounted for >4% of total SO_x emissions under air-blown conditions, while it accounted for >6% under oxy-fuel conditions. These numbers were in line with typical coal power plants where SO_3 consists of 2–5% of total SO_x concentrations with slightly higher SO_2 to SO_3 conversion under oxy-fuel conditions. For sub-bituminous coal, it can be seen that gaseous SO_3 was exceedingly low, accounting for only 0.07% of total gaseous SO_x emissions. This is due to two reasons: the sub-bituminous coal had a low sulphur content; it also produced highly alkaline ash, which resulted in nearly 14% of the input sulphur being retained in the ash deposits. The highly alkaline fly ash could also effectively capture most of the SO_3, accounting for its very low concentrations in the flue gas. It is highly probable that additional SO_2 could be neutralized by unreacted cations in the superfine fly ash trapped on the surfaces of the fabric filter.

Some studies have indeed pointed to a disproportionally higher SO_3 concentration under this scenario due to the higher partial pressure of SO_2 and O_2. For example, Maier *et al.* presented data obtained from a 500 kW pilot-scale test unit that showed a relative SO_3 concentration increase from 1.1% under air-blown conditions to 4.6% under oxy-fuel conditions.[16]

As noted above, in the case of high-sulphur fuels such as petcoke and some high-sulphur coals with 6% of sulphur content, SO_2 concentrations have been shown to reach greater than 1.5%. If we consider that SO_3 concentrations under typical combustion conditions account for 2–5% of total SO_x, this means that SO_3 concentrations could be as high as 300 to 750 ppm. Due to the corrosive properties of SO_3 under typical flue gas conditions, it is obvious that, for high-sulphur coals, consideration must be taken for build-up of SO_3 in the flue gas recirculation loop. On the other hand, corrosion due to SO_3 may not be a significant problem for low-sulphur coals and/or coals that produce highly alkaline ash deposits that can retain sulphur species.

A direct effect of potentially significantly higher SO_2-SO_3 concentrations in the flue gas concerns the flue gas recirculation options. For a low-sulphur coal, it is probably acceptable to recycle the flue gas before the FGD. This can contribute to less chance of air ingress, improved plant efficiency and lower capital and operating costs due to the scaling-down of the FGD. For high-sulphur fuels, however, it is necessary to draw recycled flue gas downstream of the FGD to avoid any concern for corrosion and hazardous flue gas egress. A more detailed discussion on this topic is presented in Chapter 3.

8.4 Mercury and trace elements

Mercury is a very toxic element that is released through coal combustion. Even though mercury content in coal is very low (typically in ppb level), it has a tendency to accumulate in fish and birds in the highly toxic form of methylmercury, which then is transmitted to humans as they consume these foods.

Similar to sulphur, mercury exists in either pyritic or organic forms. Once mercury enters the furnace with fuel, it is released as elemental mercury, Hg^0. As Hg^0 travels downstream along with the flue gas, part of it may go through a series of chemical changes to transform into the oxidized form, Hg^{2+}. This transformation depends on a number of factors, mainly temperature, chlorine content in the flue gas and surface characteristics. Typically, a high-chlorine coal will promote the transformation of Hg^0 to Hg^{2+}; while for a low-chlorine coal, most of the Hg will remain as elemental. For a complete description of the complex Hg chemistry, readers should refer to Galbreath *et al.*[19] and Zygarlicke and Pavlish.[20]

The difference in the two gaseous forms of mercury is important in that oxidized mercury is water soluble and can be removed with good efficiency in conventional wet scrubbing while elemental mercury is usually released through the stacks. This means that elemental mercury is much more difficult to control than oxidized mercury.

Not much research has been done to study mercury evolution specific to oxy-fuel conditions. One of the major consequences of oxy-fuel combustion is much higher CO_2 partial pressure in the flue gas. But, since CO_2 does not play any direct role in Hg chemistry, this should not have any noticeable impact on mercury chemistry. Recently, some very preliminary studies seem to indicate that increased oxidant effect (attributable to high O_2 concentration) may promote Hg oxidation. However, oxygen concentration is higher only in the flame zone where the temperature is usually too high for oxidized Hg to remain stable, while oxygen concentration in the flue gas under oxy-fuel conditions is generally the same as for air-blown conditions, and oxidant *per se* may have limited, if any, direct influence on Hg oxidation. On the other hand, the higher flue gas SO_2 concentration in the oxy-fuel case could theoretically promote Hg^0 oxidation through its interaction with SO_2 to produce Hg^{2+}: $Hg^0(g) + SO_2(g) + O_2(g) \leftrightarrow HgSO_4(s)$. $HgSO_4$ can then react with chlorine species to produce $HgCl_2$. Both $HgSO_4$ and $HgCl_2$ can be removed by wet FGD. However, it has not been extensively studied to see if this pathway becomes noticeably more effective with rising SO_2 concentrations. Clearly, more research is urgently needed to elucidate Hg behaviour under oxy-fuel conditions.

Mercury emissions control technology has been extensively studied for the past decade and yet no satisfactory methods have been found. One of the main problems related to mercury emissions control is its extremely small concentration in the flue gas, typically in ppm levels, which makes any current capture technology almost prohibitively expensive on a per unit of Hg captured basis. Another issue that is equally important concerns the clean-up and disposal of mercury-containing material after its capture, which is often ignored in most Hg removal studies.

Oxy-fuel combustion offers an easier way to deal with mercury emissions through the flue gas compression train. Air Products has presented its design of the CO_2 compression train that can capture mercury along with SO_x and NO_x.[21] The design is scheduled to be tested on Vattenfall's 30 MWth Schwarze Pumpe oxy-fuel demonstration power plant in Germany. A brief discussion of this approach will be presented in this chapter.

As a mined resource, coal contains a variety of minerals other than mercury that are released when combusted. Since the amounts of these minerals are usually very low, they are designated as trace elements, or trace metals. Even though their absolute emissions are low, they have damaging implications in human health. Some of them, such as arsenic, vanadium, sodium and potassium, also play a role in promoting boiler slagging and fouling.

CanmetENERGY did some extensive surveys on the emissions of trace elements in oxy-fuel combustion (35% vol. dry O_2) as well as in air-blown conditions using its VCRF. A typical example of these results is shown in Fig. 8.2. These results were obtained with a bituminous coal burning in the VCRF

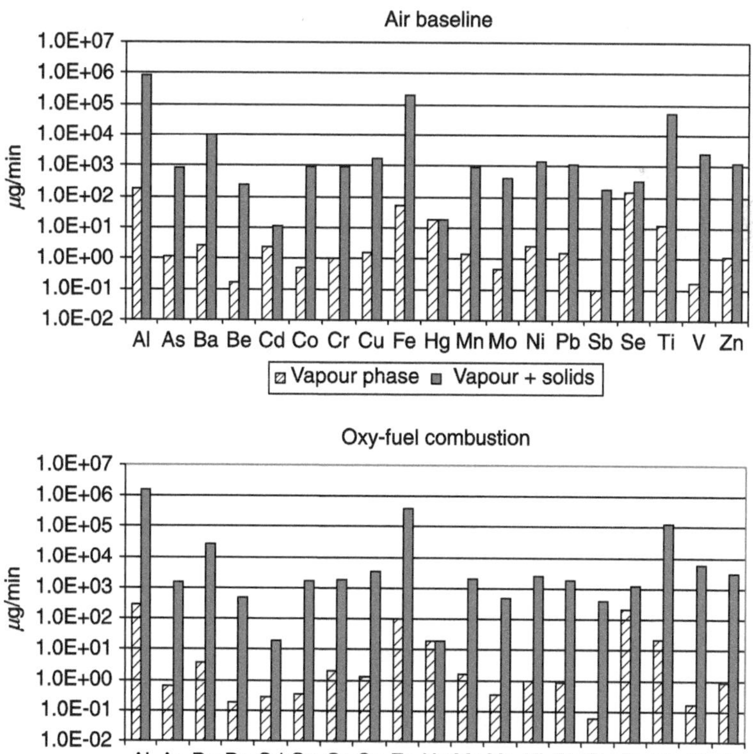

8.2 Emissions of trace metals under air-blown and oxy-fuel conditions, expressed in micrograms per minute.

at a heat input of 0.21 MW/h. Trace elements in both vapour phase and solid deposits were analyzed. Tests done with a sub-bituminous coal yielded the same results. This clearly showed that combustion modes did not affect the emissions of these trace elements. As a result, there is no need to employ extra measures to deal with emissions of trace elements under oxy-fuel conditions.

8.5 Ash formation

Ash deposition and fly ash properties are important parameters in boiler design as they exert significant influence on plant performance parameters such as heat transfer, reliability and safety. It is, therefore, important to understand how ash formation and deposition change when combustion mode switches from air-blown to oxy-fuel, especially considering changes in flue gas composition in oxy-fuel mode, which can lead to high concentrations of NO_x, SO_x and moisture. Unfortunately, so far very few studies have looked into this subject.

CanmetENERGY measured the deposits of selected trace element across its VCRF. These trace elements show enrichment and depletion in deposits at various system locations downstream of the burner and in the flue gas, indicating a strong dependency on temperature and particle size. However, partitioning of the refractory and less volatile trace elements did not vary appreciably with changes in combustion media with either coal. Accordingly, the evaluation was focused on the partitioning of the moderately and highly volatile trace elements with a high potential for emission, such as As, Se, Sb, Pb, Cd and Hg.

The following trends were observed from the trace element analyses of ash deposits collected from bituminous coal tests:

- the above-mentioned trace elements were lowest on the burner quarl and reflected the very-high-temperature environment near the flame front;
- As, Se and Sb steadily increased downstream of the burner quarl and peaked across the fabric filter;
- Cd was fairly flat downstream of the quarl and across the fabric filter;
- Pb gradually increased downstream of the quarl and peaked across the fabric filter; and
- Hg remained low downstream of the quarl, peaked steeply in the flue gas cooler by an order of magnitude and gradually declined across the fabric filter.

The close affiliation and greatly increased concentration of Hg with the halogens Cl and F at specific combustion system locations were significant. For the bituminous coal, a very high level of enrichment and accumulation of Hg with Cl/F occurred in the flue gas cooler (~60–200°C).

The elemental partitioning in ash deposits from sub-bituminous coal tests given below showed similar but slightly different trends from the bituminous coal ash deposits:

- trace elements on the burner quarl were low;
- As, Se and Sb were flat downstream of the quarl and then increased by an order of magnitude at the hot cyclone and peaked across the fabric filter;
- Cd was also flat downstream of the quarl, but increased by two orders of magnitude at the hot cyclone and peaked at this level across the fabric filter; and
- Hg was low downstream of the quarl, then increased by an order of magnitude at the flue gas cooler and remained at this level across the fabric filter.

The same close affiliation and greatly increased concentrations of Hg and Cl/F were also observed with the sub-bituminous coals, but the accumulation occurred downstream of the furnace exit before the flue gas cooler.

It is noteworthy that the high-level enrichment of Hg with Cl and F and its accumulation in a certain temperature range occurred although the coals had almost the same Hg content but different ranks, as well as Cl and F contents that differed by a factor of about 10. This observation suggests that a specific temperature window exists for optimal *in situ* reactions between Hg and Cl and F and that the best location for injection of a halogen or halide for Hg capture may be immediately upstream of this temperature window rather than through the burner.

The common denominator in the volatile trace element partitioning and sedimentation was the flue gas temperature, which ranged between 600°C and 100°C in the enrichment zones, indicating a very strong dependency on temperature and a secondary dependency on time for selective chemical reactions and/or physical bonding.

Trace element sampling of flue gas at the flue gas cooler outlet with bituminous and sub-bituminous coal trials indicated little or only a moderate change in partitioning of the more volatile trace elements with changes in combustion media. Moreover, a breakdown of these elements indicates that most of the As, Sb, Se and Pb was physically or chemically bound to the coarser particles in the filter, whereas virtually all of the Hg was in the gas phase or associated with the superfine dust.

Stam *et al.* conducted a thermodynamic study[22] on ash behaviour under oxy-fuel conditions. Their simulation showed that:

- Main ash compositions were similar for either air-blown or oxy-fuel combustion, though the latter produced a higher amount of calcium sulphate.
- For trace element speciation, there were no changes when combustion mode switched from air-blown to oxy-fuel firing.
- Slag formation was lower under oxy-fuel conditions.
- Oxy-fuel combustion presented an increase in fouling risks related with K and Na compared with air-blown combustion.

At this time, there have been no studies reported in the literature on the impacts of combustion modes on ash formation and growth. However, Imperial College in London, UK is planning detailed studies of this issue in collaboration with E.ON UK.

8.6 Integrated emissions control

A recent trend in emissions control is to adopt an integrated approach. Over the past decades, as more and more stringent emissions regulations have been implemented, the traditional piece-meal approach for emissions control has become difficult to sustain. With the advent of modern technologies, integrated emissions control has become a preferred solution and is being gradually adopted. Integrated emissions control is especially beneficial for power plants with CO_2 capture, whether by post-combustion capture or oxy-fuel processes, because both approaches either require (post-combustion capture) or prefer (oxy-fuel flue gas compression train) a clean flue gas stream.

The traditional approach for emissions control involves specific stand-alone equipment for specific pollutants (fabric filter or ESP for particulate, SCR or SNCR for NO_x control, wet or dry FGD for SO_x control and, more recently, the prospects for Hg scrubbers). Under oxy-fuel conditions, as pointed out earlier in this chapter, NO_x is usually not a problem because a large amount of recycled NO_x will be destroyed through the reburn mechanism. As such, de-NO_x units may not be required for oxy-fuel power plants. However, emissions of SO_x and mercury still need to be controlled and it is still desirable to reduce NO_x concentrations in the vent stream from flue gas compression train as much as possible. Integrated emissions control attempts to use a single piece of equipment to remove or reduce emissions of multiple pollutants simultaneously by creating an environment that induces synergies among various pollutants and facilitates their removal. The potential advantages of integrated emissions control are obvious: higher efficiency, lower capital and operating cost and better reliability; and oxy-fuel fired power plants are superior candidates for these technologies compared with air-fired plants, due mainly to the fact that the volume of the flue gas stream to be treated is considerably smaller in the oxy-fuel case.

Air Products Inc. has recently disclosed its approach for integrated emissions control for oxy-fuel power plants.[21] It is based on the well-known Lead Chamber process, developed more than two centuries ago for sulphuric acid production. The original process involves injection of SO_2, steam and NO_x into large chambers and can be summed up with the following reactions:

$$\begin{aligned} &2NO_2 + H_2O \rightarrow HNO_2 + HNO_3 \\ &SO_2(l) + HNO_3 \rightarrow NOHSO_4 \\ &NOHSO_4 + HNO_2 \rightarrow H_2SO_4 + NO_2 + NO \\ &SO_2(l) + 2HNO_2 \rightarrow H_2SO_4 + 2NO \end{aligned} \qquad [8.10]$$

The overall rate-determining step in the process is the oxidation of NO by oxygen:

$$2NO + O_2 \rightarrow 2NO_2 \quad [8.11]$$

Because nitrogen oxides are absorbed and regenerated in the process, it serves as a catalyst for the overall reaction:

$$2SO_2 + 2H_2O + O_2 \rightarrow 2H_2SO_4 \quad [8.12]$$

Under ambient pressure, the conventional Lead Chamber process is extremely slow (in the order of tens of minutes) and unsuitable for power plant situations where a large flow rate of flue gas must be accommodated. However, according to Air Products Inc., the Lead Chamber reaction rates increase with pressure to the third power. As a result, when flue gas is compressed to high pressure, as it is in the flue gas compression train, the reaction rates significantly increase to the point where this process becomes feasible even for large-scale power plants without resorting to excessively large reaction chambers. It is important to note that one of the advantages of this approach is that the pressurization process does not incur any additional loss in efficiency, as it is an integral part of the flue gas compression train, the purpose of which is to produce CO_2 in supercritical state for sequestration. An additional advantage of this process in the context of oxy-fuel combustion is that, thanks to flue gas recirculation, the gas stream to be treated is of a much lower volume, further reducing the size of the reaction chamber.

Based on the reactions involved in the Lead Chamber mechanism, one can see that both SO_x and NO_x can be removed using this process with the end products as sulphuric acid and a small amount of nitric acid formed from excess NO_x. Air Products Inc. shows an SO_2 removal rate of 100% and NO_x removal rate of 90–99%. In addition, mercury can also be completely removed through its reaction with nitric acid.

An interesting observation with the Lead Chamber process is that, in this case, NO_x is a necessary initial component for the entire process to work optimally. It is necessary to ensure that there is enough NO_x getting through to the CO_2 compression train, at least during the compression train start-up. According to the kinetic mechanism highlighted above, 2 mol NO is needed to completely convert 1.33 mol SO_2 (l) to H_2SO_4, which means that for a high-sulphur coal, a flue gas with relatively high concentrations of NO_x is needed. Since NO_x is being regenerated during the process, in theory, no fresh supply of NO_x is needed once the process is under way. As a result, some operational flexibility will be necessary to ensure enough NO_x supply to the compression train during start-up and to reduce the NO_x once stable operation is established. One way to achieve this could be through manipulating burner oxygen injection pattern and burner swirl settings.

The Lead Chamber process is an excellent illustration of the superiority of oxy-fuel power plants in terms of integrated emissions control implementation.

Due to the slow rate of reaction of this process at atmospheric pressure, it is not realistic to be deployed for air-fired units simply because the size of the reaction vessel would dwarf the rest of the power plants and, in addition, the compression cost in order to process a flue gas stream that contains >70% nitrogen would be prohibitive. With a much smaller volume of flue gas and the ability to compress this stream to an efficient operating pressure at no additional cost, the Lead Chamber process can be successfully adopted for oxy-fuel power plants. Air Products Inc. is currently planning to put this technology to test on Vattenfall's 30 MWth oxy-fuel demonstration power plant.

While Air Products Inc.'s high-pressure Lead Chamber process can be considered as an example of full integrated emissions control, other less integrated but probably more practical approaches have been tested. CanmetENERGY, for example, has been trying to use the condensing heat exchanger for simultaneous SO_x and Hg removal and showed excellent removal of SO_x but so far limited success with Hg removal.[23] The less integrated approach could prove advantageous for high-sulphur fuels that require the dry recycle option in which SO_x must be removed before being recycled to avoid corrosion problems.

8.7 Vent stream from flue gas compression train

It should be noted that there is a vent stream coming out of the flue gas compression train. The vent stream typically consists of carbon dioxide and non-condensable gases, such as NO_x, CO, O_2 and Ar, and very little, if any, moisture and other contaminants. Although the volume of this vent stream is very small, it is highly concentrated in CO_2 and other air pollutants and likely not allowed to be directly vented to the atmosphere untreated. The simplest solution is to dilute this stream with nitrogen from the air separation units (ASUs) before venting to the stack.

A better solution would be to recycle this stream back to the furnace after a simple process to separate CO_2 from the rest of the components. The separated CO_2 can then be recycled back to the furnace. This option presents an obvious advantage: by recycling the separated CO_2, it can significantly increase CO_2 capture rate of the power plant. Currently, most of the CO_2-capture power plants aim at a CO_2 capture rate of about 90%. With this option, a capture rate of almost 100% can be realized. Since the vent stream is already under pressure and is very clean, a membrane-based separation process can be implemented at limited cost.

Improving this separation step further, one can envisage a membrane-based approach to separate both CO_2 and O_2 from the rest of the components. The advantage of this is that the oxygen in the vent stream can be reused, helping improve the overall power plant efficiency by reducing power requirements for the ASU. Air Products Inc. estimates that ASU power consumption can be reduced by about 5% with this approach. It also suggests that this separation process is energy efficient and economical.[21]

8.8 Conclusion

Even though oxy-fuel combustion provides opportunities for near-zero emissions from power plants, at the current state of the art, it is still necessary to implement emissions control technologies in order to facilitate downstream flue gas compression and to minimize corrosion concerns under certain circumstances. Under oxy-fuel conditions, emissions characteristics are heavily influenced by two factors – large quantities of CO_2-rich recycled flue gas and higher wind-box oxygen concentrations.

It has been widely shown in the literature that NO_x emissions are greatly reduced under oxy-fuel conditions through the NO_x reburn mechanism and the suppression of the thermal NO_x pathway. These emissions reductions can be notably improved through conventional low-NO_x approaches, such as oxidant and fuel staging. Oxy-fuel combustion introduces another powerful pathway for NO_x reduction that was not available to conventional air-blown combustion in the form of oxygen delivery manipulation. By combining these techniques, it is possible to reduce NO_x emissions by 60–70% compared with air-blown combustion.

With flue gas recirculation, certain pollutants can accumulate in the flue gas recycle loop if they are not removed before the flue gas is recycled and if they are not destroyed in the furnace. One of the main pollutants that falls into this category is SO_x. It has been shown that SO_2 concentration can double or triple in the flue gas recycle loop if it is not removed before being recycled, as in the wet recycle approach. A coal with 4–6% sulphur can thus produce a recycled flue gas stream with SO_2 concentrations well above 1% or even 2%. Along with this high SO_2 concentration comes a proportionally higher concentration of SO_3 and its related lower acid dew point that can lead to serious potential for corrosion. However, with a low-sulphur coal, the corrosion concern will be much lower, and it is beneficial in this case to adopt the wet recycle option for improved efficiency and lower operating costs.

While experimental data show that emissions of trace metals, including mercury, do not present significant changes under oxy-fuel conditions, there is evidence that ash formation and growth pattern may be different. Research in this important area is, however, still in its early stages.

Recent developments in integrated emissions control technologies for oxy-fuel power plants promise to produce highly concentrated carbon dioxide (>99%) that is ready for transport and sequestration with limited additional cost. Some of these new technologies may also give opportunities to build oxy-fuel power plants with very high (close to 100%) CO_2 capture rates with limited additional cost.

8.9 References

1 Glarborg, P., Jensen, A.D. and Johnsson, J.E., 'Fuel nitrogen conversion in solid fuel fired systems', *Progress in Energy and Combustion Science*, **29** (2003) 89–113.

2 Okazaki, K. and Ando, T., 'NOx reduction mechanism in coal combustion with recycled CO_2', *Energy*, **22** (1997) No. 2/3, 207–215.

3 Hampartsoumian, E., Folayan, O.O., Nimmo, W. and Gibbs, B.M., 'Optimization of NO_x reduction in advanced coal burning systems and the effect of coal type', *Fuel*, **82** (2003) 373–384.

4 Chambrion, P., Orikasa, H., Suzuki, T., Kyotani, T. and Tomita, A., 'A study of the C-NO reaction by using isotopically labelled C and NO', *Fuel*, **76** (1997) 493.

5 Aarna, I. and Suuberg, E.M., 'The role of carbon monoxide in the NO-carbon reaction', *Energy and Fuels*, **13** (1999) 1145–1153.

6 Illan-Gomez, M., Linares-Solano, A., Radovic, L.R. and Salinas-Martinez de Lecea, C., 'No reduction by activated carbons. 2. Catalytic effect of potassium', *Energy & Fuels*, **9** (1995) 97–103.

7 Guo, F. and Hecker, W.C., 'Effects of CaO and burnout on the kinetics of NO reduction by beulah zap char', *Symposium (International) on Combustion*, **26**(2) (1996) 2251–2257.

8 Andersson, K., Normann, F., Johnsson, F. and Leckner, B., 'NO emission during oxy-fuel combustion of lignite', *Industrial and Engineering Chemistry Research*, **47** (2008) 1835–1845.

9 Hao, L., Zailani, R. and Gibbs, B.M., 'Comparisons of pulverized coal combustion in air and in mixtures of O_2/CO_2', *Fuel*, **84** (2005) 833–840.

10 Hao, L., Zailani, R. and Gibbs, B.M., 'Pulverized coal combustion in air and in O_2/CO_2 mixtures with NOx recycle', *Fuel*, **84** (2005) 2109–2115.

11 Farzan, H., Vecci, S.J., Châtel-Pélage, F., Pranda, P. and Bose, A.C., 'Pilot-scale evaluation of coal combustion in an oxygen-enriched recycled flue gas', *The 30th International Technical Conference on Coal Utilization and Fuel Systems*, 2005, Clearwater, Florida.

12 Tan, Y., Croiset, E., Douglas, M.A., Thambimuthu, K., 'Combustion characteristics of coal in a mixture of oxygen and recycled flue gas', *Fuel*, **85** (2006) 507–512.

13 Tan, Y., Chui, E., Douglas, M. and Thambimuthu, K., 'Oxy-fuel coal burner design: from CFD modeling to pilot scale testing', *6th International Conference On Greenhouse Gas Control Technologies*, 2002, Kyoto, Japan.

14 Kiga, T., Takano, S., Kimura, N., Omata K., Okawa, M., Mori, T. and Kato, M., 'Characteristics of pulverized-coal combustion in the system of oxygen/recycled flue gas combustion', *Energy Conversion and Management*, **38** (1997) S129–S134.

15 Croiset, E. and Thambimuthu, K., 'NO_x and SO_2 emissions from O_2/CO_2 recycle coal combustion', *Fuel* **80** (2001) 2117–2121.

16 Maier, J., Dhungel, B., Mönckert, P., Kull, R. and Scheffknecht, G., 'Impact of recycled gas species (SO_2, NO) on emission behaviour and fly ash quality during oxy-coal combustion', *The 33rd International Technical Conference on Coal Utilization and Fuel Systems*, 2008, Clearwater, Florida.

17 Wall, T., Liu, Y., Spero, C., Elliott, L., Khare, S., Rathnam, R., Zeenathal, F., Moghtaderi, B., Buhre, B., Sheng, C., Gupta, R., Yamada, T., Makino, K. and Yu, J., 'An overview on oxyfuel coal combustion – state of the art research and technology development', *Chemical Engineering Research and Design* **87** (2009) 1003–1016.

18 Chagger, H., Goddard, P., Murdoch, P. and Williams, A., 'Effect of SO_2 on the reduction of NOx by reburning with methane', *Fuel* **70** (1991) 1137–1142.

19 Galbreath, K.C., Zygarlicke, C.J. and Toman, D.L., 'Mercury-chlorine-fly ash interactions in a coal combustion flue gas', *Air and Waste Management Association's 91st Annual Meeting & Exhibition*, 14–18 June, 1998, San Diego, CA.

20 Zygarlicke, C.J. and Pavlish, J.H., 'The fate and control of mercury emissions from coal-fired systems', *13th Annual International Pittsburgh Coal Conference Proceedings: Coal-Energy and the Environment*, Vol. 1 (1996) 241–246.
21 White, V. and Fogash, K., 'Purification of oxy-fuel derived CO2: Current developments and future plans', *1st Oxy-fuel Combustion Conference*, September 2009, Cottbus, Germany.
22 Stam, A., Ploumen, P. and Brem, G., 'Ash related aspects of oxy-combustion of coal and biomass: a thermodynamic approach', *The 33rd International Technical Conference on Coal Utilization and Fuel Systems*, 2009, Clearwater, Florida.
23 Tan, Y., Lu, D., Anthony, E.J., Dureau, R., Mortazavi, R. and Douglas, M., 'Mercury removal from coal combustion by Fenton reactions. Paper B: Pilot-scale tests', *Fuel* **86** (2007), 2798–2805.

9

Oxy-fuel heat transfer characteristics and impacts on boiler design

Y. LIU, T. WALL, S. KHARE, The University of Newcastle, Australia and R. GUPTA, The University of Alberta, Canada

Abstract: An outline of the basis of radiative transfer which dominates furnace heat transfer for coal-fired oxy-fuel furnaces is presented and illustrated by comparisons of the significance of the radiating species influencing heat transfer, and also differences between air-fired and oxy-fuel fired furnaces. Heat balances and heat transfer for three different oxy-fuel furnaces of 1.2 MWt, 30 and 420 MWe are given by a simple well-stirred reactor model and also comprehensive computational fluid dynamics models. It is concluded that oxy-fuel combustion can be operated at the same heat transfer rate as air-firing, but results in operational changes – lower volumetric gas flow rate, lower adiabatic flame temperature and lower flue exit gas temperature – associated with the higher gas emissivity of oxy-fuel firing and different heat capacity of the flue gas.

Key words: oxy-fuel, heat transfer, radiation.

9.1 Introduction

Oxy-fuel combustion is considered an important technology option for carbon capture and storage (CCS) from existing and new-build power plants. The understanding and prediction of heat transfer aspects underpins oxy-fuel technology demonstration and deployment. In oxy-fuel firing furnaces fuel combusts with oxygen producing major gaseous products of carbon dioxide and water vapor, and solid soot, char and fly ash particles. The heat transfer between combustion products and the water cooling wall determines furnace thermal efficiency. In oxy-fuel technology, a recycled flue gas is used to modulate flame temperature and make up the flue gas volume for heat transfer in the convective pass section of the boiler. Depending on whether water vapor is removed from the recycled flue gas, oxy-fuel technology can be classified as dry recycle and wet recycle operations. Depending on the temperature regions at which the recycled flue gas is extracted and whether recycled flue gas is heated, oxy-fuel technology can be classified as hot recycle and cold recycle.

Understanding radiative heat transfer, which is a dominant mechanism of heat transfer in the combustion chamber, is critical. Experimental studies of heat transfer in oxy-fuel combustion have been carried out at laboratory, pilot and demonstration scale furnaces. The types of oxy-fuel combustion furnaces include vertical down-firing furnace,[1–3] horizontal furnace,[4] wall firing furnace,[5] and tangential firing furnace.[6] Various types of fuels have been tested in oxy-fuel combustion conditions, including gaseous fuel such as natural gas (CH_4),[7] propane (C_3H_8),[8,9] solid fuels such as lignite,[1,10–12] sub-bituminous coal[11–13] and bituminous coal.[11,12]

For theoretical prediction of radiative heat transfer, it is necessary to understand the complex interaction between radiative heat transfer and several different phenomena in the combustion process, as illustrated in Fig. 9.1. Combustion products' properties and distribution are required by the radiative transport equation to predict radiative heat transfer. In industrial applications, temperatures and combustion products are unevenly spatially distributed inside the reactor; thus a spatial dependence of temperature and radiative properties should be taken into account. Coal drying, devolatilization, gaseous volatile matter turbulent combustion, heterogeneous char reactions, ash/slag formation and particle dispersion determine the species concentration/size distribution and temperature distribution inside the three dimensional chamber, which in turn impact radiative properties of combustion products. The emissivity of combustion products is determined by contributions of gaseous and particulate matter. The gaseous contributions include those from CO_2 and H_2O. Particulate matter contributions include soot, char and fly ash. Emissivity of the furnace wall depends on the extent of ash deposition present on the wall, which strongly depends on the mineral matter in the coal. For radiative heat transfer in oxy-fuel firing, one also needs to consider the total exchange area, which is related to furnace geometry.

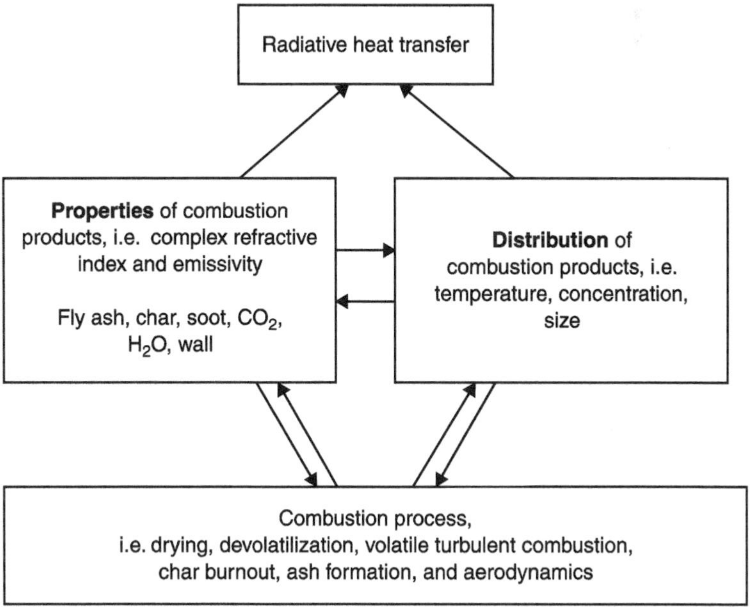

9.1 Radiative heat transfer prediction and its dependence on combustion processes.

Computational fluid dynamics (CFD) based comprehensive combustion models couple heat transfer, fluid dynamics and coal reaction kinetics, and are capable of predicting species and temperature distribution. By incorporating gas/solid radiative properties into radiation heat transfer models, temperature distribution and heat flux distribution on the wall can be determined. Several commercial codes such as FLUENT[14] and CFX,[13] and proprietary code such as VEGA-3[4] have been adapted to oxy-fuel combustion.

In this chapter, we first introduce the design criteria for retrofitting an existing air-firing furnace into an oxy-fuel firing furnace and then a simplified theoretical model is presented to illustrate the impacts of major factors on radiative heat transfer; the temperature dependent radiation properties of oxy-fuel combustion properties are discussed in detail. The chapter concludes with radiation heat transfer modeling by a comprehensive CFD model.

One type of coal fired in three different furnaces is simulated. The properties of the coal used are listed in Table 9.1, and detailed information on the three furnaces (called F-A, F-B, F-C) is listed in Table 9.2. F-A is a pilot scale coal-testing furnace, F-B is a unit of a scale considered suitable for a retrofit demonstration and F-C is a unit of practical scale. All the furnaces have water cooling walls. Also given in Table 9.2 are factors influencing radiative transfer, the mean beam length being an average distance traveled by a radiative (gas) beam in the furnace to the walls, which determines the 'average' emissivity of the furnace gases. Radiative heat transfer in furnaces is furnace dependent; thus some general trends will be presented to indicate the impact of major factors.

Table 9.1 Analysis data of coal used in calculations

	Sub-bituminous coal
Proximate analysis, as received %	
Moisture	8.8
Ash	17.6
Char	50.2
Volatile	23.4
Ultimate analysis, dry ash free %	
C	78.6
H	3.5
N	0.9
S	0.3
O	16.7
High heating value, MJ/kg, dry base	23.7
SiO_2	47.6
Al_2O_3	28.6
Fe_2O_3	16
CaO	1.64
MgO	1.12
TiO_2	2.22

Table 9.2 Furnace details

Furnace details	F-A	F-B	F-C
Unit output, MWe	1.2 MWt	30	420
Coal feed rate, kg/s	0.033	5.45	54.17
Furnace volume, m^3	8.9	670.3	12031.9
Total furnace wall area, m^2	29.6	501.6	3482.8
Wall fraction	0.94	0.96	0.96
Effective furnace wall area, m^2	27.7	481.1	3349.8
Mean beam length, m	1.1	4.68	12.1
Wall temperature, K	1200	1000	1000
Wall emissivity	0.7	0.7	0.7

9.2 Heat transfer criteria for oxy-fuel combustion

Compared with conventional air combustion, oxy-fuel combustion has several differences. Removal of nitrogen from air in oxy-fuel combustion leads to lower volumetric flue gas flow rate and different inlet mass flow conditions through burner inlets compared with air combustion. In oxy-fuel combustion, nitrogen is removed in the air separation unit used to generate oxygen and combustion gas species is dominated by CO_2/H_2O compared with $N_2/CO_2/H_2O$ in air combustion. CO_2 and N_2 have intrinsic differences in physical–chemical properties in heat capacity, density, thermal diffusivity and emissivity. To design a new oxy-fuel furnace chamber or retrofit an existing furnace to oxy-fuel technology, it is critical to understand the impacts of changes of systematical variables and intrinsic variables on temperature distribution and heat transfer properties. The following sections will discuss the impact of systematical variables and intrinsic variables on heat transfer.

9.2.1 Adiabatic flame temperature and combustion stoichiometry for oxy-fuel furnaces with recycled flue gas

The adiabatic flame temperature (AFT) is defined as the temperature attained when all of the chemical reaction heat released heats combustion products. In practical coal combustion systems, excess oxygen is usually applied to achieve complete coal burnout. In this chapter we set O_2 concentration in the flue gas at 3.3% v/v. AFT in pure oxygen combustion is high so recycled flue gas is introduced to moderate AFT. The AFT depends on the heating value of the fuel, the temperature and amount of oxidizer, and final combustion product composition. To equate the sum of the heating value of fuel and sensible heat of oxidizers to the sensible heat of the combustion product, AFT can be calculated. The information on standard heats of formation and heat capacity which are used to calculate sensible heat of combustion products can be found in compiled thermodynamic data such as the JANAF Table.

A calculated result of AFT vs. recycled flue gas ratio and dependence of AFT on fuel heating value and recycled flue gas temperatures is illustrated in Fig. 9.2.

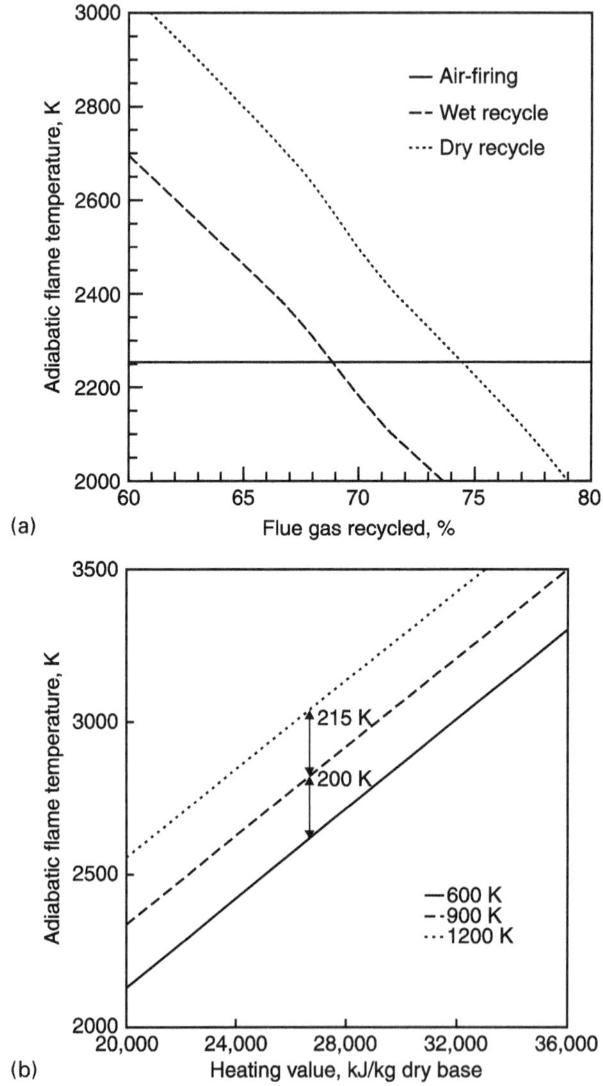

9.2 Adiabatic flame temperature in oxy-fuel combustion. (a) Adiabatic flame temperature in oxy-fuel combustion under various flue gas recycle ratio and dry/wet recycle modes, compared with air-firing adiabatic flame temperature. (b) Adiabatic flame temperature as a function of fuel heating values and recycle flue gas temperatures for 67% wet recycle oxy-fuel combustion, assuming fuel heating value and fuel chemistry are independent.

Under the same recycled flue gas ratio, dry recycle has higher adiabatic flame temperature than wet recycle oxy-fuel combustion. With increasing recycled flue gas from 60% to 80% recycle ratio, AFT monotonically decreases from above 3000 K to below 2000 K. For the cases considered, 69% recycle ratio for wet recycle oxy-fuel has the same AFT as that of air-firing. At 74.5% recycle ratio, dry recycle oxy-fuel obtains the identical AFT to that of air-firing. AFT increases with increasing fuel heating values and recycled flue gas temperatures. In wet recycle oxy-fuel combustion under 67% recycle rate, increasing recycled flue gas temperatures by 300 K makes AFT increase about 200 K. AFT represents total energy input into a furnace for specific fuel rate and combustion stoichiometry, and thus is generally used as a design criterion for air-firing combustion chambers. In oxy-fuel combustion, warm recycled flue gas can carry energy into the furnace chamber. At 71% recycle ratio in wet recycle oxy-fuel, about 8.5% energy is carried over by recycled flue gas into the oxy-fuel combustion chamber if the temperature of the flue gas is the same as that of the preheated air, in air-firing, that carries about 8.5% energy into the combustion system. Although the recycled flue gas contribution to energy input is minor, increased flue gas can favor heat transfer in a convective pass downstream of the furnace chamber.

9.2.2 The relationship between recycle ratio and oxygen (O_2) concentration

The absence of nitrogen in oxidant reduces the flue gas volume and recycled flue gas is used to compensate for the absence of nitrogen. By increasing recycled flue gas to about 77% in wet recycle oxy-fuel, the same volumetric flue gas flow rate as in air-firing is achieved. At the same recycle ratio, dry recycle oxy-fuel results in a lower gas flow rate than wet recycle because water vapor is removed.

By increasing proportions of flue gas recycled from 50% to 80%, oxygen concentrations at burner inlet decrease from 53% to 24% and 45% to 19% for dry recycle and wet recycle conditions respectively, as indicated in Fig. 9.3. The higher oxygen concentration for dry recycle is due to the absence of the water vapor dilution that exists in the wet recycle. The oxygen required to completely combust coal is defined as theoretical oxygen required. In practice the oxygen supply is greater than this theoretical value to achieve complete burnout; thus excess oxygen is defined as the ratio of oxygen excessively supplied to the theoretical oxygen requirement. Excess oxygen comes partly from the pure oxygen stream from the air separation unit and partly from recycled flue gas, the contribution from the recycled flue gas being the greater. Excess oxygen at burner inlets increases with increasing recycled flue gas for recycle flue gas containing 3.3% oxygen. When 71% of flue gas is recycled, 15.9% and 12.6% excess oxygen at burner inlets is obtained for wet and dry recycle oxy-fuel combustion respectively. The higher excess oxygen in wet recycle than in dry

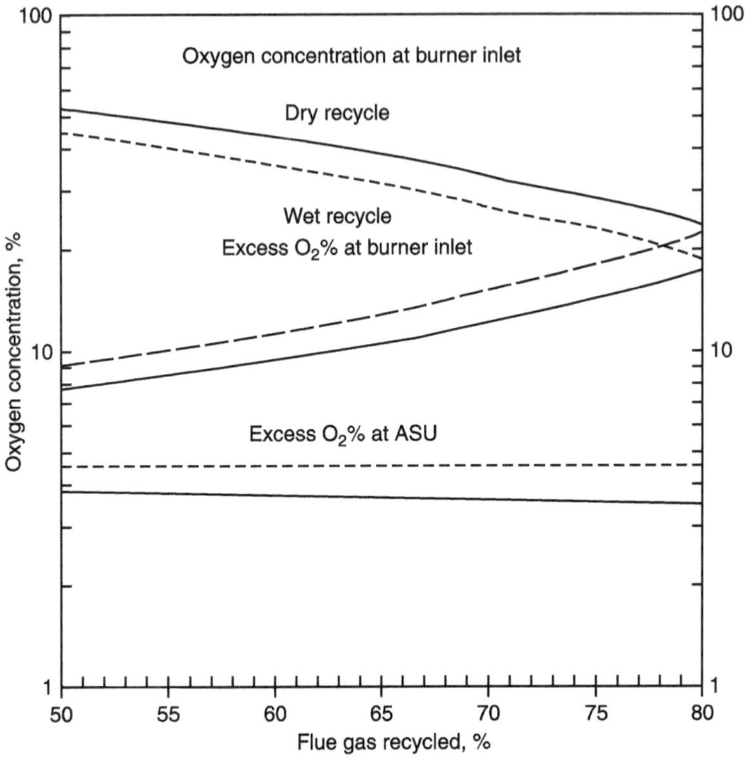

9.3 Oxygen concentrations (% v/v) and excess oxygen (oxygen at burner/stoichiometric oxygen for fuel) at burner inlet, excess oxygen for air separation unit (oxygen supplied by ASU/stoichiometric oxygen for fuel) as a function of recycled flue gas ratio in oxy-fuel combustion.

recycle oxy-fuel is due to higher excess oxygen from the air separation unit outlet in wet recycle, when both recycle modes achieve 3.3% oxygen in the flue gas.

9.2.3 Technical methods in estimation of radiant heat transfer

For retrofitting an existing furnace into an oxy-fuel furnace, the same matched heat transfer is required. The most important factors impacting heat transfer in oxy-fuel combustion are temperature and emissivity of the combustion products (gases and particles) and walls. In practical oxy-fuel combustion, both temperature and emissivity vary through the furnace. The heat balance can be monitored by several means. First, by monitoring the flue gas temperature and composition at furnace exit, the flue gas enthalpy can be obtained. By subtracting flue gas enthalpy from total energy input into the furnace, total heat absorbed by the furnace walls can be estimated. Secondly, to monitor wall heat flux at various locations along the furnace walls, the total heat absorbed can be approximated by summation of all the heat flux measured. Thirdly, to monitor the water flow rate and temperature/pressure increase when the water passes through water cooling walls, the heat absorbed by the water can be calculated.

Flame temperature, radiation intensities or radiative heat flux have been measured in several oxy-fuel combustion experimental rigs. Gas temperatures can be measured with a water-cooled suction pyrometer, in which a thermocouple is protected by a ceramic tube. Radiation intensity measurements were performed with a radiometer[1,8–10,13] or computer-controlled scanning monochrometer for spectral radiation intensity.[15] The radiation detected by the radiometer consists of radiation emitted by the flame, and radiation emitted and reflected by any furnace wall in view, thus the Schmidt method can be used to estimate gas temperature.[14]

9.3 Theoretical heat transfer analysis

9.3.1 Heat transfer properties comparison between air and oxy-fuel combustions

Coal combustion products include gaseous CO_2 and H_2O and fly ash particles, with some intermediate solid products including char and soot. The contribution to the radiative transfer in flames is due to luminous radiation and non-luminous radiation. Non-luminous flame radiation is concentrated in gaseous absorption lines and bands in the infrared spectrum with windows existing between them, resulting from the transitions between the molecular energy states, particularly the vibration-rotation energy states. In hydrocarbon combustion, carbon dioxide and water vapor are the dominant contributors to the non-luminous radiation. In luminous flames a continuum radiation in the visible spectrum and infrared is also emitted by the soot, char and fly ash, which contribute greatly to the luminosity of the flames.[16] To achieve a better understanding on emissivity, the relative contributions to the total emissivity by fly ash, char, soot and gases need to be assessed. In air-firing, it is probable that the continuum radiation from fly ash particles is more important because these particles exist in almost the entire furnace volume while char and soot particles are usually present in a relatively small fraction of the entire furnace volume.[17]

The radiative properties, i.e. emissivity, absorptivity, reflectivity etc., are spectral dependent. The fractional blackbody emissive power indicates over 90% of radiative heat transfer exists in the wavelength range between 1 and 10 microns when gas temperature is above 1500 K. Thus in the following discussion on the radiative properties of gas, soot, fly ash and char, we will concentrate on the wavelength range between 1 and 10 microns. For engineering applications and simulations, the total properties which are integrated from the whole spectrum determine the overall heat transfer efficiency.

Compared with air-firing, the oxy-fuel combustion chamber is filled by a medium with higher concentrations of CO_2 and H_2O, and particulate matter of char, soot and fly ash. The gas composition and ash concentrations in oxy-fuel combustion at 67% recycle ratio are presented in Table 9.3; in such a case air leakage is neglected. Figure 9.4 further shows gas compositions from oxy-fuel combustion flue gas reported in the literature. The total inert value in the flue gas is usually less than 15%, but the ratios between water vapor content to carbon

Table 9.3 Gas compositions in wet and dry oxy-fuel combustion at 67% recycle ratio; air leakage which may take place in the practical process is disregarded; the ash concentration is calculated at the flue gas temperature of 1600 K

	Air-firing	Oxy-fuel combustion	
		Wet recycle	Dry recycle
O_2, %	3.3	3.3	3.3
N_2, %	74.7	0.3	0.3
CO_2, %	15.6	68.5	84.7
H_2O, %	6.3	27.7	11.4
Ash concentrations, m^3/m^3	2.4E-06	3.1E-06	3.8E-06

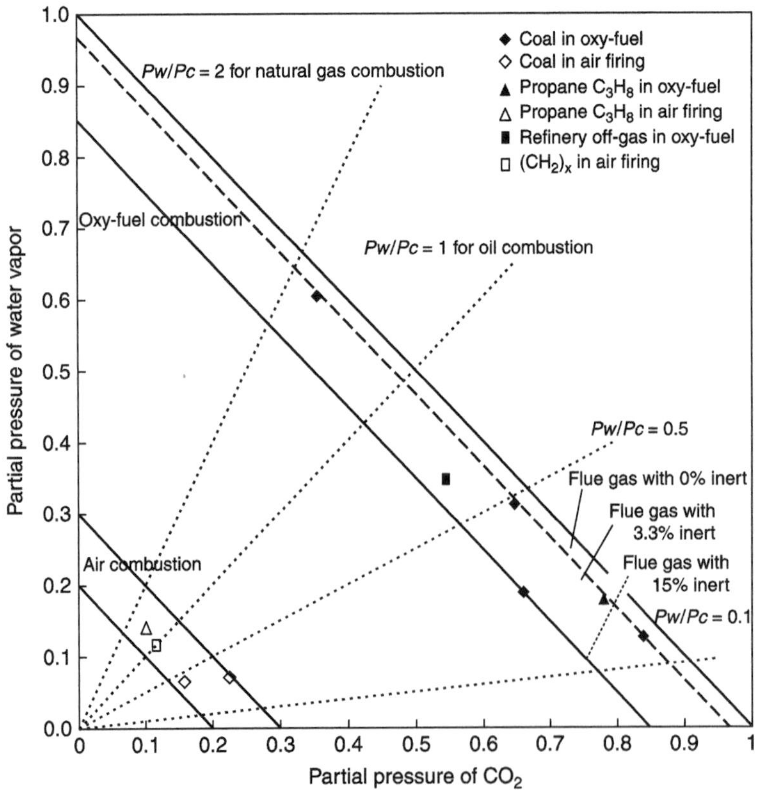

9.4 Gas compositions from oxy-fuel combustion compared with those from air-firing.

dioxide content vary greatly. The major inert gas originates from excess O_2, Ar and N_2 from the air separation unit, N_2 from air leakage, and impurities from coal combustion.

The emissivity of CO_2 and H_2O depends on several parameters such as gas temperature, total pressure, absorbing gas partial pressures, path lengths and wavelength. The total emissivity is defined as the Planck mean emissivity over the whole wavelength spectrum. There are several existing models which can be applied to predict total radiative properties of combustion gaseous products, including Hottel's emissivity charts and regression correlation,[18,19] Weighted Sum of Gray Gas Model (WSGGM),[20–26] Total Emissivity Model,[27,28] Wide Band Model (WBM),[29,30] Narrow Band Model (NBM) and Line-By-Line Model (LBLM). These models are developed based on physical spectroscopic measurement of gas absorption.

WSGGM was first presented by Hottel for the zonal method for the radiative heat transfer analysis;[18] later it was demonstrated by Modest that WSGGM also can be applied to any solution method for the equation for transfer provided all boundaries are black and the medium is non-scattering.[31] In the model total emissivity is approximated by weighted sum of gray gases:

$$\varepsilon(T, PpL) = \sum_{i=0}^{I} a_{\varepsilon,i}(T)\left[1 - e^{-k_i PpL}\right] \qquad [9.1]$$

where $a_{\varepsilon,i}$ is the emissivity weighting factor for ith gray gas as based on gas temperature T, $[1 - e^{-k_i PS}]$ is ith gray gas emissivity, k_i is the absorption coefficient related to ith gray gas, PpL is the partial pressure-path length product (atm-m), Pp is the sum of the partial pressures of the absorbing gases (atm), L is the path length (m), k_i is the absorption coefficient of ith gray gas (atm-m)-1, and I is the number of gray gases.

In this way, WSGGM represents the radiation properties of total emissivity and absorptivity by a summation of a number of terms, each given by the multiplication of a weighting factor and a gray emissivity. WSGGMs are limited to specific gas compositions, i.e. the ratio of the partial pressures of water (Pw) and CO_2 (Pc), Pw/Pc = 1[21,23] or 2[21,23,26] for oil or natural gas combustion. For fuel oil, CO_2 and H_2O both have concentrations of 10%, for natural gas 10% and 20%. These WSGGM predict a maximum emissivity value ranging between 0.5 and 0.6 when path length is greater than 5 atm-m. The characteristics of emissivity increasing with path length are present only in the small path length range; after that region it reaches an asymptotic limit.

The Exponential Wide-band Model (EWBM) first developed by Edwards and Menard is by far the most common of the wide-band models. In the model, three physical parameters are necessary to describe total gas band absorptances over a wide range of temperature, path length and pressure conditions, i.e. the integrated band intensity α, the line-width-to spacing parameter β, and the bandwidth parameter ω. The parameters α, β, ω are functions of temperature, pressure and wave number. In the EWBM, it is assumed that the three parameters vary solely with temperature over each vibration-rotation band, the influence of pressure through the pressure broadening parameter, while the parameter mass absorption coefficient accounts for the spectral dependence.[29] The integrated band intensity parameter α represents the area under the mass absorptance curve:

$$\alpha = \int_0^\infty k_\eta d_\eta = \int_0^\infty \left(\frac{S}{d}\right)_\eta d_\eta \qquad [9.2]$$

The dependence of α on temperature is given by the relation with a function ψ, which depends on v_k, the vibrational quantum number, g_k the degeneracy of the fundamental band v_k, and δ_k the vibrational transition number.[32] The parameter β is defined as β times the mean line-width to spacing ratio for a dilute mixture at one atmosphere total pressure. The dependence of β on temperature is described by function Φ, which in turn can be approximated by the formula: $\sum_{k=0}^{N} \alpha_k T^k$.

The emissivity of pure water vapor and pure carbon dioxide as a function of path length are presented in Fig. 9.5. The temperature used in the calculation is fixed at 1500 K. Emissivities were calculated by Hottel's chart and WSGGM methods. Emissivities of pure CO_2 and H_2O initially increase with path length, approaching an asymptotic limit at large path length. The asymptotic limits are about 0.25 and 0.6 for CO_2 and H_2O respectively. The differences in emissivities calculated by different models are within 0.1. The original Hottel's chart provided CO_2 emissivity data in a range with path length less than 150 atm-cm, and H_2O emissivity data in a range with path length less than 600 atm-cm.

Radiation emitted by the particulate matter including char, fly ash and soot is distributed continuously over the wavelength spectrum, unlike the discrete band nature of gas emission. The radiative properties of these particulates depend on their size distribution, particle shape, volume fraction concentration and chemical composition as well as temperature and wavelength. During the course of coal combustion, conversion from coal to char/soot and fly ash changes volumetric fraction and radiative properties of particles.

The radiative properties of particles required for typical radiative transfer calculations are absorption and scattering coefficients, and scattering phase function. The material structure of the particulate matter determines its complex index of refraction. The real part of the complex refractive index, which is called the refractive index, is the ratio of the speed of light in a vacuum to that within the particle for light at normal incidence, controlling scattering. The imaginary part, which is also called the absorption index, is directly related to the rate of attenuation of radiation with depth within the material, controlling absorption. The size parameter $x = \pi D/\lambda$ determines the interaction between electromagnetic field and particle, where D is particle diameter and λ is wavelength. When $x \ll 1$, which is applicable for soot, Rayleigh scattering applies; when $x \approx 1$, which is the case for char and fly ash particles, Lorenz-Mie theory determines how the incident wave is absorbed and scattered by an isolated homogeneous spherical obstacle; when $x \gg 1$, which is the case for coarse coal/char and fly-ash particles, geometric optics applies.[33]

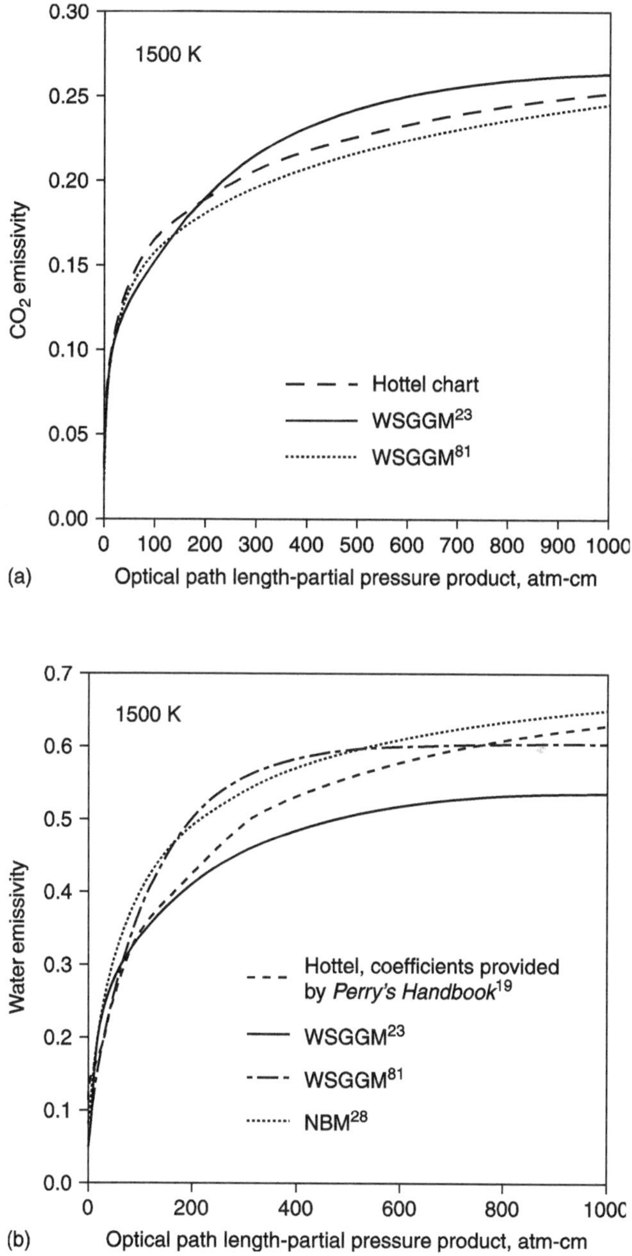

9.5 Emissivity of CO_2 and H_2O at 1500 K, from different methodologies. (a) CO_2; (b) H_2O.

The ratio of absorbed and scattered energy to the incident radiation gives the absorption and scattering cross sections, which have units of area. The absorption and scattering efficiency factors are the ratio of absorption/scattering cross section to the geometric area cross section of a particle. The extinction efficiency factor approaches a constant asymptotic value of 2 with increasing size parameter, while the efficiency factors of absorption and scattering approach limits dependent on their complex index of refraction $m = n - ik$. In the following sections, we will discuss the optical properties of char, soot, fly ash and ash deposits on the wall.

The optical properties of char are considered to be more significant than those of coal, as coal devolatilization time is much shorter than the char burnout time. An in situ measurement technique, such as FTIR emission-transmission technique, is applied to measure char spectral emissivity at flame temperature; however, the complex refractive index has not been provided.[34–37] Ex situ measurement, such as transmission technique, also has been reported.[38]

Soot is released at the stage of coal devolatilization, and is considered an important emitter in the oxy-fuel flame.[8,9] The various optical properties of soot – volumetric fraction, size distribution and shape, and temperature – affect its heat transfer properties. Soot optical properties, i.e. the complex index of refraction, have been measured from various hydrocarbon fuels in air combustion.[39–44] In this chapter, the optical constant of soot is estimated by experimental measurements made by Chang and Charalampopoulos[43] for a wavelength range of 0.4 to 30 microns. It has been found that chemical composition affects soot optical properties, especially the H/C ratio.[45] The absorption coefficient of soot depends on wavelength according to the relationship $1/\lambda^a$, $\alpha \approx 1$. Soot particles are typically 0.03–0.1 microns in diameter. Compared with the dominant wavelength of the thermal radiation, the size parameter is in the order of 0.1; therefore the scattering effect by individual soot particles can be disregarded. In such a Rayleigh limit of thermal radiation wavelength range, α close to unity has been observed from experiment.[46] The volumetric concentration of soot lies in the magnitude between 10^{-5} and 10^{-8}. Soot absorption coefficient is dependent on soot concentration and independent of particle size:

$$k_\lambda = \frac{36\pi nk}{(n^2 - k^2 + 2)^2 + 4n^2k^2} \frac{f_v}{\lambda} \qquad [9.3]$$

The total emissivity of soot is expressed as:

$$\varepsilon = \frac{1}{\sigma T^4} \int_0^\infty (1 - e^{-k_\lambda L}) E_{b\lambda} d\lambda \qquad [9.4]$$

Soot emissivity is independent of particle size, but dependent on the complex index of refraction (assuming it is wavelength independent), volume fraction, temperature and path length. Sensitivity study indicates soot concentration is more significant than its optical constants.

Emissivity of a single fly ash particle depends on optical constants, particle size and wavelength. Emissivity of a fly ash cloud depends on emissivity of single particles, number concentration, particle size distribution and path length.[47]

$$\varepsilon = 1 - e^{-k_{a,\lambda}L} = 1 - e^{-\int_a \int_\lambda Q_{a,\lambda} B_a A_p d\lambda da L} \quad [9.5]$$

where $Q_{a,\lambda}$ is the spectral cloud absorption efficiency, B_a is dust loading or number concentration, A_p is the projected surface area of cloud and L is the path length.

The gas temperature and combustion stoichiometry determine the flue gas volume, which impacts number concentration. The ash particle concentrations were measured from several Australian power plants by Gupta and Wall,[48] giving concentrations at the level of a few grams per cubic meter, which is consistent with mass balance calculation.

Optical properties of fly ash have been measured at laboratory, pilot and full scale combustion systems,[49] by reflectance technique,[50] transmission technique,[48,50] light angular scattering technique[51,52] and FTIR emission technique.[53] The optical constant depends on wavelength,[50] chemical composition,[50] unburnt carbon[54] and temperature.[50] The most frequently referenced optical constants of fly ash are provided by Goodwin,[50] presented in the wavelength range from 0.5 to 12 microns. The measured experimental data also has been related to ash composition and formulated by mathematical model.[50,17] The real part of the refractive index does not vary greatly for different ashes. In the 1–8 microns wavelength range, the imaginary part of the refractive index depends primarily on the iron, silica and hydroxide contents of the slag. Iron is the primary absorber for wavelengths shorter than 4 microns. In the 4–8 micron range, the absorption index is dominated by vibration absorption due to Si-O-Si and Si-O. The optical constants in the wavelength range 8–13 microns were observed to be dominated by three absorption bands. Optical constant for the long wavelength region > 13 microns are assumed to be dominated by the Reststrahlen bands of Al_2O_3 and MgO.[17]

The dependence of single ash particle emissivity on the iron content in ash and particle size is presented in Fig. 9.6, based on a Mie solution to the radiation and particle interaction. Mie theory gives an analytical solution to the interaction between radiative electromagnetic wave and particle matter, based on the size parameter and complex index of refraction. The single particle scattering code can be found from Bohren and Huffman[55] and Wiscombe.[56] With increasing particle size, the emissivity of single fly ash particle increases. Figure 9.6 also indicates higher iron ash has higher emissivity than lower iron ash. Dependence of emissivity on temperature has been calculated by Wall, who indicated emissivity decreases with increasing temperature.[57] By integrating spectral emissivity over the whole wavelength band and over the whole size distribution range, the total emissivity of fly ash can be obtained.

The boundary condition, i.e. wall temperature distribution and emissivity/reflectivity, must be known before the radiative transfer equation can be solved.

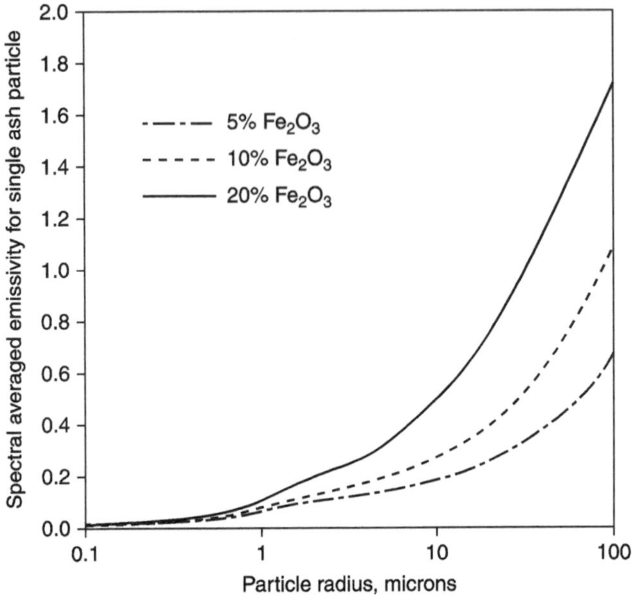

9.6 Single ash particle emissivity as a function of ash particle size and ash chemistry.

The radiative properties of the furnace walls are strongly dependent on whether ash deposit is present or not, the ash deposition characteristics being dependent on the furnace operation condition and the mineral matter composition of coal. When there is no ash deposit on the furnace wall, the emissivity of the wall can be selected as oxidized iron, with the emissivity expression $\varepsilon = 0.72 + 2 \times 10^{-4}T$ for highly oxidized iron obtained from Hampartsoumian *et al.*[58] Similar equations also have been provided for coal ashes.[59]

When there is ash deposit present on the furnace wall, the emissivity of the wall depends on the surface roughness and porosity, both depending on the structure of the ash deposit, i.e. whether it is particulate/sintered/molten. Other factors such as temperature, wavelength and chemical–physical properties which depend on coal rank and type, mineral matter in coal, and deposit history are also involved.[57,59,60–64] The emissivity of slag is measured with optical reflectance technique,[50] or Diffuse Reflectance Infrared Fourier Transform Spectrometry (DRIFTS).[65] General observations are that high iron and unburnt carbon result in higher emissivity. Thus emissivity of western US coals is lower than that of eastern coals; emissivity decreases with increasing temperature; coarse ash particles give higher emissivity than fine ash. The emissivity of a rapidly quenching amorphous surface is also higher than a slowly cooled crystalline sample. Measured spectral emissivity of deposits was integrated to obtain total emissivity with values between 0.4 and 0.7.[59,65–68] Molten slag deposit can be considered a smooth surface, thus the emissivity can be predicted as:[69]

$$\varepsilon = 1 - \frac{(n_\lambda - 1)^2 + k_\lambda^2}{(n_\lambda + 1)^2 + k_\lambda^2} \quad [9.6]$$

The predicted ash deposit emissivity is about 0.95, independent of ash composition.[70] This reported data is higher than emissivity measured from ash deposits.

When ash deposit is present on the wall surfaces, the surface temperature is controlled by deposit heat conductivity, emissivity and deposit thickness.[64]

9.3.2 Heat transfer under a Well-stirred Model assumption for a small furnace

The Well-stirred Model (WSM) is a zero-dimensional model and the predicted overall heat transfer is estimated from a single gas temperature at the furnace exit.[71] Well-stirred or zero-dimensional models calculate the heat transfer rate for a given load. In this model, the combustion products are assumed to be gray and at a uniform single temperature. Also, the temperature and the radiative properties of the furnace wall are assumed to be constant. Therefore, at steady-state conditions, the energy entering the system delivers a part of the heat to the furnace wall by radiative heat transfer and the remainder of the heat in the flue gas can be absorbed in the convective section. This model can provide an approximate performance prediction for different furnace sizes and can help in establishing the effects of changes in operating parameters on heat transfer trends. It can also examine the pilot-scale experimental data against its predictions and for knowing the model capability for large scale units.

In the furnace, heat transfer Q between furnace gases and wall is approximately proportional to the difference in gas and wall temperatures raised to the fourth power and an effective emissivity:

$$Q = \varepsilon_f \sigma A \left(T_g^4 - T_w^4\right) \quad [9.7]$$

For the furnaces with the cold or sink fraction of the total wall surface area close to unity, effective emissivity is expressed as:

$$\varepsilon_f = \frac{1}{\frac{1}{\varepsilon_g} + \frac{1}{\varepsilon_w} - 1} \quad [9.8]$$

Total heat released by flue gas is calculated as:

$$Q = H_F \frac{T_{aft} - T_g}{T_{aft} - T_{ref}} \quad [9.9]$$

By equating equations [9.7] and [9.9], the gas temperature T_g is obtained. T_{aft} and T_{ref} are adiabatic flame temperature and reference temperature respectively.

Conclusions are presented for retrofitting an existing boiler with oxy-fuel combustion.

Some constraints were set as:

- To maintain 3.3% oxygen in the flue gas as commonly met in air-firing conditions.
- To match total heat absorption from combustion chamber to that in air-firing conditions, by varying the recycle flue gas ratio.

Compared with the air-firing condition oxy-fuel combustion operates at:

- Lower AFT by about 50 K.
- Lower exit flue gas temperature by about 30 K.
- Similar energy input and furnace thermal efficiency are achieved.

From the conclusions, for oxy-fuel firing, furnace heat transfer efficiency is about 49% for Furnace-A, which is on the same level as air-firing. Equation [9.8] indicates the efficiency in oxy-fuel firing is determined by lower AFT and lower flue exit gas temperature (FEGT) in oxy-fuel firing, implying small temperature differences between AFT and FEGT in oxy-fuel firing. Mass balance calculation shows reduced flue gas in oxy-fuel firing, while higher heat capacity of the flue gas can compensate for reduced mass flow.

9.3.3 Heat transfer characteristics in larger furnaces

For heat transfer scaling in oxy-fuel combustion, the factors of furnace geometry, thermal heat loading, temperature and emissivity should be taken into account. With increasing furnace size, the ratio of the furnace volume to surface heat transfer area decreases and thermal heat loading increases; thus the firing density, defined as heat release in unit furnace volume or furnace surface, increases. Gas temperature and gas emissivity usually increase with increasing furnace scale, which is characterized by geometrical dimensions of furnace. The mean geometrical path length ranges from 1.1 to 4.7 to 12.1 meters for Furnaces A, B and C listed in Table 9.2.

The mass balance calculation and AFT calculation are scale independent. The same phenomena have been observed as in the pilot-scale furnace heat transfer: higher AFT and lower gas temperature are met in oxy-fuel combustion; thus higher heat transfer efficiency is achieved in the same scale furnace.

The emissivity as a function of furnace scale is qualitatively presented in Fig. 9.7 for air-firing, wet recycle and dry recycle oxy-fuel combustion, with the respective contributions by fly ash and gas highlighted. Here the soot and char contribution have been disregarded. The calculation conditions are gas temperature of 1600 K, pressure 1 atm, water vapor and carbon dioxide partial pressures 0.27 and 0.69 for wet recycle and 0.11 and 0.85 for dry recycle, as shown in Table 9.3. For Fig. 9.7 the emissivity calculation for gas emissivity is based on a Wide Band

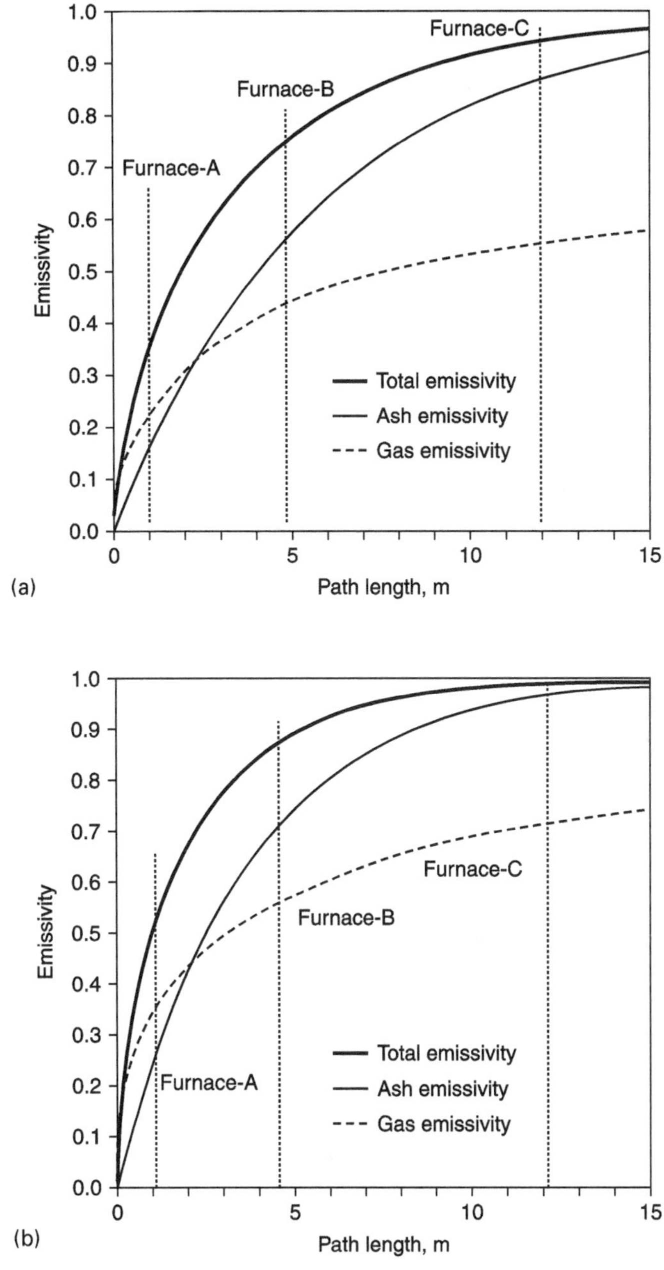

9.7 Emissivity for furnace heat transfer. (a) Air case; (b) oxy dry recycle; (c) oxy wet recycle.

(Continued)

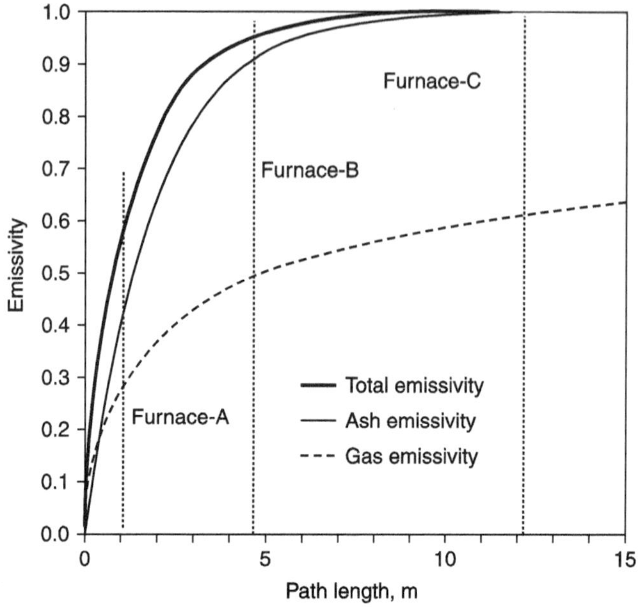

9.7 Continued.

Model and overlapped bands between water vapor and carbon dioxide are ignored. The ash emissivity calculation is based on experimental measured ash concentration and particle size distribution from a pilot-scale oxy-fuel combustion program, with the ash optical property data obtained from the ash database maintained at the University of Newcastle. A total of ten size bins and 61 spectral bins were integrated to obtain total size and spectral averaged ash emissivity data. In detailed engineering analysis, dependence of emissivity on gas temperature is taken into account by the Well-stirred Model and explained in detail in the following text. For air-firing, predicted emissivity in Furnace A is 0.35, while emissivity increases with furnace size to 0.75 for Furnace B and 0.94 for Furnace C. Ash emissivity increases more rapidly against path length than gas emissivity. For furnace size greater than Furnace B, gas emissivity approaches an asymptotic limit while ash emissivity continuously increases. For Furnace A, the gas emissivity contribution is higher than that for fly ash; however, for larger Furnaces B and C, the ash contribution is dominant.

Both wet recycle and dry recycle oxy-fuel combustion exhibit higher emissivity than air-firing due to higher concentrations of H_2O, CO_2 and solid particles, in which reduced flue gas volumetric flow rate plays an important role. Furnace A emissivity is higher in the wet recycle case compared with the dry recycle case. When furnace size increases to Furnace B and C, there is no significant difference in emissivity between wet recycle and dry recycle and differences in emissivity between oxy-fuel firing and air-firing become smaller.

The ratio of surface area to volume of a combustion chamber is higher in small furnaces than in larger furnaces. Thus the size of a large furnace is governed more by the surface area necessary for heat transfer than by the volume necessary to burn the coal in the small furnace. As shown in Fig. 9.8, with increasing in furnace size, FEGT increases for both air-firing and oxy-fuel firing. Oxy-fuel firing has lower AFTs and FEGTs than air-firing.

For large scale furnaces, the value of flue gas emissivity approaches unit so that the characteristics of water cooling wall and ash deposit emissivity play a more significant role.

9.4 Computational fluid dynamics (CFD) radiation heat transfer models

The previous sections present radiative heat transfer characteristics in oxy-fuel combustion based on a homogeneous furnace temperature and species concentration distribution assumption. The predicted result established that key factors impacting scaling radiative heat transfer are emissivity and furnace geometry. As illustrated by Fig. 9.1, the heat transfer in practical furnaces is strongly interacted with other processes simultaneously taking place with heat transfer, including coal drying, devolatilization, volatile matter combustion, char combustion, ash formation and fluid aerodynamics (momentum transfer); thus an

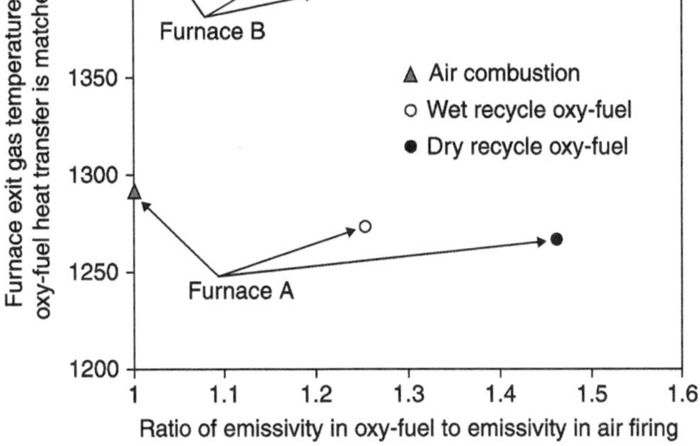

9.8 Emissivity ratios (oxy/air) and furnace exit gas temperatures for the three furnaces.

inhomogeneous temperature and concentration distribution exists inside the furnace. To match heat transfer in retrofitted oxy-fuel combustion, momentum transfer and coal reactivity also need to be taken into account for furnace scaling. A CFD-based combustion model is capable of such work. CFD models have been used for oxy-fuel experimental measurement data interpretation[72,73] oxy-fuel combustion burner modeling,[13,74] and oxy-fuel combustion.[14,75–77]

Using a combustion model to predict radiative heat transfer is based on two assumptions: first there is a predictable relationship between coal combustion rate and spatially distributed combustion product temperature and concentrations; secondly, there is a predictable relationship between combustion product spatial distribution and its radiative heat transfer. The combustion product radiative properties presented in the previous section are coupled with a CFD-based comprehensive combustion model to study radiative heat transfer in oxy-fuel furnaces. The radiative transfer equation (RTE) in a three-dimensional and heterogeneous medium can be solved by several methods, such as P1, DO, DT and Monte Carlo methods.

Coal and oxidizer are injected into the furnace at different temperatures and velocities through a burner which is of a concentric annulus structure. The inflow conditions are shown in Table 9.4. Momentum transfer results in recirculation in the near burner zone and thus various flame types can form.[14] Coal particles are ignited by a hot flame by convective and radiative heat transfer. Coal particles disperse and react in the turbulent fluid and this determines the temperature and species concentration distributions.

The fundamental laws for a combustion model are energy conservation, mass conservation and momentum conservation in a solid-gas chemical reacting system.[78] Radiative heat transfer, turbulent flow, particle dispersion and coal reactions are simulated with corresponding models. Selection of model and model parameters are critical and should be validated against experimental measurements.

Table 9.4 Mass input conditions in oxy-fuel combustion compared with air-firing for a pilot-scale test furnace[14]

	Air-firing	Oxy-fuel combustion
Coal flow rate, kg/h	120	120
Primary air flow rate, kg/h	195	240
Secondary air flow rate, kg/h	780	625
Primary air temperature, °C	80	130
Secondary air temperature, °C	280	220
Primary air velocity, m/s	20	23
Secondary air velocity, m/s	35	21
Primary momentum flux, kg/s.m^2/s^2	35.7	54.1
Secondary momentum flux, kg/s.m^2/s^2	270.2	74.1
Momentum flux ratio (primary/secondary)	0.13	0.73

In the work by Khare *et al.*,[14] the P1 model is chosen for radiative heat transfer, the standard $k - \varepsilon$ model is chosen for turbulence, the DPM model is chosen for solid–gas flow and empirical single kinetic model and kinetic/diffusion models are chosen for devolatilization and char combustion. The coal reaction kinetic data was measured from a drop tube furnace and different reactivity in oxy-fuel and air combustion was observed.[79]

As mentioned in the previous section, matching heat transfer in oxy-fuel combustion results in reduced air flow through the burner. In practice, primary air flow rate is fixed to transport coal particles; thus primary air in oxy-fuel is comparable to that in air-firing while secondary air flow rate is reduced dramatically.[14] This changes the momentum ratio of primary air to secondary air, changes the fluid aerodynamics, and impacts coal ignition, coal reaction and finally heat transfer. The illustrative numbers for such changes are shown in Table 9.4 for a pilot-scale test furnace.

The impact of momentum transfer difference on ignition in oxy-fuel combustion has been studied by Khare *et al.*[14] for a pilot-scale furnace. The model predicts delayed flame ignition due to changes in the momentum ratio of primary/secondary air. In the pilot-scale test rig, flame ignition delay was observed only for partial loading operation oxy-fuel combustion[80] and the measured temperature profile cannot indicate whether or not there is significant ignition delay due to discontinuous measurements in the axial direction, as shown in Fig. 9.9.[14]

Heat transfer in a demonstration scale furnace is presented in Wall *et al.*[5] The flame shape and wall heat absorption are shown in Fig. 9.10 and Fig. 9.11. The type-2 flame was predicted with internal and external recirculation in both

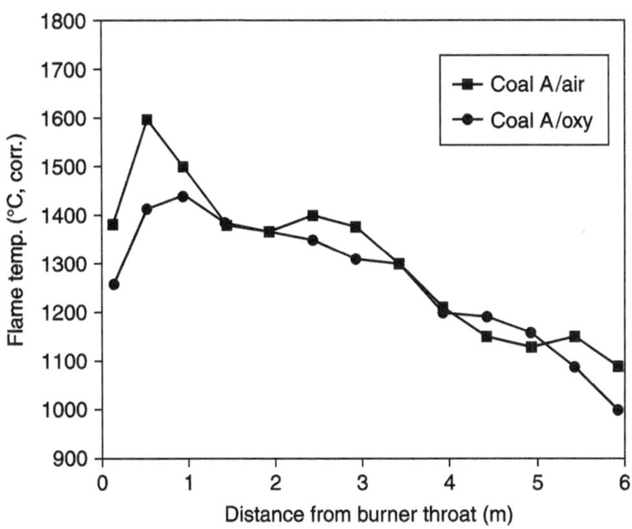

9.9 Gas temperature profiles measured by radiation pyrometer.[5]

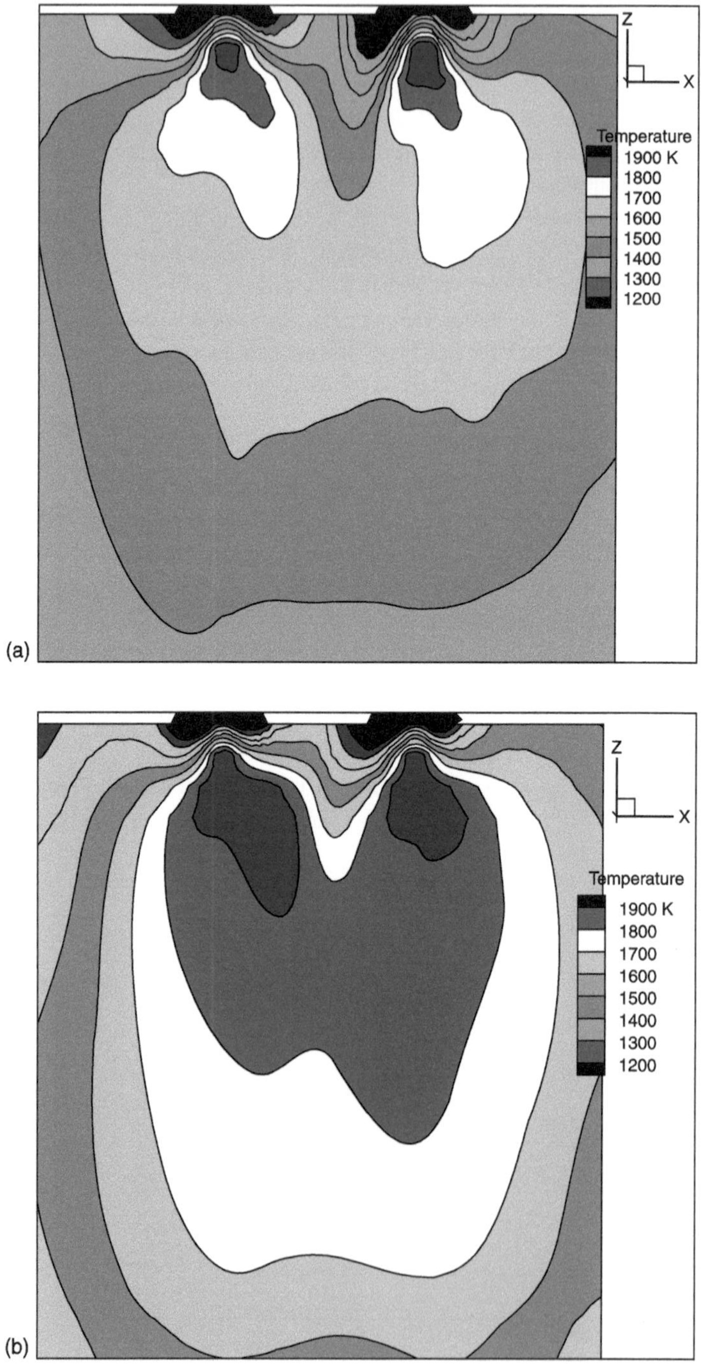

9.10 Predicted horizontal burner plane for the air case (a) compared with the oxy case (b) for a 30 MWe furnace.[5]

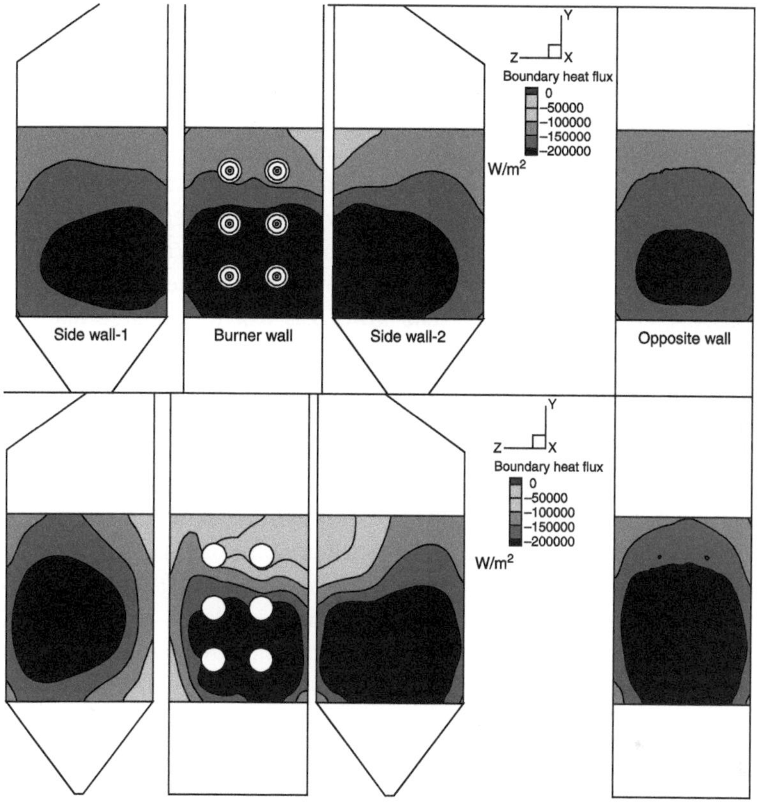

9.11 Folded furnace wall heat flux contours for air (top) and oxy (bottom) cases, showing higher burner wall heat flux for the air case and higher rear wall heat flux for the oxy case.[5]

air-firing and oxy-fuel firing, but the oxy-fuel flame was more elongated, extending to the rear wall, as shown in Fig. 9.10. The overall furnace heat transfer between air combustion and oxy-fuel combustion in a retrofitted furnace can be matched; however, the wall heat transfer distribution differs for air and oxy-fuel combustion, as shown in Fig. 9.11. Higher heat absorption on the front wall compared with the rear wall for the air combustion was observed, while in oxy-fuel combustion, more heat absorption in the rear wall was observed due to the longer flame.

9.5 Conclusions

The accuracy of heat transfer prediction is dependent on the accuracy of models and input parameters to the models, as well as boundary conditions. The radiative properties including gas, fly ash, char and soot are critical properties for heat transfer consideration in oxy-fuel combustion, as the optical path length in oxy-fuel combustion is greater than in the same furnace operated in air-combustion. For large furnaces which are oxy-fired, the emissivity data is beyond the Hottel data range due to high CO_2 and H_2O partial pressure and the Wide Band Model is used here for emissivity prediction.

Heat transfer is one aspect of a complicated combustion process and its consideration cannot be isolated from other processes taking place. Based on a Well-stirred Model assumption, radiative heat transfer theory combined with radiative properties developed in this chapter predict a recycled flue gas ratio to match heat transfer in air-firing for a retrofitted furnace. Oxy-fuel combustion then operates at lower AFT and lower gas exit temperature due to high gas emissivity met in oxy-fuel combustion. When radiative heat transfer is coupled with momentum transfer, mass transfer and coal reactivity by a CFD combustion model, the predicted heat transfer pattern can differ greatly from that in air-firing due to changes in burner flows, changing flame shape and wall heat flux distribution. For such a situation, section 9.5 suggests an elongated flame in oxy-fuel combustion can change heat transfer distribution.

Other research challenges remain which influence heat transfer. Laboratory scale rig-determined coal reactivity data should be validated against data obtained from larger scale furnaces. Currently few validated predictions have been published. Data validation and model refinement will continuously develop with accumulation of data, and model prediction uncertainty will reduce.

9.6 Acknowledgements

The authors wish to acknowledge the financial support provided by the Australian Coal Association Research Program.

9.7 References

1 Andersson, K., *et al.*, *Radiation intensity of lignite-fired oxy-fuel flames*. Experimental Thermal and Fluid Science, 2008. **33**: pp. 67–76.

2 Wall, T. and J. Yu, *Coal-fired oxyfuel technology status and progress to deployment*. In *Proceedings of the 34th International Technical Conference on Clean Coal and Fuel Systems*, B.A. Sakkestad (ed.), 2009, Clearwater, Florida, USA: Coal Technology Association. pp. 1–12.

3 Stromberg, L., G. Lindgren, and J. Jacoby, *First test results from Vattenfall's 30 MWth Oxyfuel Pilot Plant in Schwarze Pumpe*. In *Proceedings of the 34th International Technical Conference on Clean Coal and Fuel Systems*, B.A. Sakkestad (ed.), 2009, Clearwater, Florida, USA: Coal Technology Association.

4 Kiga, T., *O_2/RFG Combustion-applied pulverised coal fired plant for CO_2 recovery*. In *Advanced Coal Combustion*, T. Miura (ed.), 2001, Huntington, New York: Nova Science Publishers Inc. pp. 185–241.

5 Wall, T., *et al.*, *An overview on oxyfuel coal combustion. State of the art research and technology development*. Chemical Engineering Research and Design, 2009. **87**: pp. 1003–1016.

6 Chae, T., *et al. A numerical study on characteristics of oxy pulverized coal combustion in a corner-firing boiler of 125 MWth scale*. In *Proceedings of the 34th International Technical Conference on Clean Coal and Fuel Systems*, B.A. Sakkestad (ed.), 2009. Clearwater, Florida, USA: Coal Technology Association.

7 Tan, Y., M.A. Douglas, and K.V. Thambimuthu, *CO_2 capture using oxygen enhanced combustion strategies for natural gas power plants*. Fuel, 2002. **81**: pp. 1007–1016.

8 Andersson, K. and F. Johnsson, *Flame and radiation characteristics of gas-fired O_2/CO_2 combustion*. Fuel, 2007. **86**: pp. 656–668.

9 Andersson, K., *et al.*, *Radiation intensity of propane-fired oxy-fuel flames: implications for soot formation*. Energy & Fuels, 2008. **22**: pp. 1535–1541.

10 Hjartstam, S., *et al.*, *Combustion characteristics of lignite-fired oxy-fuel flames*. Fuel, 2009. **88**: pp. 2216–2224.

11 Tan, Y., *et al.*, *Combustion characteristics of coal in a mixture of oxygen and recycled flue gas*. Fuel, 2006. **85**: pp. 507–512.

12 McCauley, K., *et al.*, *Commerialization of oxy-coal combustion: applying results of a large 30 MWth pilot project*. Energy Procedia, 2009. **1**: pp. 439–446.

13 Chui, E.H., M.A. Douglas, and Y. Tan, *Modelling of oxy-fuel combustion for a Western Canadian sub-bituminous coal*. Fuel, 2003. **82**: pp. 1201–1201.

14 Khare, S.P., *et al.*, *Factors influencing the ignition of flames from air-fired swirl pf burners retrofitted to oxy-fuel*. Fuel, 2008. **87**: pp. 1042–1049.

15 Farzan, H., *et al. Pilot-scale evaluation of coal combustion in an oxygen-enriched recycled flue gas*. In *The 30th International Technical Conference on Coal Utilization and Fuel Systems. Coal Technology: Yesterday – Today – Tomorrow*. 2005. Clearwater, Florida, USA: Coal Technology Association.

16 Tien, C.L. and S.C. Lee, *Flame radiation*. Progress in Energy and Combustion Science, 1982. **8**: pp. 41–59.

17 Im, K.H. and R.K. Ahluwalia, *Radiation properties of coal combustion products*. International Journal of Heat and Mass Transfer, 1993. **36**(2): pp. 293–302.

18 Hottel, H.C. and A.F. Sarofim, *Radiative Transfer*. 1967, New York: McGraw-Hill. pp. 520.

19 Perry, R.H., D.W. Green, and J.O. Maloney, *Perry's Chemical Engineers' Handbook*. 7th ed. 1998, New York: McGraw-Hill.

20 Hottel, H.C. and A.F. Sarofim, *The effect of gas flow patterns on radiative transfer in cylindrical furnaces*. International Journal of Heat and Mass Transfer, 1965. **8**: pp. 1153–1169.

21 Taylor, P.B. and P.J. Foster, *The total emissivities of luminous and non-luminous flames*. International Journal of Heat and Mass Transfer, 1974. **17**: pp. 1591–1605.

22 Truelove, J.S., *The Zone Method for Radiative Heat Transfer Calculations in Cylindrical Geometries*. 1975; Harwell, UK: Atomic Energy Authority.

23 Smith, T.F., Z.F. Shen, and J.N. Friedman, *Evaluation of coefficients for the Weighted Sum of Gray Gases Model*. Journal of Heat Transfer, 1982. **104**: pp. 602–608.

24 Edwards, D.K., *Molecular gas band radiation*. In *Advances in Heat Transfer*. 1976, New York: Academic Press. pp. 115–193.

25 Coppalle, A. and P. Vervisch, *The total emissivities of high-temperature flames*. Combustion and Flame, 1983. **49**: pp. 101–108.

26 Soufiani, A., *A comparison between Weighted Sum of Gray Gases and Statistical Narrow-Band Radiation Models for combustion applications*. Combustion and Flame, 1994. **97**: pp. 240–250.

27 Modak, A.T., *Exponential wide band parameters for the pure rotational band of water vapor*. Journal of Quantitative Spectroscopy and Radiative Transfer, 1979. **21**(2): pp. 131–142.
28 Leckner, B., *Spectral and total emissivity of water vapor and carbon dioxide*. Combustion and Flame, 1972. **19**: pp. 33–48.
29 Edwards, D.K. and A. Balakrishnan, *Thermal radiation by combustion gases*. International Journal of Heat and Mass Transfer, 1973. **16**: pp. 25–40.
30 Modak, A.T., *Radiation from products of combustion*. Fire Research, 1978. **30**: pp. 339–361.
31 Modest, M.F., *The weighted-sum-of-gray-gases model for arbitrary solution methods in radiative transfer*. Journal of Heat Transfer, 1991. **113**(3): pp. 650–656.
32 Lallemant, N. and R. Weber, *A computationally efficient procedure for calculating gas radiative properties using the exponential wide band model*. International Journal of Heat and Mass Transfer, 1996. **39**(15): pp. 3273–3286.
33 Kim, C. and N. Lior, *Easily computable good approximations for spectral radiative properties of particle-gas components and mixture in pulverized coal combustors*. Fuel, 1995. **74**(12): pp. 1891–1902.
34 Solomon, P.R., *et al*. *The spectral emittance of pulverized coal and char*. In *Symposium (International) on Combustion*. 1986, Pittsburgh, PA, USA: The Combustion Institute.
35 Solomon, P.R., *et al*., *Analysis of particle emittance, composition, size and temperature by FTIR emission/transmission spectroscopy*. Fuel, 1987. **66**: pp. 897–908.
36 Best, P.E., *et al*., *Extension of emission-transmission technique to particulate samples using FTIR*. Combustion and Flame, 1986. **66**: pp. 47–66.
37 Bhattacharya, S.P. and T.F. Wall, *Development of emittance of coal particles during devolatilisation and burnoff*. Fuel, 1999. **78**: pp. 511–519.
38 Brewster, M.Q. and T. Kunitomo, *The optical constants of coal, char, and limestone*. Journal of Heat Transfer, 1974. **106**: pp. 678–683.
39 Dalzell, W.H. and A.F. Sarofim, *Optical constants of soot and their application to heat-flux calculations*. ASME Journal of Heat Transfer, 1969. **91**(1): pp. 100–104.
40 Stull, V.R. and G.N. Plass, *Emissivity of dispersed carbon particles*. Journal of the Optical Society of America, 1960. **50**(2): pp. 121–129.
41 Howarth, C.R., P.J. Foster, and M.W. Thring, *The effect of temperature on the extinction of radiation by soot particles*. In *Proceedings of the Third International Heat Transfer Conference*, 1966. Washington D.C.: Hemisphere.
42 Lee, S.C. and C.L. Tien, *Optical constants of soot in hydrocarbon flames*. In *Eighteenth Symposium (International) on Combustion*. 1980, Pittsburgh, PA, USA: The Combustion Institute.
43 Chang, H. and T.T. Charalampopoulos, *Determination of the wavelength dependence of refractive indices of flame soot*. In *Proceedings of the Royal Society (London) A*. 1990.
44 Felske, J.D., T.T. Charalampopoulos, and H.S. Hura, *Determination of refractive indices of soot particles from the reflectivities of compressed soot particles*. Combustion Science and Technology, 1984. **37**: pp. 263–284.
45 Millikan, R.C., *Optical properties of soot*. Journal of the Optical Society of America, 1961. **51**: pp. 698–699.
46 Yuen, W.W. and C.L. Tien, *A simple calculation scheme for the luminous-flame emissivity*. In *Symposium (International) on Combustion*. 1977, Pittsburgh, PA, USA: The Combustion Institute.

47 Liu, F. and J. Swithenbank, *The effects of particle size distribution and refractive index on fly-ash radiative properties using a simplified approach.* International Journal of Heat and Mass Transfer, 1993. **36**(7): pp. 1905–1912.

48 Gupta, R.P. and T.F. Wall, *The optical properties of fly ash in coal fired furnaces.* Combustion and Flame, 1985. **61**: pp. 145–151.

49 Lowe, A., I.M. Stewart, and T.F. Wall, *The measurement and interpretation of radiation from fly ash particles in large pulverized coal flames.* In *Symposium on Combustion.* 1979, Pittsburgh, PA, USA: The Combustion Institute.

50 Goodwin, D.G. and M. Mitchner, *Flyash radiative properties and effects on radiative heat transfer in coal-fired systems.* International Journal of Heat and Mass Transfer, 1989. **32**(4): pp. 627–638.

51 Bhattacharya, S.P., T.F. Wall, and R.P. Gupta, *An analysis of the angular scatter measurement to determine the optical constants of coal and ashy materials.* International Communication on Heat and Mass Transfer, 1996. **23**(6): pp. 809–821.

52 Wyatt, P.J., *Some chemical, physical, and optical properties of fly ash particles.* Applied Optics, 1980. **19**: pp. 975–983.

53 Zygarlicke, C.J., D.P. McCollor, and C.R. Crocker, *Task 3.2 – Ash emissivity characterization and prediction. Final report.* 1999, Grand Forks, North Dakota: EERC. pp. 28.

54 Ouazzane, A.K., *et al.*, *Design of an optical instrument to measure the carbon content of fly ash.* Fuel, 2002. **81**: pp. 1907–1911.

55 Bohren, C. and D. Huffman, *Absorption and Scattering of Light by Small Particles.* 1983, New York: John Wiley & Sons.

56 Wiscombe, W.J., *Mie scattering calculations: advances in technique and fast, vector-speed computer codes.* In *NCAR Technical Note.* 1979, Boulder, CO, USA: NCAR.

57 Wall, T.F., *et al.*, *The properties and thermal effects of ash deposits in coal-fired furnaces.* Progress in Energy and Combustion Science, 1993. **19**: pp. 487–504.

58 Hampartsoumian, E., *et al.*, *The radiant emissivity of some materials at high temperatures – review.* Journal of the Institute of Energy, 2001. **74**(500): pp. 91–99.

59 Zbogar, A. and F. Frandsen, *Surface emissivity of coal ashes.* IFRF Combustion Journal, 2003.

60 Mulcahy, M.F.R., J. Boow, and P.R. Goard, *Fireside deposits and their effect on heat transfer in a pulverized-fuel-fired boiler. Part I: the radiant emittance and effective thermal conductance of the deposits.* Journal of the Institute of Fuel, 1966. **39**: pp. 385–394.

61 Boow, J. and P.R. Goard, *Fireside deposits and their effect on heat transfer in a pulverized-fuelfired boiler. Part III: the influence of the physical characteristics of the deposit on its radiant emittance and effective thermal conductance.* Journal of the Institute of Fuel, 1969. **42**: pp. 412–419.

62 Markham, J.R., *et al.*, *Measurement of radiative properties of ash and slag by FT-IR emission and reflection spectroscopy.* Journal of Heat Transfer, 1992. **114**(2): pp. 458–464.

63 Richards, C.H., *et al. Radiative heat transfer in pulverized-coal-fired boilers-development of the absorbptive/reflective character of inital ash deposits.* In *Proceedings of the 25th Symposium (International) on Combustion.* 1994, Pittsburgh, PA, USA: The Combustion Institute.

64 Baxter, L.L., *Influence of ash deposit chemistry and structure on physical and transport properties.* Fuel Processing Technology, 1998. **56**: pp. 81–88.

65 Reid, R., J. Hoskisson, and L. Baxter. *Radiative properties of coal and biomass deposits and slags*. In *Proceedings of the 34th International Technical Conference on Clean Coal and Fuel Systems*. 2009, Clearwater, Florida, USA: Coal Technology Association.
66 Wall, T.F. and H.B. Becker, *Total absorptivities and emissivities of particulate coal ash from spectral band emissivity measurements*. Journal of Engineering for Gas Turbines and Power, 1984. **106**: pp. 771–776.
67 Best, P.E., *et al.*, *Extension of emission-transmission technique to particulate samples using FT-IR*. Combustion and Flame, 1986. **66**: pp. 47–66.
68 Bhattacharya, S.P., T.F. Wall, and M. Arduini-Schuster, *A study on the importance of dependent radiative effects in determining the spectral and total emittance of particulate ash deposits in pulverized fuel fired furnaces*. Chemical Engineering and Processing, 1997. **36**: pp. 423–432.
69 Siegel, R. and J.R. Howell, *Thermal Radiation Heat Transfer*. 2002, New York: Taylor & Francis.
70 Bhattacharya, S.P., *A theoretical investigation of the influence of optical constants and particle size on the radiative properties and heat transfer involving ash clouds and deposits*. Chemical Engineering and Processing, 2000. **39**: pp. 471–483.
71 Viskanta, R. and M.P. Menguc, *Radiation heat transfer in combustion systems*. Progress in Energy and Combustion Science, 1987. **13**(2): pp. 97–160.
72 Richter, W., W. Li, and R. Payne, *Two Dimensional Modeling of Fossil Fueled Power Plant Behavior when Using CO_2-O_2 or CO_2-H_2O-O_2 Instead of Air to Support Combustion*. 1987, Argonne, Illinois: Argonne National Laboratory.
73 Nozaki, T., *et al.*, *Analysis of the flame formed during oxidation of pulverised coal by an O_2-CO_2 mixture*. Energy, 1997. **22**(2/3): pp. 199–205.
74 Chui, E.H., *et al.*, *Numerical investigation of oxy-coal combustion to evaluate burner and combustor design concepts*. Energy, 2004. **29**(9–10): pp. 1285–1296.
75 Toporov, D., *et al.*, *Detailed investigation of a pulverized fuel swirl flame in CO_2/O_2 atmosphere*. Combustion and Flame, 2008. **155**: pp. 605–618.
76 Chae, T., *et al.*, *A numerical study on characteristics of oxy pulverized coal combustion in a corner-firing boiler of 125 MWth scale*. In *Proceedings of the 34th International Technical Conference on Clean Coal and Fuel Systems*. B.A. Sakkestad, (ed.) 2009, Clearwater, Florida, USA: Coal Technology Association.
77 Eriksson, K., *CFD Modelling for oxy-combustion process*. In *2nd Workshop of the Oxy-fuel Combustion Network*, 2007, Hilton Garden Inn, Windsor, Connecticut.
78 Truelove, J.S. and R.G. Williams. *Coal combustion models for flame scaling*. In *Symposium (International) on Combustion*. 1988, Pittsburgh, PA, USA: The Combustion Institute.
79 Rathnam, R.K., *et al.*, *Differences in reactivity of pulverised coal in air (O_2/N_2) and oxy-fuel (O_2/CO_2) conditions*. Fuel Processing Technology, 2009. **90**: pp. 797–802.
80 Yamada, T., *et al.*, *Comparison of combustion characteristics of between oxy-fuel and air combustion*. In *The Proceedings of the 31st International Technical Conference on Coal Utilization and Fuel Systems*. 2006. Sheraton Sand Key, Clearwater, Florida, USA: Coal Technology Association.
81 Farag, I.H. and T.A. Allam, *Carbon dioxide standard emissivity by mixed gray-gases model*. Chemical Engineering Communications, 1982. **14**: pp. 123–131.

10
Current and future oxygen (O_2) supply technologies for oxy-fuel combustion

N. M. PROSSER and M. M. SHAH, Praxair, Inc., USA

Abstract: This chapter describes air separation technologies for supplying oxygen for oxy-fuel combustion. Cryogenic air separation processes and equipment technologies are described, including a historical as well as a forward looking perspective on those technologies. Specific attention is given to the oxygen requirements for oxy-fuel power plants and processes suitable for this application. An advanced technology under development that is based on oxygen transport membranes (OTMs) is also described.

Key words: cryogenic air separation, oxygen, low purity oxygen, oxy-fuel combustion, oxygen transport membranes, CO_2 capture.

10.1 Introduction

10.1.1 A historical perspective of air separation technology

In the late 19th century oxygen was produced primarily by chemical means. However, the industrial revolution, and especially iron and steel production and fabrication, created the need for higher volumes of oxygen. A number of inventors/entrepreneurs contributed to the emergence of cryogenic oxygen production and its ultimate superiority over previous methods. Karl von Linde was the first to appreciate the industrial implications of the production of liquid air. His air liquefier used Joule–Thomson expansion of high pressure gas to produce cold. Later, an ammonia pre-cooler was added, which gave an efficiency improvement. In 1902, Georges Claude and his colleagues developed a practical piston expander to let down the pressure of air while decreasing enthalpy by extracting work (Royal, 2009). The Claude cycle was more efficient, and its principle still provides the basis for the modern refrigeration cycles employed in most air separation and liquefaction installations today. Paulus Heylandt further improved liquefaction efficiency and ran the expansion engine at a higher temperature, which avoided Claude's problems with solidifying lubricant (Ruhemann, 1949, pp. 123–134).

Linde produced oxygen in a single column, which was first operated in 1902. This was greatly improved by his invention of the double column in 1910, which was the fruition of intense competition during the first decade of the 20th century (Royal, 2009). The double column enabled high recovery of high purity oxygen for the first time. By the late 1930s industrial needs led to the improvement of the double column to enable economic production of large quantities of both high purity oxygen and nitrogen. The double column makes effective use of nature's air composition, with four parts nitrogen to one part oxygen by mole such that the condensation needs of the first column (high pressure column) match the boil-up needs of the second column (low pressure column). This is provided by a single condenser/reboiler, usually called the main condenser. The amount of vapor reboil and liquid reflux needed for each distillation column is matched perfectly by the process. The efficiency, effectiveness, and elegance of the double column are evidenced by its continued use today for most high purity oxygen installations, albeit with many improvements and variations. Various methods to recover argon, which constitutes about 1% of the air, have been added to the double column (Ruhemann, 1949, pp. 141–221). An argon side-arm column became accepted as the best method by the industry.

Industrial demands, primarily of steel and chemical production, led to the need for greater production rates from air separation plants through much of the 20th century. In turn, increased capability for higher production rates led to both broader and more intense usage by industry. The heavy industrial expansion and the innovation spurred by World War II greatly increased demand. On-site gas plants, serving local customers by pipeline, became prevalent beginning in the 1940s in place of small liquid-only plants. By the 1960s enclaves of large, multi-plant facilities served customers via systems of pipelines and liquid distribution. Plant production capacities of today are orders of magnitude greater than those of the 1940s.

The suite of applications for the industrial gases is vast. Ultra-pure products are delivered to electronics and microchip companies. Oxygen is used in many operations to intensify production rates by replacing or augmenting air. Such use can also reduce waste streams and facilitate pollution control. Nitrogen's stability enables its use for purging and blanketing to preserve the quality and integrity of reactive or degradable products. In the coming years, energy applications are expected to be a key driver for growth in demand. Large supplies of coal in Asia and North America and low value energy sources such as the Canadian oil sands and waste products from petroleum refining are leading to a prevalence of gasification projects. Gasification is being used for chemical production, fuel conversion and power generation. Gas to liquids technology is being used for monetizing remote gas reserves.

Oxy-coal combustion and IGCC (integrated gasification combined cycle) are modern, clean power generation technologies that enable CO_2 capture and sequestration. Most recent large oxygen plants have been built to serve energy applications. Figure 10.1 shows the enormous increase in oxygen production in the US for the last 100 years (US Bureau of Commerce, 2008). In 2007, US oxygen production reached 730,000 million cuft/year (~75,000 tonnes/day).

10.1.2 Oxygen (O_2) supply methods

Atmospheric gas components are supplied to customers via a number of different commercial methods. Liquid products from cryogenic air separation plants are supplied via truck tanker to smaller users, where cryogenic storage vessels with

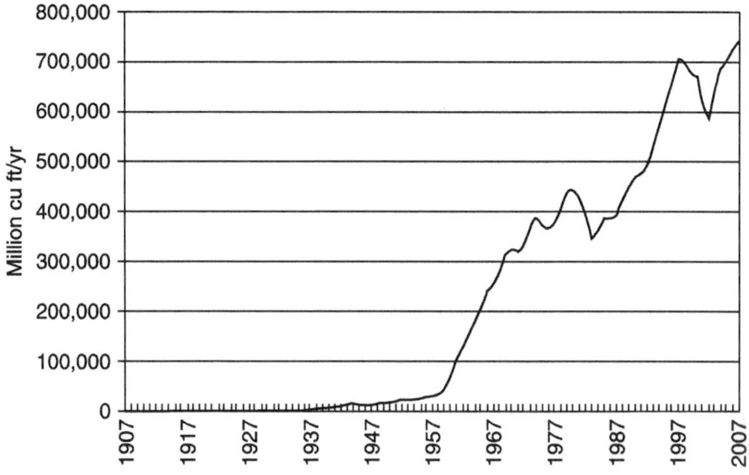

10.1 100 years of US oxygen production (US Bureau of Commerce, 2008).

pumps and vaporizers provide gaseous product on demand. Small, relatively modest purity nitrogen needs can be supplied using polymeric membrane systems (up to 10 tonnes/day, 97–99.9% nitrogen by volume). Here, bundles of hollow fiber membranes allow oxygen, carbon dioxide, and water vapor to permeate faster than nitrogen from a pressurized air feed. Polymeric membrane systems are not used for oxygen supply. Vacuum pressure swing adsorption (VPSA) systems provide more economical oxygen supply than liquid delivery for customers with larger demand, so long as lower purity is acceptable. For even larger demand, or for high purity or multiple product requirements, cryogenic air separation systems are needed. An overview of VPSA technology is given in section 10.3. Cryogenic air separation, particularly as it applies to oxy-coal power generation needs, is discussed in section 10.4. An advanced oxygen transport membrane (OTM) based technology with a potential for significant reduction in parasitic power consumption for oxygen supply is discussed in section 10.5. This OTM technology is not yet available commercially.

10.2 Oxygen supply needs for oxy-coal power plants

While the majority of worldwide oxygen plants produce oxygen at high purity (>99.5% by vol.), there are many applications where lower purity oxygen (90–98% by vol.) is satisfactory. In such cases, it is usually more cost effective to use low purity oxygen. Examples of such applications are pig iron manufacture from blast furnaces, glass furnaces, primary metal refining (such as copper and gold), and gasification (where high purity products are not made, such as IGCC).

For oxy-coal power plants, low purity oxygen supply is adequate. The purpose of the air separation plant is to reject nitrogen from air so that the flue gas is concentrated in CO_2 to enable its economical purification. Typically, the flue gas exiting the boiler will be about 80% by vol. CO_2 (on a dry basis). The bulk of the impurities in the flue gas are atmospheric gases: nitrogen, oxygen, and argon. Three factors affect the level of these impurities: excess oxygen in combustion, air in-leakage, and impurities in oxygen product. Purification of the flue gas to at least 95% CO_2 will be required by the downstream CO_2 processing unit (CPU). The selection of the optimum oxygen purity from the air separation unit (ASU) involves a trade-off between the cost of higher purity oxygen production and the savings from the CPU for lower impurity levels in the flue gas. Figure 10.2 shows the effect of increasing oxygen purity on the overall cost of purified CO_2, factoring in the additional power and cost of higher purity oxygen from the ASU and the savings of higher quality flue gas on the downstream CPU. The optimal oxygen purity, considering the overall system, is in the range of 95% to 97.5%. Above about 97.5% purity, the air separation plant power trend increases much more rapidly (see section 10.4.2), which explains the increase in cost per tonne of CO_2 captured.

The power demands of air separation and CO_2 purification and compression reduce the net output of power plant by 20–25%. In addition, the ASU and CPU significantly increase the capital cost of power plant. Figure 10.3 illustrates the impact of the ASU and CPU on net power output and the cost of electricity. The cost and power consumption of the air separation unit and the CO_2 processing unit raise the cost of electricity by 60% for the oxy-coal installation. Clearly, this illustrates the vital need for improvements in efficiency and capital for the ASU and CPU for the ultimate competitiveness and viability of oxy-coal option for capturing CO_2 from power plants.

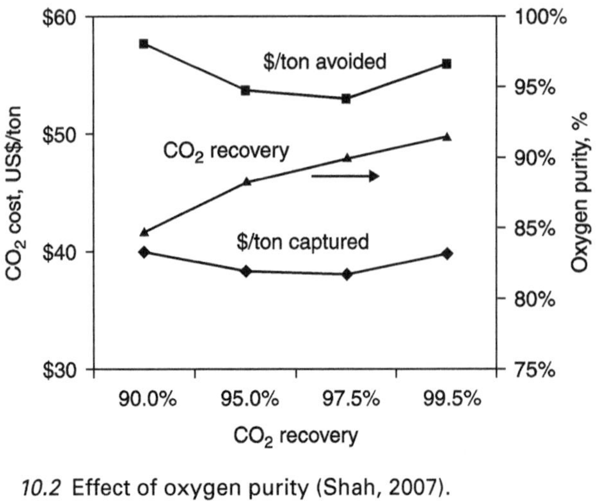

10.2 Effect of oxygen purity (Shah, 2007).

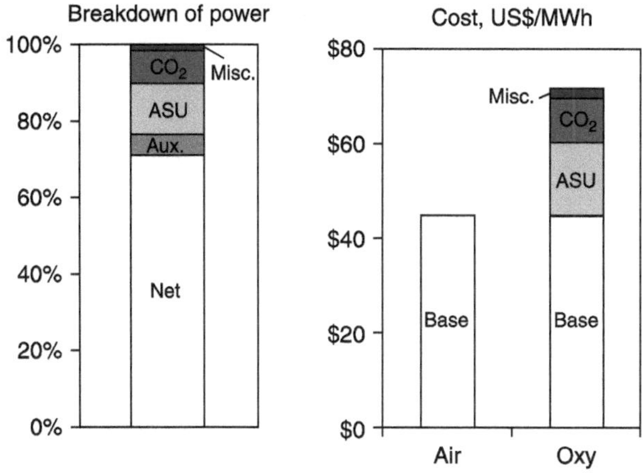

10.3 Impacts of ASU and CPU on power output and cost of electricity (Shah, 2006). ASU, air separation unit; CO_2, CO_2 compression and purification.

Very large quantities of oxygen will be needed for oxy-coal power plants. Commercial scale power plants with gross output of 500 MW will require 7000–9000 tonnes/day of oxygen. It is expected that this need will best be met with two air separation plants. Plants of this size offer more opportunity for efficiency gains compared with the more conventional capacity plants. Oxygen pressure at the point of injection with the recycled flue gas stream is at most 0.2 barg (3 psig). Rapid load following may be necessary for the air separation plant to efficiently supply the power plant, as it responds to the needs of the power grid.

10.3 Vacuum pressure swing adsorption technology

Pressure swing adsorption systems for oxygen production were first used in the 1970s. Since then, the technology has rapidly improved. The early systems operated the entire operational sequence above atmospheric pressure. However, incorporation of a sub-atmospheric regeneration step much more fully utilized the characteristics of the adsorbent. This vacuum pressure swing adsorption method resulted in a significant decrease in energy consumption (Smolarek *et al.*, 2000). In 1989, the first vacuum pressure swing adsorption installation was started up. Today's VPSA systems use less than 50% of the power of the earlier PSA designs. One-bed systems and two-bed systems are employed. The one-bed systems offer capital savings, but are limited to about one-half of the production of a two-bed system because a single machine provides both feed as well as vacuum functions as opposed to the individual feed and vacuum blowers of the two-bed system.

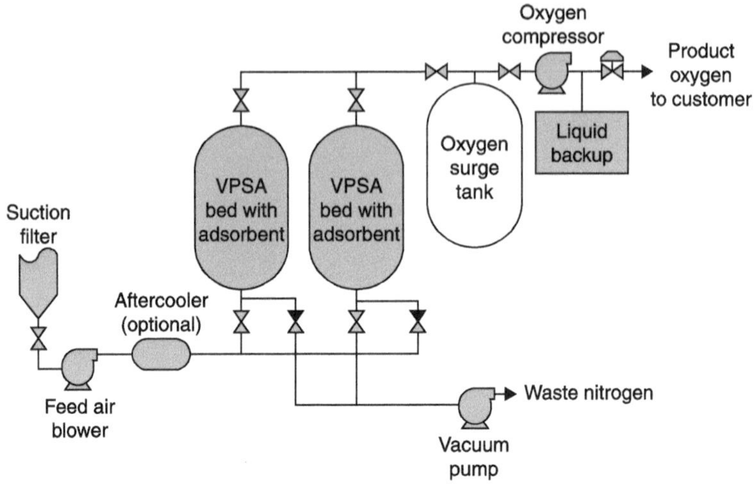

10.4 Two-bed oxygen VPSA system.

A simplified two-bed schematic is shown in Fig. 10.4. The feed air blower supplies air to the on-stream adsorbent vessel. Nitrogen, water vapor, carbon dioxide, and atmospheric hydrocarbons are preferentially adsorbed, while most of the oxygen and argon pass through. The product oxygen stream is available at a natural pressure of 0.2–0.35 barg (3–5 psig) from the system. An oxygen compressor is used when higher pressures are needed. The oxygen purity is 90–93% by volume. System capacity and efficiency suffer dramatically above 93% oxygen purity. The off-stream adsorbent vessel is regenerated under vacuum (0.3–0.7 bara, 4–10 psia) using the vacuum pump (blower). The nitrogen, carbon dioxide, and moisture desorb during this time. At the end of the desorption period, a small quantity of product oxygen is used to purge the remaining desorbed contaminants prior to repressurizing the bed for the adsorption step. Today's advanced systems have a cycle time of less than one minute. There is a period of time during each cycle in which no oxygen is produced. The surge tank, which is a low pressure storage vessel downstream of the adsorbent bed, enables a smooth supply of gas oxygen to the customer.

In addition to the step improvement that the introduction of VPSA enabled, innovations in every area of system and component design and operation have led to continued, large improvements in capital cost and energy efficiency. Development can be divided into four subject areas: adsorbent materials, process design, system design, and component packaging. Adsorbent advances have been the key to substantially reducing the cost of VPSA oxygen. Today's VPSA bed sizes are less than one-tenth that of the early systems. In addition to the reduction in vessel size and adsorbent requirements, improved adsorbent has enabled a reduction in the size of the machinery and all other components, and has led to lower power consumption. Lithium exchanged Type X (LiX) zeolites are the most cost effective for air separation (Chao, 1989). They provide a large increase in nitrogen working capacity. Of the many process design improvements, an example is the incorporation of performance enhancing steps such as pressure equalization and product purge (Baksh *et al.*, 1996). System design improvements over the years have greatly simplified VPSAs. Today's systems use fewer adsorbent beds and valves, and significantly less adsorbent than the earlier commercial units. Component packaging improvements have enabled compact, skid mounted equipment layouts. This reduces overall cost by reducing the engineering and field installation costs through efficient shop assembly. Also, this has led to shorter project schedules and increased equipment recoverability. Figure 10.5 illustrates the marked improvement in total VPSA oxygen cost since its early commercialization, and the improvement with respect to capacity.

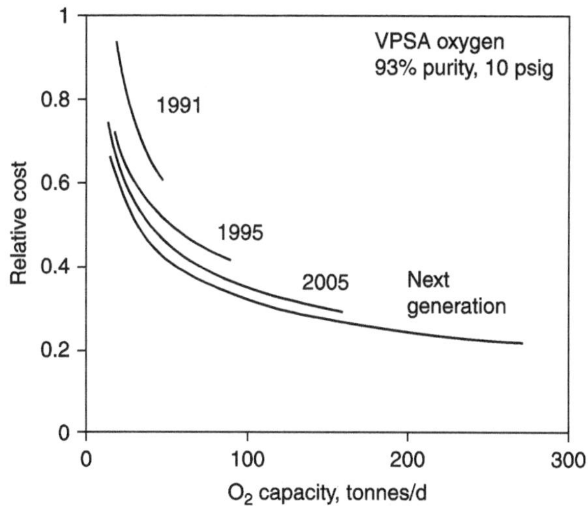

10.5 VPSA oxygen cost and capacity improvement (Shah, 2005).

The maximum capacity available from a VPSA plant is dictated by the capacity of commercially available blowers used for feed and vacuum service, and by vessel size constraints. Higher capacities are generally achieved with the installation of multiple VPSA plants; however, due to economies of scale there is a practical limit above which cryogenic supply becomes more economically attractive. Figure 10.6 illustrates the typical range of economic application for VPSA systems. In addition to this, VPSA systems are best suited to applications with modest oxygen pressure requirements due to the large expense of an oxygen compressor to deliver higher pressures. For application in an oxy-coal installation VPSA is not the economical choice primarily because of the large oxygen production rates needed. The lower purity of VPSA also would add cost for the CO_2 purification system.

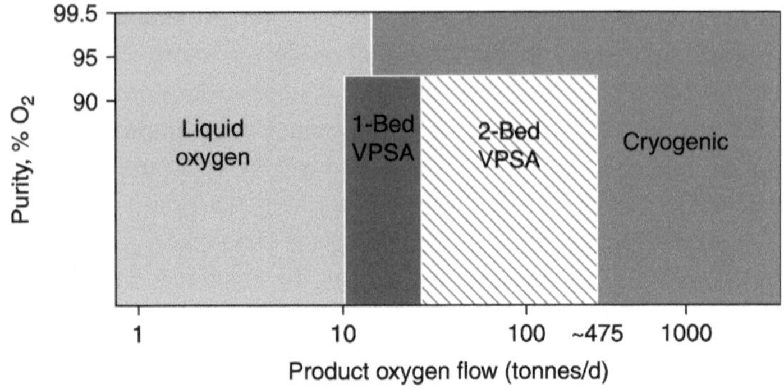

10.6 Oxygen supply system typical economic ranges.

10.4 Cryogenic air separation technology

10.4.1 Air separation process description

The air separation process can be best understood by examining the double column process, which is illustrated in Fig. 10.7. This process is the workhorse of the industry; it is widely used, albeit with many variations, in well over 90% of oxygen plants. For simplicity, the example process of Fig. 10.7 produces only

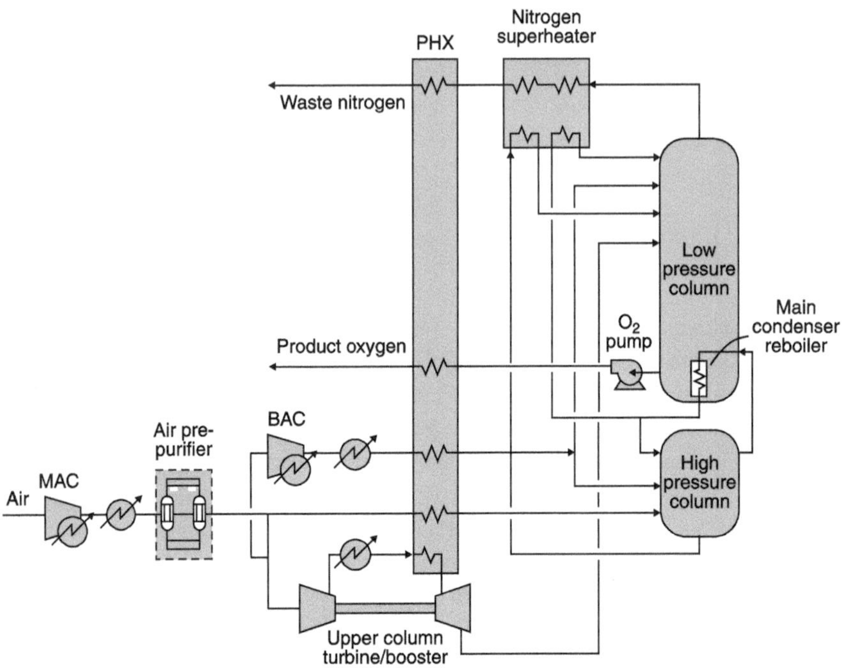

10.7 Double column process.

gaseous oxygen. In most installations, other products such as gaseous nitrogen, liquid oxygen and nitrogen, and argon are manufactured. Even rare gases such as neon, krypton, and xenon are made in some very large plants. The widely used argon making process adds a third column to the double column process, and is aptly named the triple column process. Oxygen product is required at a minimum of 99.5% purity in many industrial applications. In the double column process of Fig. 10.7, oxygen from the low pressure column is withdrawn as a liquid and pumped before it is vaporized and warmed. This 'liquid oxygen pumping process' is widely used in the industry today, primarily to avoid the use of an oxygen compressor to raise the pressure of the gaseous oxygen after it is warmed.

Referring again to Fig. 10.7, filtered air is compressed in the main air compressor (MAC), a multistage, intercooled compressor. The discharge air pressure of this compressor is set by the pressure of the high pressure column, which is the destination of most of the air. Air is then cooled, usually using water, to remove the heat of compression, either in a conventional shell-and-tube heat exchanger or a direct contact heat exchanger. Often an air chilling step is also employed. The air pre-purifier removes moisture and carbon dioxide from the feed air to avoid freezing in the cryogenic portion of the process. In addition, other air contaminants such as nitrous oxide and heavy hydrocarbons are removed, which greatly enhances process safety. The air pre-purifier contains adsorbent material such as alumina and molecular sieve. It usually consists of two vessels; one is on stream while the other is regenerated by nitrogen exiting the cryogenic process. The purified air is now ready to be fed to the cryogenic portion of the process. All of the equipment in the cryogenic portion of the process is housed in an enclosure called a cold box that is filled with an insulating material, such as perlite.

The largest portion of the air is fed directly to the primary heat exchanger (PHX). This is a large, multi-stream heat exchanger where the air feeds are cooled against the warming products. Brazed aluminum plate-fin heat exchanger technology is used for the PHX. This provides a very large heat transfer surface area in a compact volume. Also, because power efficiency is very important, stream pressure losses are low and very small temperature differences are achievable. This technology allows for multiple streams (six, or more, if desired) to be combined in the same heat exchanger. In addition, streams can be fed and removed at optimal locations in the PHX for improved efficiency. A single heat exchanger also can combine the functionality of more than one service. Air separation plant suppliers have extensive capabilities and experience in brazed aluminum heat exchanger design, enabling the achievement of the optimum combination of heat transfer efficiency versus pressure loss, while also practically minimizing the heat exchanger capital cost based on an overall cost/benefit analysis.

Often the PHX consists of multiple heat exchanger cores in parallel in order to handle the flow capacity. Another portion of the air leaving the air pre-purifier is fed to the booster air compressor (BAC). This may be a single stage compressor, or a multistage, intercooled compressor, depending on its discharge pressure. The elevated pressure air from the BAC provides most of the heat duty to boil the pumped liquid oxygen in the PHX. In order to do this, it normally must be about 2.5 times greater in pressure than the boiling oxygen, and it condenses against the boiling oxygen in the PHX. The remainder of the air provides the refrigeration needs of the process. This is vital in order to keep the equipment cold in the cold box and to provide sufficient liquid volumes in the system for the distillation. A high efficiency turbo-expander extracts energy from the system to provide refrigeration. By using highly efficient aerodynamics and drive mechanisms, the penalty to the process for providing the refrigeration is reduced. In Fig. 10.7, the turbo-expander powers a booster compressor. This raises the pressure of the feed to the turbine and reduces the turbine flow needed. The air to the turbine is cooled to an intermediate temperature in the PHX, turbo-expanded, and fed to the low pressure column.

The high pressure distillation column and the low pressure distillation column make up the double column system. Mass transfer in the distillation columns is accomplished using sieve trays or structured packings. Structured packing consists of thin aluminum sheets formed in an advanced, corrugated pattern. Down-flowing liquid in the distillation column wets the structured packings, and affords a large surface area for contact against up-flowing vapor.

In the distillation column, the light (more volatile) components concentrate in the vapor, while the heavy (less volatile) components concentrate in the liquid. The continuous contact of up-flowing vapor and down-flowing liquid provides many equilibrium stages of separation. Air can be considered a ternary (three component) system for our purposes. Nitrogen is the lightest component, oxygen is the heaviest component, while argon is slightly lighter than oxygen. Hence, oxygen concentrates in the liquid and in the bottom of the distillation columns, while nitrogen concentrates in the vapor and in the top of the distillation columns. Argon is distributed between the top and bottom of the distillation columns. In order to provide vapor and liquid for the functioning of the distillation columns, reboilers and condensers are needed. For cryogenic air separation, these are driven by process streams of different pressure. The different pressure of the driving stream means it will condense or vaporize at a different temperature than the column fluid, which affords a temperature difference for heat transfer.

The main air feed exiting the cold end of the PHX is slightly above its dew point. It is fed to the base of the high pressure column, where the vapor is enriched in nitrogen as it flows upward through the column. Nearly pure nitrogen vapor is withdrawn at the top of the high pressure column. It condenses in the main condenser/reboiler while liquid oxygen in the bottom of the low pressure column boils. A portion of the condensed nitrogen is returned to the high pressure column to provide liquid reflux. The remainder is sent to the top of the low pressure column to provide reflux after it is subcooled in the nitrogen superheater.

One point of thermal integration (i.e. the main condenser/reboiler) characterizes the double column, and the requirement to boil lower volatility oxygen with higher volatility nitrogen in the main condenser/reboiler sets the pressure of the high pressure column. The prevalent use of the double column over the last century, and its continuing use is a testament to its utility, efficiency, and simplicity. The condensed air from the PHX is distributed between the high pressure and low pressure columns. A liquid stream enriched in oxygen is withdrawn from the sump of the high pressure column. It is fed to an intermediate point of the low pressure column after it is subcooled in the nitrogen superheater. Product oxygen liquid is withdrawn from the sump of the low pressure column. After pumping to the desired product pressure, it is boiled and warmed to ambient temperature, against the feed air streams. It is then delivered, usually via pipeline, to the customer(s).

Vapor nitrogen from the top of the low pressure column is warmed in the nitrogen superheater and the PHX, to ambient temperature. A portion of this stream is used for regeneration of the air pre-purifiers. Waste nitrogen can also be used for evaporative water cooling. The exhaust of the waste nitrogen to atmosphere sets the pressure of the air separation system, and ultimately the pressure at which air is discharged from the MAC. Nitrogen flow circuit pressure drops through piping, heat exchangers, valves, etc. set the pressure of the low pressure column. However, it is the pressure difference required to drive the main condenser/reboiler that most raises the MAC pressure, much more than flow induced pressure drops. In fact, the main condenser/reboiler has a multiplicative effect, raising the pressure of the high pressure column to about 3.5 times that of the low pressure column.

10.4.2 Air separation processes for low purity oxygen

In the quest for power savings significant gains can be made when oxygen is produced at less than 98% purity. However, to take advantage of this, a change from the double column process must be made. At the lower oxygen product purities, oxygen no longer must be separated from argon, which is a difficult separation. Only separation of oxygen and nitrogen is now required, which is relatively easy. The vapor boil-up and the liquid reflux generated in the double column process provide much more driving force for the distillation in the low pressure column than necessary; this manifests itself in high power. Instead, for low purity oxygen, another type of air separation process can be used which generates less boil-up and reflux for the distillation, and requires less power. A good example is the dual reboiler process (Cheung, 1995) illustrated in Fig. 10.8. A separate side column is shown in addition to the medium pressure column and low pressure column. However, the side column can be considered part of the low pressure column and can be built either separately or in combination with the low pressure column.

The major difference compared with the double column is the two condenser/reboilers. The use of two points of thermal integration is a key to the power savings this process enables. Oxygen is produced at the base of the side column at

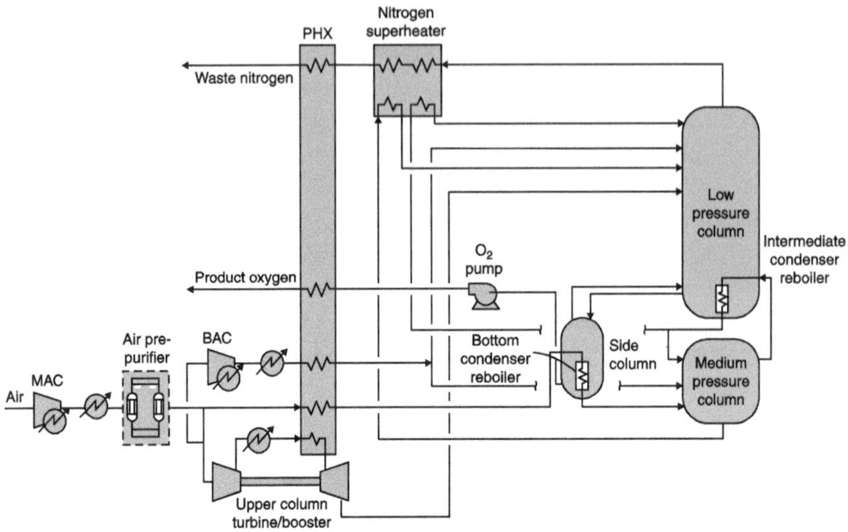

10.8 Dual reboiler process.

the outlet of the bottom condenser/reboiler. Feed air from the PHX is the condensing fluid here. This differs from the double column where nitrogen condenses against product purity oxygen in the main condenser/reboiler. Since air is a mixture with a higher dew point than nitrogen, it can drive the boiling in the bottom condenser/reboiler at a lower pressure. The pressure multiplier is reduced to about 2.5. This means that to produce 95% purity oxygen the MAC discharge pressure need only be about 3.8 bara (55 psia) for the dual reboiler process rather than the about 5.0 bara (73 psia) required for the double column process. This reduced pressure provides the power savings. The partially condensed air is now passed to the bottom of the medium pressure column. The medium pressure column produces nitrogen vapor at the top, as the high pressure column does for the double column process.

The difference now is that the boiling fluid in the intermediate condenser/reboiler is much lower in oxygen content compared with the product purity oxygen of the main condenser/reboiler in the double column process. As a result the medium pressure column operates at the same lower pressure as the air driving the bottom condenser/reboiler. The boiling for the low pressure column takes place in two steps for the dual reboiler process, and so the boil-up is provided in two steps. Only the boil-up that is needed for low purity oxygen production is provided. In the meantime, much more of the feed air is condensed, by virtue of its use to drive the bottom condenser/reboiler. As a result, the liquid nitrogen reflux for the top of the low pressure column is also reduced to a level appropriate for low purity oxygen production. So, the reduced driving forces for distillation in the dual reboiler process are manifested in a power savings. Classic examples of

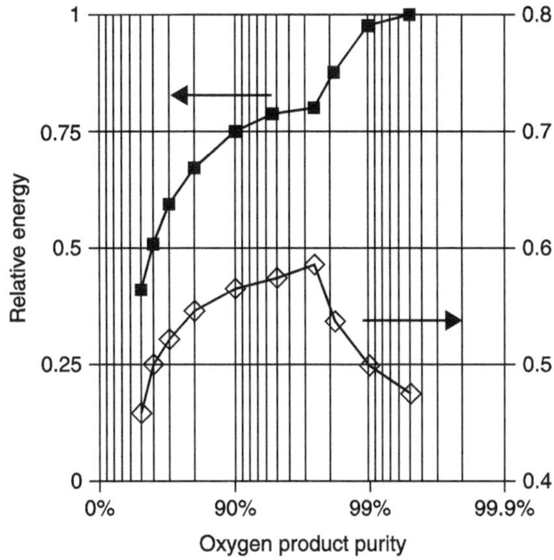

10.9 Effect of oxygen purity on ASU separation energy (Shah, 2005). Note: only separation energy for distillation is shown here.

earlier dual reboiler process patents are listed for reference (Jakob, 1958; Gaumer, 1965; Kleinberg, 1987).

Figure 10.9 illustrates the trend in separation energy (power) and separation efficiency as a function of product oxygen purity (Shah, 2005). Standard merchant and industrial grade purity is 99.5% oxygen. A rapid decrease in separation energy required takes place as the product oxygen purity is decreased from 99.5% to 97.5%, corresponding to the change from an oxygen–argon separation to an oxygen – nitrogen separation and the switch to an improved low purity oxygen process. Below 97.5% the energy reduction is much more gradual. The sudden decrease in separation efficiency above 97.5% oxygen purity is indicative of the much higher reboil and reflux requirements for the oxygen–argon separation.

A McCabe-Thiele diagram (Treybal, 1980, pp. 339–352), while inadequate as a design tool, provides an excellent illustrative representation of the distillation process for our purposes. Figures 10.10–10.12 show McCabe-Thiele diagrams for the low pressure column in the double column process, the dual reboiler process, and a further enhanced low purity oxygen process, respectively. In the diagrams, the upper curved line represents the equilibrium line. This is the equilibrium composition of nitrogen in the liquid phase compared with the equilibrium composition of nitrogen in the vapor phase at any point in the distillation column (balance is oxygen and argon). The lower straight lines are the operating lines. They represent nitrogen compositions of the down-flowing liquid and the up-flowing vapor phases as they pass at any level of the distillation column. The

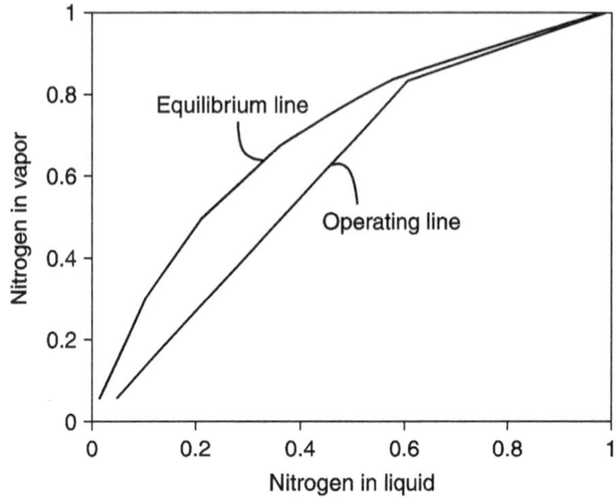

10.10 McCabe-Thiele diagram for low pressure column of double column process.

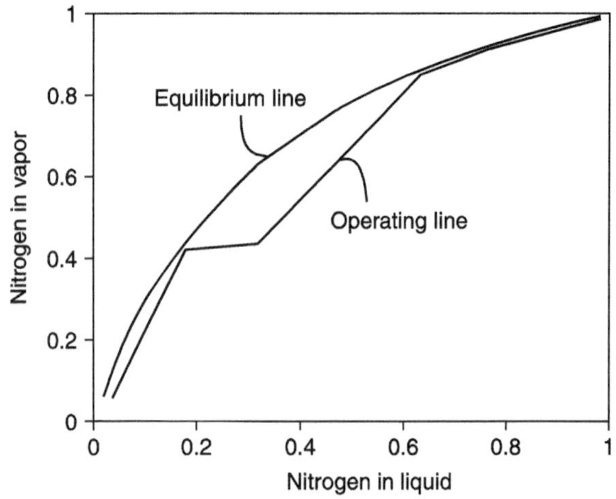

10.11 McCabe-Thiele diagram for low pressure column of dual reboiler process.

space between the operating line and the equilibrium line on the diagrams represents the driving force for the distillation: if the operating line is close to the equilibrium line then there is little distillation driving force; if the operating line is distant from the equilibrium line the distillation driving force is large. If the operating and equilibrium lines are coincident, then there is no driving force for distillation and no composition change will take place.

The largest distillation driving force occurs when the operating lines fall along the diagonal of the McCabe-Thiele diagram (45° line); this represents a condition

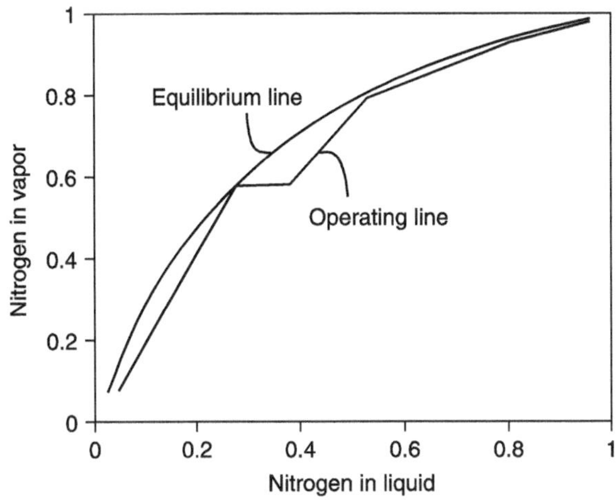

10.12 McCabe-Thiele diagram for low pressure column of further enhanced process.

of 'infinite reflux,' where no vapor or liquid products are withdrawn from the column. As this condition is approached, the number of stages needed to accomplish a given separation is reduced, but the energy input per unit of product increases. We can consider the McCabe-Thiele diagram analogous to a temperature profile in a heat exchanger; larger temperature differences mean higher quality (higher temperature) energy is needed, but the heat transfer area is reduced. The slope of the operating line represents the internal reflux ratio (liquid molar flow/ vapor molar flow) in any given section of the distillation column.

The McCabe-Thiele diagrams of Fig. 10.10 to Fig. 10.12 illustrate the driving force for distillation in the low pressure column. A large driving force is indicative of excess energy needed to drive the distillation (more duty than is necessary for reboiling and condensation). The large driving forces for distillation are readily apparent for the low pressure column of the double column process, particularly in the bottom section (Fig. 10.10). Reduced driving forces for distillation in the McCabe-Thiele diagram for the dual reboiler process are apparent (Fig. 10.11); this has been successfully converted into a power savings. The bottom reboiling step provides only the boil-up needed for the bottom portion of the distillation column, evidenced by the tight approach between the operating line and the equilibrium line there. Additional reboil is provided at the intermediate condenser/reboiler, now somewhat in excess of what is needed (a large driving force is now apparent). In the top section of the low pressure column in the dual reboiler process a tighter coincidence of the operating line and equilibrium line occurs, which is evidence of less liquid nitrogen reflux. The closer coincidence of the operating lines and the equilibrium line also requires more stages to accomplish the separation, but for a low purity oxygen product, total stage requirements remain very manageable.

The McCabe-Thiele diagram of Fig. 10.12 is that of a further enhanced process (Prosser and Shah, 2003), where the location of the intermediate condenser/reboiler in the low pressure column can be optimized. The reduced distillation driving force over the whole of the low pressure column is evident, and this process reduces the separation power by an additional 5% compared with the dual reboiler process. The separation unit power to produce 95% purity oxygen at low pressure from each of these processes is provided for comparison in Table 10.1. Additional improvements are possible, as can be imagined, by using additional condenser/reboilers to further attenuate the slope of the operating lines. Several three-reboiler processes have been published, of which examples are provided (Rathbone, 1996; Howard *et al.*, 2009).

10.4.3 Impact of elevated pressure oxygen production

Oxygen production at pressures significantly above atmospheric is commonly needed. Even relatively low pressure applications, such as glass furnaces and primary metal ore smelters, often require modestly elevated pressure oxygen to accommodate system controls and burners. Steel manufacture is a large traditional user of oxygen, and final refinement of steel in the basic oxygen furnace requires oxygen at about 30 bar. Oxygen supply for gasifiers is usually needed at 40–90 bar. Prior to about 1990, elevated pressure oxygen supply was usually generated by means of an oxygen compressor. Unlike the liquid oxygen pumped processes of Fig. 10.7 and Fig. 10.8, oxygen was withdrawn from the base of the low pressure column as a vapor and warmed as it passed through the primary heat exchanger. The oxygen was available at slightly above atmospheric pressure at the warm end of the PHX, from which it was fed to the suction of the product oxygen compressor. However, very high compressor costs and significant safety

Table 10.1 Power consumption comparison. Air separation processes producing 95% purity oxygen

Process	Unit separation power, kW-h/tonne[1,2,3,4]	Relative power
Double column process	230	1.00
Dual reboiler process	200	0.87
Further enhanced process	190	0.83

[1] Tonne = 1000 kg = 2204.6 lb.
[2] Power for production of oxygen at pressure deducted, so that separation power is based on oxygen produced at 1.1 bara
[3] Includes major power consumers, no auxiliary consumers such as pumps, cooling tower fans, etc.
[4] Assumes ambient $T = 15°C$, cooling water $T = 21°C$, expected design values for efficiencies, pressure drops, temperature differences.

concerns associated with oxygen compressors have led to the prevalence of liquid oxygen pumped air separation processes. In these processes, oxygen pressure is generated from the liquid pump. A high pressure stream such as air from the BAC is needed at sufficient pressure and flow to vaporize the oxygen. The additional power for the high pressure oxygen production is primarily from the BAC. Capital cost for the high pressure oxygen is due to the high pressure BAC, the oxygen pump, and the cost differences for the PHX.

While power plants are operated at low pressures, the efficiency benefits of operating an oxy-coal power plant at high pressure have been studied in Hong *et al.* (2009) and Zheng *et al.* (2007). Hong *et al.* (2009) considered boiler pressures up to 10 bara, in Zheng *et al.* (2007) pressures up to 80 bara were considered. Figure 10.13 illustrates the increased power and cost on a normalized basis to deliver oxygen to higher pressure boilers. The total capital of the air separation plant is relatively unaffected by oxygen pressure. The increased capital cost of the BAC and liquid pump is mitigated by savings in the PHX. The PHX cost tends to decrease as a result of more effective heat transfer and smaller cross-sectional flow areas, even though higher working pressures are required. The cost of oxygen increases by about 21% for an 80 bara boiler with a power cost of US$0.10/kW-h, and by about 26% with a power cost of US$0.15/kW-h, as illustrated by the curves in Fig. 10.13.

10.4.4 Technology advancements in cryogenic oxygen production

Although cryogenic oxygen production is more than a century old, innovation continues to drive down the real cost of products. Global competition in the industry, high product energy intensity and high capital investment intensity

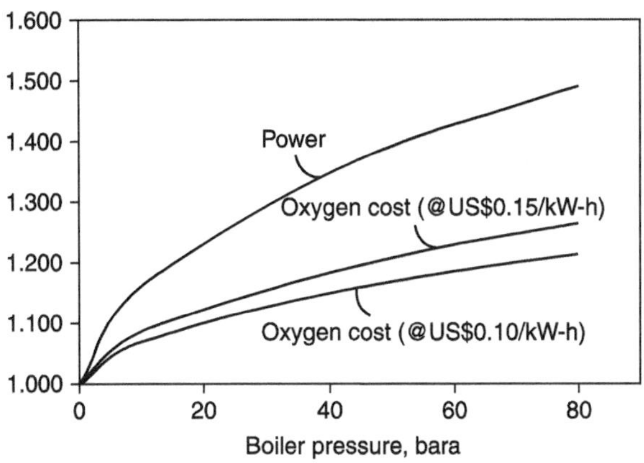

10.13 ASU power and oxygen cost related to delivered pressure.

provide continued motivation for improvement. Markedly increased energy and plant construction costs in recent years have further incentivized the industrial gas industry. An ever-widening suite of applications, many of which enhance efficiency, reduce waste streams and pollution, or improve product quality, contributes to growth in industrial gas sales that exceeds GDP growth rates. The industry players attribute a high value to technology as a result. This is demonstrated by the fact that well over 100 patents for cryogenic air separation and liquefaction were issued in the United States over the five year period 2004–2008. In Fig. 10.14, aggregate technology-driven power improvements are itemized over the last two decades.

Even while these large improvements in energy efficiency were developed and implemented, large capital cost savings were realized for cryogenic air separation plants. Figure 10.15 shows the cost trend compared with the inflation adjusted cost from 1980 (Drnevich *et al.*, 1981, Table 7-1). These savings resulted partly from technology advances, and also from improved project execution methods such as plant standardization and improved fabrication and construction strategies.

Also of particular interest regarding oxygen supply for oxy-fuel combustion is the development of large plants. Over the history of the air separation industry, the

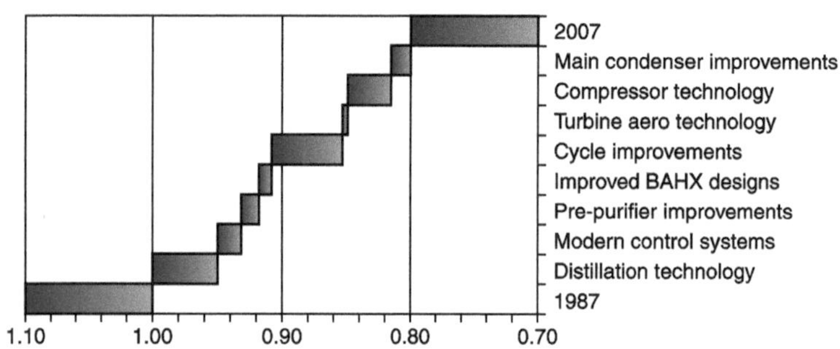

10.14 Indexed technology-driven oxygen separation power changes.

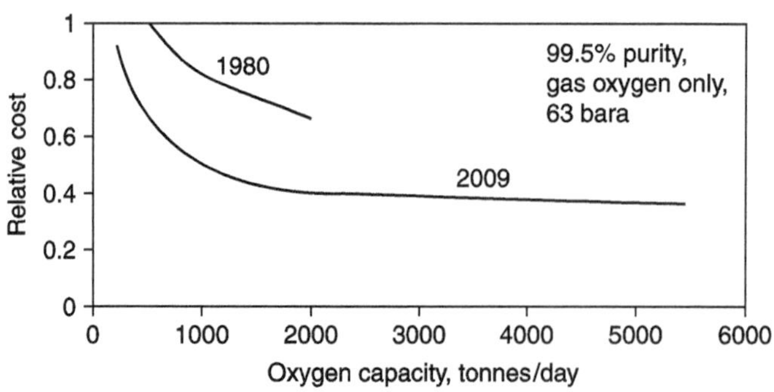

10.15 Air separation plant cost trend.

maximum plant capacity has increased dramatically. This was largely driven by market developments, such as the shift from liquid only plants to on-site gas plants, beginning in the 1940s. Technology enabled the change. In 1960 the largest plant produced 200 tonnes of oxygen per day; by 1980 single train plants of 2000 tonnes per day were operating. Today, the largest operating plant produces 3900 tonnes of oxygen per day. Rapid growth in large energy applications is expected to lead to even larger plants soon, and the industry currently is proposing single train plants with oxygen capacities exceeding 5000 tonnes per day.

While significant improvement has been achieved through improved process configurations, particularly as the requirements for products and co-products change, much of the source of the improvement has been led by equipment innovation.

The use of structured packing as mass transfer media in distillation columns represents one of the most notable advances in air separation technology in recent years (Lockett, 1989; Victor and Lockett, 1989; Lockett and Victor, 1990). It was first deployed in the late 1980s, and it is now widely used in almost all new air separation plants. Chiefly, its benefit is a large reduction in the pressure drop of the up-flowing vapor in the columns compared with trays. This is especially important in the low pressure column of the double column because of the 3.5 multiplicative factor of pressure from the low pressure column to the high pressure column and the main air compressor discharge. That is, a trayed low pressure column typically may incur 0.4 bar (6 psi) in pressure drop, which is reduced to 0.1 bar (1.5 psi) when packed. This pressure drop reduction of 0.3 bar lowers the MAC discharge pressure by about 1.1 bar (16 psi).

The low pressure drop has also facilitated increased staging to optimize power savings or enable higher purity products with structured packing. This has made the manufacture of product purity argon by purely cryogenic means possible (Bianchi *et al.*, 1992), which has saved cost and eliminated the need for hydrogen in argon manufacture, an important advance that has allowed broad scale production in emerging markets. Improved capacity utilization has been successfully achieved in air separation plants and remains a continued focus of structured packing development (Billingham and Lockett, 1997, 1999, 2000; Ender *et al.*, 1997). Structured packing density in each section of each distillation column is optimized, depending on the different flow-rates in each section, in order to maximize the effective utilization of each section (Lockett *et al.*, 1992b). Structured packing also provides a large turndown capability, to 40% of design capacity or less, while maintaining efficient mass transfer (Lockett *et al.*, 1992a).

Special heat transfer equipment has always been used in the cryogenic process industry, driven by specialized needs (Timmerhaus and Flynn, 1989, pp. 189–215):

- Low temperature materials
- Multi-stream capability
- Small temperature differences (from Second Law of Thermodynamics, magnitude of temperature difference becomes more important as absolute temperature falls)

- Low pressure drop for reduced compression requirements
- Large surface area to volume ratio to minimize heat leak
- High heat transfer to reduce surface area
- Low mass to minimize start-up time
- High pressure capability to provide design flexibility
- High reliability with minimal maintenance to reduce shut-downs.

Early air separation plants used shell-and-tube, concentric tube and coiled tube heat exchangers. Later, regenerators enabled larger plants, providing large surface areas and low pressure drops. In these devices, always built in at least pairs, the feed air and product stream flows would be switched, and the warming product stream would remove frozen carbon dioxide and water deposited by the air. The addition of an imbedded coil allowed the oxygen to be withdrawn without contamination. Ultimately, the need to produce more clean products limited the utility of regenerators.

Brazed aluminum plate-fin heat exchangers (BAHX), which are constructed from alternating layers of corrugated, die-formed aluminum fins, between flat aluminum parting sheets to form individual flow passages, were introduced commercially in the mid-1950s and became the industry standard almost immediately (Baker and Fisher, 1992). The fins provide extended surfaces for heat transfer, and the stream layer pattern and fin selection are optimized for the best trade-off between heat transfer and pressure drop. They enable multiple product streams; today's designs commonly cool and warm six or more streams.

Reversing heat exchangers (RHX) were commonly used as the main heat exchanger in air separation plants. In these designs, the main air feed and waste nitrogen switched passages at about 10 minute intervals so that the nitrogen swept out solid carbon dioxide and moisture. Unlike the regenerator, the RHX enabled a portion of the nitrogen to be withdrawn as clean product. However, the large volume of nitrogen required as a sweep stream limited the amount of clean nitrogen product to about 30% of the air (50% total clean products). Advances in zeolite (molecular sieve) technology led to the prevalence of air pre-purified plants, beginning in the 1970s. A nitrogen flow of only about 10% of the air was required to regenerate a temperature-swing pre-purifier, and the remainder of the products could be withdrawn without contamination (90% total clean products). Improved plant reliability resulted from the elimination of the switching valves and pressure cycling associated with RHXs, and air pre-purified plants provided an improved safety profile due to more complete hydrocarbon and contaminant removal. Plant operability improved due to much less frequent switching of the air pre-purifiers.

Advances in BAHX technology since then have largely been due to improved fabrication technology. Early designs were limited to 17 × 21 inches in cross-section and 108 inches long, with design pressures less than 14 bara (200 psia). These were brazed in salt baths (Gregory, 1987). Today, large vacuum furnaces enable block sizes of greater than 4 × 5 feet cross-section and 25 feet length, with design pressures exceeding 100 bara (1450 psia), and tremendous design flexibility (Markussen, 2004; Markussen and Lewis, 2005). Today's PHX designs feature gas-to-gas and gas-to-liquid heat transfer, and portions of the product oxygen and product nitrogen are often vaporized at very high pressures. The function of the nitrogen superheater is frequently combined with the PHX in the same brazement (Pahade *et al.*, 2000). Current development has much to do with optimization of the complex design features of BAHXs (Reneaume and Niclout, 2003).

Prior to the mid-1950s, reboilers in air separation plants used traditional shell-and-tube configurations. The need for small temperature differences led to very large surface requirements due to poor heat transfer coefficients on smooth surfaces. The advent of BAHXs led to much more effective designs. A modularized shell-and-tube design utilizing an enhanced heat transfer surface on the boiling side, invented in 1947 (Baker and Fisher, 1992), and a fluted surface on the condensing side is still the basis of effective reboiler designs for Praxair. Recently, development has focused around minimizing or eliminating the internal pinch in pool boiling (thermosiphon) configurations by reducing the liquid head (Schweigert *et al.*, 2004), or eliminating the liquid head through the use of falling-film (or downflow) configurations (Lockett and Srinivasan, 1997). To maintain satisfactory surface wetting, which is necessary for safe operation, a recirculation pump is usually required for a falling-film reboiler. However, the pump adds cost and an energy penalty, as well as a reliability concern. Pumpless apparatus has recently been developed (Satchell *et al.*, 1998; Chakravarthy *et al.*, 2005, 2008).

The advent of air pre-purification technology clearly led to a shift in the capabilities of air separation plants. Continued development since the 1970s has reduced the energy cost and capital cost of pre-purification, while improving reliability and capacity. The prevalent type of air pre-purifier uses temperature-swing adsorption (TSA), where regeneration of the off-stream bed is accomplished by a combination of heat and reduced pressure. The low pressure waste nitrogen stream is heated to 300–600°F prior to introduction in the bed for regeneration. Molecular sieve only beds were first used. However, it was found that dual-layer beds, with a first layer of alumina followed by molecular sieve, enabled a 35% reduction in regeneration energy and a reduction in regeneration temperature (Wilson *et al.*, 1984). This is due to the reduced energy of desorption of water on alumina and to the reduction of waste heat loss. The dual-layer design was also more robust and less susceptible to damage by free moisture (Grenier *et al.*, 1984).

Further energy was saved by improving the method of thermal regeneration such that a minimum of waste heat is lost by using an efficient thermal pulse (Oliker, 1982). Air pre-purifiers conventionally consist of two vertical, cylindrical beds, in parallel. As the plant capacity grows, however, this configuration becomes impractical because the bed diameter becomes too large. As an alternative, three-bed systems have been used, where two beds are on-stream while one is regenerated. At very large capacities, the beds may be reconfigured in the horizontal direction. Here, each cylindrical vessel is laid on its side and the air flows vertically through the horizontal bed. This allows a very large flow area, which can be increased by increasing the length of the horizontal bed. This system provides a number of difficult design challenges, and flow maldistribution can be problematic as the bed length increases. This also can lead to significant operational issues.

An alternative that is widely used is the radial flow bed system (Bosquain *et al.*, 1985; Kerry, 1991; Libal *et al.*, 1998; Tentarelli, 2000). Here the bed is configured in an annular ring, and air flows through the annulus from the outside to the center. This design is more robust, and more amenable to large capacities, without flow maldistribution problems. The bed is fixed and cannot fluidize. Screens are free to expand and contract during thermal cycles, unlike the other configurations.

Adsorbent materials represent another area of considerable improvement in recent years. New molecular sieves have higher carbon dioxide working capacities, enabling smaller beds with concurrent cost and energy savings. Also, compound beds have been employed, where a third layer of specialty adsorbent is added for complete removal of specific impurities, especially hydrogen, carbon monoxide, and nitrous oxide.

The application of pressure-swing adsorption (PSA) is yet another area of recent innovation in air pre-purification. Here, the beds are cleaned with a flow of unheated nitrogen, and regeneration heating is completely eliminated, leading to significant cost and regeneration energy savings. However, preferred adsorbent materials are considerably weaker to enable effective regeneration and satisfactory desorption of impurities. This means the adsorbent working capacity is less. The PSA air pre-purifier requires a much higher regeneration nitrogen flow, and the cycle time is much shorter than for TSA, resulting in higher blowdown losses. All of these factors tend to limit the size of PSA air pre-purifiers so that they are not used for very large plants.

Compressors provide the gas flow and pressurization needs to operate the ASU. They are very large power consumers and use about 95% of the total plant energy. Compressors typically limit the efficient turndown capability of the air separation plant. As such, the efficiency and rangeability of the compressors directly impact the efficiency and rangeability of the air separation plant. Furthermore, compressors contribute to a large portion of the capital cost (Drnevich *et al.*, 1981, Fig. 7-2). Turbo-machinery technology development in air separation has led to improvements in compressor efficiency, rangeability, and cost (Mahoney *et al.*, 1999; Abdelwahab *et al.*, 2008). Turbo-expanders are an essential component in the reverse Brayton cycles used in air separation plants and liquefiers for refrigeration generation. Turbine efficiency improvements have been a very significant source of efficiency improvement in the industry, both for new plants and retrofits of existing plants (Wulf, 1991).

The tightly integrated air separation process exhibits a high degree of dependence among process variables. In addition to the primary control loop, a supervisory control system is used to optimize overall performance and move the process from one operating point to another. Model predictive control (MPC) is an advanced, multivariable control system that has resulted in a step change improvement in robustness and efficiency of air separation plants since its implementation (Canney, 1996). The capability of rapid, direct, on-spec changes of production levels of multiple plant products is enabled with MPC. Effective, remote operation of plants is also facilitated with MPC. Special equipment configurations and methods have been devised to facilitate especially rapid capacity modulation (Darredeau *et al.*, 1999), which may be important for the oxygen supply to an oxy-fuel power plant. Advances in controls for air separation have been wide in breadth, from the primary control level (Howard *et al.*, 1995), to the overall plant level, to the multi-plant level (Megan *et al.*, 2004).

10.5 Oxygen transport membrane (OTM) technology

With the current state of the art, the parasitic power associated with the cryogenic ASU can amount to 12–15% of the gross output of the power plant depending on the steam cycle efficiency. For a mature technology like cryogenic air separation, it will be difficult to reduce its parasitic load below 10% of power plant output. A completely novel approach for supplying oxygen to oxy-combustion will be required to achieve a step change in efficiency improvement.

Praxair is developing an oxygen transport membrane (OTM) technology that could revolutionize the way oxy-combustion is carried out. Instead of separating a pure stream of oxygen in an air separation unit and then delivering it to a boiler for combustion, the OTM technology integrates oxygen separation and combustion in one unit. In the novel OTM boiler concept, ceramic membrane tubes keep the air and fuel streams apart. Oxygen from the air side is transported through OTM to the fuel side and it reacts with the fuel on the membrane surface itself. Building on this concept, Praxair has defined a power cycle concept that integrates OTM-based oxidation units and other well-known unit operations for a step change in efficiency improvement of a power plant with CO_2 capture.

10.5.1 Principle of operation of oxygen transport membrane (OTM)

An OTM consists of a robust inert porous support coated with a dense gas separation layer (Nagabhushana *et al.*, 2009), as illustrated in Fig. 10.16. The dense layer is made of dual phase materials capable of transporting oxygen ions and electrons. The ionic phase is made from fully stabilized zirconium oxide and the electronic phase is made from a metallic oxide containing lanthanum, strontium, chromium, and manganese. The dense layer is kept as thin as practical in order to minimize the resistance to oxygen transport. The porous support layer is made from partially stabilized zirconium oxide. The composite membrane can be formed as a planner element or as a tube.

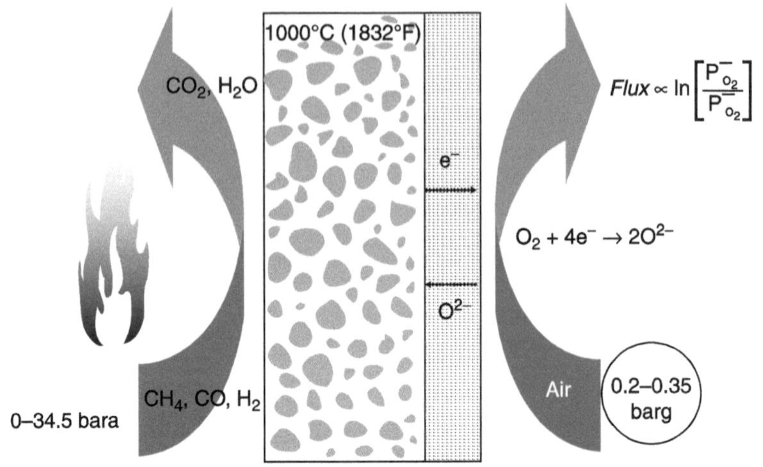

10.16 Oxy-combustion using oxygen transport membranes.

Air flows on the cathode side of the membrane where molecular oxygen reacts with oxygen vacancies and electrons on the membrane surface to form oxygen ions, which transport through the dense layer using a chemical potential difference as the driving force. Fuel species (CO, H_2, CH_4 etc.) are fed to the anode side of the membrane where they diffuse through the porous support and react with oxygen ions at the dense layer surface to form oxidation products (H_2O, CO_2) and oxygen vacancies and electrons in the crystalline lattice structure of the separation layer. The liberated electrons traverse back through the dense layer to the cathode side for recombining with oxygen again. The oxidation products diffuse through the porous support into the fuel stream.

Since oxygen is consumed as soon as it passes through the membrane, the oxygen partial pressure on the fuel side is always close to zero. As a result, it is possible to transport oxygen from the air side at near ambient pressure to the fuel side kept at pressures as high 34.5 bara (500 psia).

10.5.2 OTM boiler

Praxair's approach uses a tubular architecture for OTM modules. These modules can be integrated into an OTM boiler. In one concept (Bool and Kobayashi, 2002), lance tubes are inserted inside the OTM tubes for supplying the air inside the tubes. Fuel is supplied on the outside of the tubes. The OTM boiler consists of a large array of the OTM tubes interspersed with steam tubes as shown schematically in Fig. 10.17. Fuel is introduced at one end of the boiler and the oxidation products, CO_2 and H_2O, are withdrawn from the other end. The oxy-combustion reaction takes place on the outside of the tubes. The thermal energy released from the reaction is transferred to steam tubes for generating and superheating steam. The temperature on the surface of the OTM tube is maintained at 900 to 1100°C.

10.5.3 Advanced power cycle

To adopt the concept of an OTM boiler for a coal-based power plant, the advanced power cycle concept shown in Fig. 10.18 has been developed (Shah *et al.*, 2008). Major steps in the advanced power cycle are gasification, OTM-based partial oxidation, syngas expansion, OTM-based oxyfuel combustion for generating steam, steam expansion for power generation, and CO_2 compression and purification.

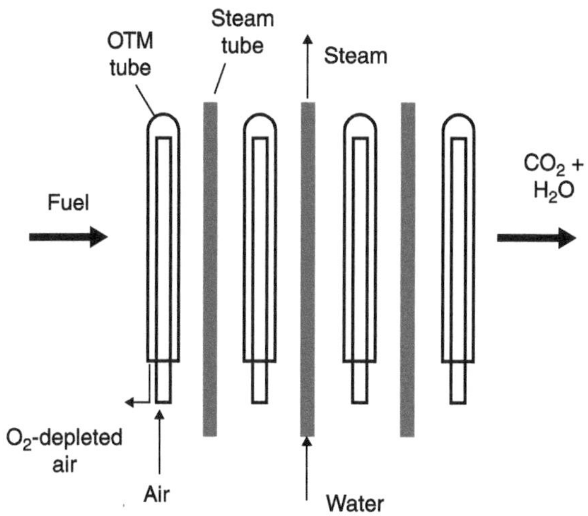

10.17 OTM boiler.

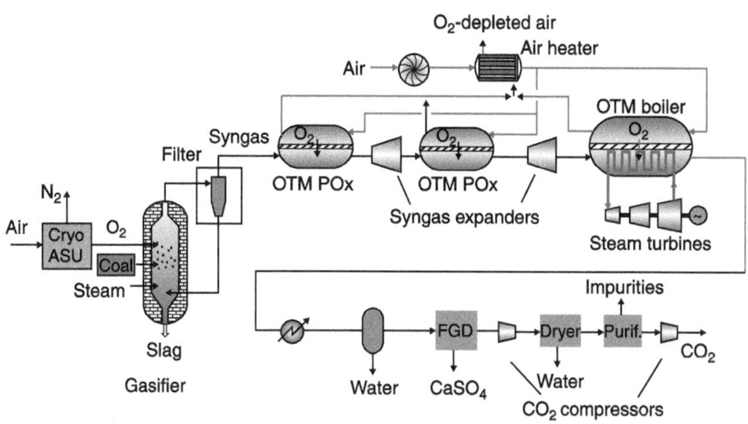

10.18 Advanced power cycle concept based on OTM boiler. FGD = flue gas desulphurization; POx = partial oxidation.

The main premise behind this concept is to oxidize only gaseous fuels using OTM tubes. Therefore, an oxygen-blown gasifier is used first to convert coal into syngas at 20 to 35 bara (290–510 psia). Oxygen for gasification is supplied from a cryogenic ASU. It is desirable to select a gasifier that will minimize the consumption of cryogenic oxygen. Since the boiler is envisioned to be operating at atmospheric pressure, syngas is expanded in two stages to recover power. To maximize power recovery, syngas is heated between two stages of expansion and optionally before the first stage of expansion. The syngas heating is accomplished using the OTM-based partial oxidation units. If syngas from the gasifier is at 537°C (1000°F) such as from a BGL gasifier, it is first heated to 982°C (1800°F). The heated syngas is expanded in a turbine to 4.5–6 bara (65–87 psia) to recover power. The expanded syngas from the first syngas expander is then passed through an OTM partial oxidation unit for raising its temperature back up to 982°C and then expanded in a second syngas expander to slightly above the ambient pressure for more power recovery. The expanded syngas is sent to the OTM boiler. Air used in the OTM POx units and in the OTM boiler is compressed to 1.35 bara (20 psia). It is then preheated close to the operating temperature of OTM by heat exchange with the oxygen-depleted air exiting the OTM units.

Within the OTM boiler, syngas is first passed over an array of OTM tubes that are interspersed with steam tubes. The preheated air is passed inside the OTM tubes. Oxygen from the air transports across the membrane and reacts with the syngas on the outside of the tubes. The thermal energy from the oxidation reaction will be transferred to the steam tubes by radiation and by convection through flow of the fuel stream that passes alternately over OTM and steam tubes. In the OTM zone, the steam and OTM tubes will be placed such that the temperature is maintained at the optimum level for membrane performance. Since the rate of oxygen transport is limited by the availability of the membrane area, the oxidation of syngas will take place over a large area (the OTM zone) within the boiler. As the syngas is oxidized, the driving force for oxygen transport will decrease and the required membrane area will increase. For practical reasons, the OTM will be used to supply oxygen to the fuel side until 80–90% fuel utilization is achieved. The remaining fuel will be combusted using oxygen supplied from the cryogenic ASU. After the fuel is completely oxidized with externally supplied oxygen, the flue gas will pass through a convective section of the boiler for further steam generation and boiler feed water preheating. The flue gas exiting the boiler is processed according to a purification process proposed for a conventional oxy-fuel technology (see Chapter 11) to produce a compressed and purified CO_2 stream for sequestration.

10.5.4 Cost and performance projections

The main motivation behind the development of the advanced power cycle is the possibility of achieving very high efficiency, even with CO_2 capture. High efficiency results from the use of OTM for oxygen supply, syngas expanders for additional power generation and low grade heat recovery. In the advanced OTM power cycle, 65–70 % of the total oxygen needed in the process is supplied through OTM and the remainder is supplied from a cryogenic ASU. Since the OTM requires air to be compressed to only 1.35 bara (20 psia), the power consumption per unit volume of oxygen is ~70% lower for OTM units than that for cryogenic air separation. A comparison presented by Shah and Christie (2007) showed that the OTM process of Fig. 10.18 will result in 4.5 percentage points higher efficiency (HHV) compared with a conventional oxy-combustion based power plant.

High efficiency of the process results in higher power output from a given investment, which in turn leads to lower specific capital cost (US$/kW). The specific capital cost (US$/kW) for the OTM process is projected to be 50% higher than the air-fired PC (pulverized coal) power plant. High efficiency also reduces the fuel cost. These dual benefits of high efficiency lead to lower cost of electricity in the OTM process. According to a recent analysis, shown in Table 10.2 (Shah *et al.*, 2009), the increase in cost of electricity for the OTM power cycle with CO_2 capture and storage over a reference air-fired PC power plant is less than 35%. In comparison, the cost of electricity for an oxygen-fired PC boiler with oxygen supplied from cryogenic air separation increases by 50–60% (DOE/NETL, 2008; see also Fig. 10.3). The cost of electricity for the OTM power cycle is not very sensitive to the cost of OTM (Shah *et al.*, 2009) because the installed cost of OTM elements accounts for less than 10% of total capital cost of the power plant.

10.5.5 Comparison with integrated gasification combined cycle (IGCC)

Although the OTM power cycle uses a gasifier at the front end of the process, its CO_2 capture characteristics are similar to an oxy-combustion process. Table 10.3 shows a comparison of the key features of the OTM power cycle and IGCC. If CO_2 capture is required from an IGCC-based power plant, the syngas from the gasifier is subjected to water gas shift reaction to convert carbon monoxide and steam into carbon dioxide and hydrogen. The shifted syngas is cooled to near

Table 10.2 Cost comparison of air-fired PC and advanced power cycle (Shah *et al.*, 2009)

	Air-fired PC	OTM process
Coal price US$3.0/MMBtu (HHV)		
Cost of electricity, US$/MW-h	US$83	US$110
Cost of CO_2 capture, US$/ton*		US$30
Cost of CO_2 avoided, US$/ton		US$32

* Ton = 0.907 tonne.

Table 10.3 OTM power cycle vs. IGCC

	OTM power cycle	IGCC
Gasifier	Yes	Yes
Syngas expander	Yes	Yes/No
Shift reactor	No	Yes
Acid gas recovery (e.g. selexol)	No	Yes
H_2 combustion turbine	No	Yes
Steam generation unit	OTM boiler	HRSG
Oxidant for combustion	Oxygen	Air
Pre-combustion CO_2 capture	No	Yes
Oxy-combustion CO_2 capture	Yes	No
Power production from major units		
Combustion gas turbine	0	62%
Syngas expander	18%	1%
Steam cycle	82%	37%

ambient temperature. An acid gas recovery unit removes sulfur compounds and CO_2 from the cooled syngas stream and hydrogen is sent to a combined cycle section for power generation. A significant amount of fuel energy is lost as the coal is transformed into hydrogen. The report published by DOE/NETL (2007) shows that the efficiency of IGCC plant with CO_2 capture will range from 31.7% (HHV) to 32.5% (HHV) depending on the type of gasifier used.

In the OTM process, there is no loss of energy as syngas does not have to be transformed into hydrogen and syngas does not have to be cooled to ambient temperature prior to combustion for CO_2 separation. Overall, the OTM process with ultrasupercritical steam cycle is projected to achieve 37.2% (HHV) efficiency (Shah *et al.*, 2009), which is 4.7 to 5.5 percentage points higher than the efficiency of IGCC plant with CO_2 capture (DOE/NETL, 2007). In the IGCC process, CO_2 must be separated by using solvents such as selexol, rectisol, or amine. In the OTM process, the CO_2-rich stream is generated in the OTM boiler during the combustion just like in a conventional oxygen-fired PC boiler. In IGCC, about two-thirds of the power is generated by the combustion turbines and one-third of power is generated in the steam cycle. In the OTM power cycle, more than 80% of power is generated in the steam cycle and the balance of power is generated by syngas expanders. Final purification of CO_2 in the OTM process is similar to that of a conventional oxy-combustion process. Overall, the OTM power cycle shares more features with the conventional oxy-combustion process than with IGCC.

10.5.6 Technology status

All of the unit operations of the advanced power cycle shown in Fig. 10.18 can be carried out on a commercial scale using equipment available today, except for the OTM units. Both OTM-based partial oxidation and OTM boiler equipment need significant further development. Therefore, the efforts are mainly focused on advancing the OTM technology.

After several years of development, Praxair has developed membrane tubes with resistance to thermal and chemical shock. These tubes show chemical stability in the presence of high concentrations of CO_2 and H_2O. The OTM technology is currently being tested in a lab-scale tubular reactors. The lab-scale reactors have been operated under conditions anticipated in OTM POx reactor, and OTM boiler and oxygen fluxes needed for commercial viability have been achieved using the simulated syngas mixtures. Membranes are also shown to have tolerance to 1% of H_2S and COS in the syngas mixture. Among the next steps in technology development are operation of pilot scale OTM partial oxidation and boiler units with coal derived syngas and conceptual engineering design and costing of a commercial scale OTM boiler.

10.6 Future trends

Cryogenic air separation has demonstrated remarkable advances over the last two to three decades, despite the view that it is a mature technology. Advances have occurred in all sub-technology areas and there is no indication of that subsiding. A number of process and equipment technology improvements specifically benefit oxy-fuel firing of power generation plants, by reducing the parasitic load of the ASU and by reducing its cost. The very large scale of these air separation plants is near the edge of current capabilities for single trains and much is being learned currently about how to most effectively design and construct these plants. Historically, technology advances have led to increased equipment capacity capabilities in air separation. Plant outputs have grown dramatically with little apparent change in the size of the plants. The prospects for a large number of energy projects, many exploiting lower value fuel sources, with the drive for energy efficiency and process intensification portends well for the future of large air separation plants. Air separation plants are vital in many modern energy technologies. They enable improved efficiency, greatly increased production levels, and much reduced byproducts, and facilitate pollution control. Emerging markets with rapid growth in infrastructure, steel production, and chemical manufacture based on gasification will also lead to large increases in worldwide oxygen production. Further, the drive towards greener, more efficient, processes, with higher quality products in many applications points to continued growth in demand for all the major industrial gases.

The OTM-based oxy-combustion technology has achieved several significant milestones at a bench-scale. These include improved membrane tube mechanical strength, robustness, and resistance to chemical and thermal shock. Demonstration of extended operation in single- and multi-tube reactors with oxygen fluxes needed for commercial viability has been achieved. Significant challenges remain for the OTM technology. Engineering design of a large scale boiler that integrates OTM and steam tubes is a major challenge. Although durability has been demonstrated in a controlled lab environment, achieving reliability and robustness in a real operating environment will also be a major challenge. Currently, plans for a pilot scale demonstration of the OTM technology are being developed.

10.7 Acknowledgements

The authors would like to thank colleagues Mark Ackley, Paul Belanger, Max Christie, and James Handley, all of Praxair, Inc., for providing assistance in preparation of this chapter.

10.8 References

Abdelwahab, A, Baker, R L, and Gerber, G J (2008), 'Leaned Centrifugal Compressor Airfoil Diffuser', US Patent 7,448,852. 11 Nov., 2008.

Baker, C R, and Fisher, T F (1992), 'Industrial Cryogenic Engineering in the USA', in Scurlock, R G, *History and Origins of Cryogenics*, Oxford University Press, 217–254.

Baksh, M S A, Kibler, V J, and Schaub, H R (1996), 'Pressure Swing Adsorption Process', US Patent 5,518,526. 21 May, 1996.

Bianchi, J R, Bonaquist, D P, and Victor, R A (1992), 'Cryogenic Rectification Method for Producing Refined Argon', US Patent 5,133,790. 28 Jul., 1992.

Billingham, J F, and Lockett, M J (1997), 'Packing with Improved Capacity for Rectification Systems', US Patent 5,632,934. 27 May, 1997.

Billingham, J F, and Lockett, M J (1999), 'Development of a New Generation of Structured Packings for Distillation', *Trans. IChemE*, Vol. 77, Part A, Oct. 1999, 583–587.

Billingham, J F, and Lockett, M J (2000), 'Cryogenic Rectification System with High Strength and High Capacity Packing', US Patent 6,101,841. 15 Aug., 2000.

Bool, L E, and Kobayashi, H (2002), 'Oxygen Separation and Combustion Apparatus and Method', US Patent 6,394,043 B1. 28 May, 2002.

Bosquain, M, Grenier, M, Hay, L, Lehman, J-Y, Petit, P, and Sauty, P (1985), 'Reactor and Apparatus for Purifying by Adsorption', US Patent 4,541,851. 17 Sept., 1985.

Canney, W M (1996), 'Model Predictive Control Method for an Air-Separation System', US Patent 5,522,224. 4 June, 1996.

Chakravarthy, V S, Jibb, R J, Royal, J H, and Lockett, M J (2005), 'Developments in Falling Film Type (Downflow) Reboilers in the Air Separation Industry', Proceedings of the Fifth International Conference on Enhanced, Compact and Ultra-Compact Heat Exchangers: Science, Engineering and Technology, Sept. 2005, 264–272.

Chakravarthy, V S, Jibb, R J, Lockett, M J, and Royal, J H (2008), 'Cryogenic Air Separation with Once-through Main Condenser', US Patent 7,421,856. 9 Sept., 2008.

Chao, C C (1989), 'Process for Separating Nitrogen from Mixtures Thereof with Less Polar Substances', US Patent 4,859,217. 22 Aug., 1989.

Cheung, H (1995), 'Side Column Cryogenic Rectification System for Producing Low Purity Oxygen', US Patent 5,463,871. 7 Nov., 1995.

Darredeau, B, Lehman, J-Y, and Peyron, J-M (1999), 'Process and Installation for the Separation of Air', US Patent 5,901,580. 11 May, 1999.

DOE/NETL (2007), 'Cost and Performance Baseline for Fossil Energy Plants', DOE/NETL-2007/1281, Revision 1, August 2007, US Department of Energy, National Energy Technology Laboratory.

DOE/NETL (2008), 'Pulverized Coal Oxycombustion Power Plants', DOE/NETL-2007/1291, Revision 2, August 2008, US Department of Energy, National Energy Technology Laboratory.

Drnevich, R F, Ecelbarger, E J, and Portzer, J W (1981), 'Industrial Oxygen Plants – A Technology Overview for Users of Coal Gasification – Combined-Cycle Systems', EPRI AP-1674, Jan. 1981, Electric Power Research Institute.

Ender, C, Billingham, J F, Lockett, M J, Yeoman, N, Kuratle, R, Walztoni, K A, and Kallenberger, D L (1997), 'Packing for Mass Transfer Column', US Patent 6,578,829 B2. 27 May, 1997.

Gaumer, L S, Jr. (1965), 'Method for Low Temperature Separation of Gaseous Mixtures', US Patent 3,210,951. 12 Oct., 1965.

Gregory, E (1987), 'Plate and Fin Heat Exchangers', *The Chemical Engineer*, Sept. 1987, 34–39.

Grenier, M, Lehman, J-Y, Petit, P, and Eyre, D V (1984), 'Adsorption Purification for Air Separation Plants', in Kerney, P J *et al.*, *Cryogenic Processes and Equipment*, book no. G00283, ASME, 143–148.

Hong, J, Chaudhry, G, Brisson, J G, Field, R, Gazzino, M, and Ghoniem, A F (2009), 'Analysis of Oxy-Fuel Combustion Power Cycle Utilizing a Pressurized Coal Combustor', *Energy*, 34(9), Sept. 2009, 1332–1340.

Howard, H E, Bonaquist, D P, Canney, W M, and Nash, W A (1995), 'Process for Maximizing the Recovery of Argon from an Air Separation System at High Argon Recovery Rates', US Patent 5,448,893. 12 Sept., 1995.

Howard, H E, Jibb, R J, and Larson, K F (2009), 'Method and Apparatus for Separating Air', US Patent Application 2009/0277220 A1. 12 Nov., 2009.

Jakob, F (1958), 'Process and Apparatus for the Separation of Compressed Air', US Patent 2,850,880. 9 Sept., 1958.

Kerry, F G (1991), 'Front-Ends for Air Separation Plants – The Cold Facts', *Chemical Engineering Progress*, August 1991, 48–54.

Kleinberg, W T (1987), 'Cycle to Produce Low Purity Oxygen', US Patent 4,704,148. 3 Nov., 1987.

Libal, K, Fierlbeck, W, and Von Gemmingen, U (1998), 'Reactor', US Patent 5,827,485. 27 Oct., 1998.

Lockett, M J (1989), 'Gas Separation by Distillation', Proceedings of the Fifth BOC Priestley Conference, Special Publication No. 80, *Separation of Gases*, Royal Society of Chemistry, pp. 19–34.

Lockett, M J, and Srinivasan, V (1997), 'Downflow Shell and Tube Reboiler-Condenser Heat Exchanger for Cryogenic Rectification', US Patent 5,699,671. 23 Dec., 1997.

Lockett, M J, and Victor, R A (1990), 'Structured Column Packing with Liquid Holdup', US Patent 4,929,399. 20 May, 1990.

Lockett, M J, Victor, R A, and Augustyniak, J D (1992a), 'Structured Column Packing with Improved Turndown and Method', US Patent 5,132,056. 21 July, 1992.

Lockett, M J, Victor, R A, Zawierucha, R A, McIlroy, K, and Cooper, S L (1992b), 'Variable Density Structured Packing Cryogenic Distillation System', US Patent 5,100,448. 31 March, 1992.

Mahoney, K W, Allen-Hayes, C B, Leo, J M, Henry, P A, Skare, T A, Bonaquist, D P, and Handley, J R (1999), 'Cryogenic Air Separation System with Integrated Machine Compression', US Patent 5,901,579. 11 May, 1999.

Markussen, D (2004), 'All Heat Exchangers are Not Created Equal', *The Process Engineer*, Sept. 2004.

Markussen, D, and Lewis, L (2005), 'Brazed Aluminum Plate Fin Heat Exchangers – Construction, Uses and Advantages in Cryogenic Refrigeration Systems', paper presented at Spring AIChE Meeting, April 2005.

Megan, L, Lennox, D F, Sharf, P F, Adebekun, D, and Zhu, M (2004), 'Control for Pipeline Gas Distribution System', US Patent 6,697,713. 24 Feb., 2004.

Nagabhushsana, N, Lane, J M, Christie, G M, and VanHassel, B A (2009), 'Composite Oxygen Ion Transport Membrane', US Patent 7,556,676. 7 July, 2009.

Oliker, M D (1982), 'Adsorption Beds and Method of Operation Thereof', US Patent 4,324,564. 13 April, 1982.

Pahade, R F, Arman, B, and Lockett, M J (2000), 'Heat Exchange Unit for a Cryogenic Air Separation System', US Patent 6,044,902. 4 April, 2000.

Prosser, N M, and Shah, M (2003), 'Cold Compression Cryogenic Rectification System for Producing Low Purity Oxygen', US Patent 6,626,008. 30 Sept., 2003.

Rathbone, T (1996), 'Air Separation', US Patent 5,551,258. 3 Sept., 1996.

Reneaume, J.-M, and Niclout, N (2003), 'MINLP Optimization of Plate Fin Heat Exchangers', *Chemical and Biochemistry Engineering Quarterly* 17 (1), 65–76.

Royal, J (2009), 'Innovation and Enterprise: The Industrial Gases Industry in the United States', in Flank, W H, Abraham, M A, and Matthews, M A, *Innovation in Industrial and Engineering Chemistry, a Century of Achievements and Prospects for the New Millennium*, American Chemical Society.

Ruhemann, M (1949), *The Separation of Gases*, Oxford University Press, Oxford, UK.

Satchell, D P, Jr, Natarajan, V, and Clarke, R H (1998), 'Heat Exchanger', US Patent 5,775,129. 7 July, 1998.

Schweigert, K H, Wanner, A, Hecht, T, and Sotzek, M (2004), 'Multistoreyed Bath Condenser', US Patent 6,748,763. 15 June, 2004.

Shah, M M (2005), 'Capturing CO_2 from Oxy-Fuel Combustion Flue Gas', paper presented at 1st Workshop of the International Oxy-Fuel Combustion Research Network, Cottbus, Germany, 29–30 Nov., 2005, IEA GHG.

Shah, M M (2006), 'Oxy-Fuel Combustion for CO_2 Capture from PC Boilers', paper presented at 31st International Technical Conference on Coal Utilization and Fuel Systems, Clearwater, FL, USA, 21–25 May, 2006, Coal Technology Association.

Shah, M M (2007), 'Oxy-Fuel Combustion for CO_2 Capture from New and Existing PC Boilers', paper presented at Electric Power 2007 Conference, Chicago, IL, USA, 1–3 May, 2007, Electric Power.

Shah, M M and Christie, G M (2007), 'Oxy-Fuel Combustion Using OTM For CO_2 Capture from Coal Power Plants', paper presented at 2nd Workshop of the International Oxy-Combustion Research Network, Windsor, CT, USA, 25–26 January, 2007, IEA GHG.

Shah, M M, Jamal, A, Drnevich, R F, VanHassel, B A, Christie, G M, Kobayashi, H, and Bool, L E (2008), 'Electrical Power Generation Method', US Patent Application 2008/0141672 A1. 19 June, 2008.

Shah, M M, Christie, G M, Degenstein, N and Wilson, J (2009), 'Oxy-Combustion on Oxygen Transport Membranes (OTM), presented at the 1st Oxyfuel Combustion Conference, Cottbus, Germany, 7–11 Sept., IEA GHG.

Smolarek, J, Fassbaugh, J H, Rogan, M K, and Schaub, H R (2000), 'Vacuum Pressure Swing Adsorption System and Method', US Patent 6,010,555. 4 Jan., 2000.

Tentarelli, S C (2000), 'Radial Flow Adsorption Vessel', US Patent 6,086,659. 11 July, 2000.

Timmerhaus, K D, and Flynn, T M (1989), *Cryogenic Process Engineering*, Plenum Press.

Treybal, R E (1980), *Mass-Transfer Operations*, McGraw-Hill Book Company.
US Bureau of Commerce (2008), *Current Industrial Reports: Industrial Gases*, MQ325C and predecessor series; *US Statistical Abstract*, various years; available from: http://www.census.gov/manufacturing/cir/index.html [accessed 2008].
Victor, R A, and Lockett, M J (1989), 'Double Column Air Separation Process with Hybrid Upper Column', US Patent 4,838,913. 13 June, 1989.
Wilson, K B, Smith, A R, and Theobald, A (1984), 'Air Purification for Cryogenic Air Separation Units', *IOMA Broadcaster*, Jan.–Feb. 1984, 15–20.
Wulf, J B (1991), 'High Efficiency Turboexpander', US Patent 5,046,919. 10 Sept., 1991.
Zheng, L, Pomalis, R, and Clements, B (2007), 'Technical and Economic Feasibility Study of TIPS Process and Comparison with Other CO_2 Capture Power Generation Processes', paper presented at 32nd International Technical Conference on Coal Utilization and Fuel Systems, Clearwater, FL, USA, 10–15 June, 2007, Coal Technology Association.

11

Carbon dioxide (CO_2) compression and purification technology for oxy-fuel combustion

M. M. SHAH, Praxair, Inc., USA

Abstract: The chapter begins with a description of the industrial CO_2 production process. The remainder of the chapter is focused on oxy-combustion flue gas CO_2 purification, including process description, energy consumption, CO_2 recovery, capture costs, emissions, recent advances and future outlook.

Key words: CO_2 purification, oxy-fuel combustion, near-zero emissions, flue gas, CO_2 capture.

11.1 Introduction

Carbon dioxide (CO_2) is produced for use in a variety of applications in the food and beverage, chemical, electronics, energy and other industries. Depending on the application, the purity requirements could vary. Minimum purity of 99% (by vol.) is generally required with more stringent specifications for food, beverage and pharmaceutical applications. CO_2 for industrial use is produced as a liquid and is distributed by trucks, rail cars or ships as a merchant product by industrial gas companies. According to Garvey (2009), the liquid CO_2 production capacity in the US is 37,000 tons/day (33,560 tonnes/day; 1 tonne = 1,000 kg). A much larger volume of CO_2 is used in the enhanced oil recovery (EOR) operations. The total volume of fresh CO_2 supplied for EOR in the US is greater than 100,000 tonnes/day (World Resources Institute, 2008). About 80% of CO_2 for EOR is extracted from natural underground reserves at high pressure. The purity of CO_2 extracted from natural reserves is usually more than 95% (by volume). Natural CO_2 is treated to meet EOR specifications, compressed to supercritical pressure and then transported through pipelines to the oil fields. In most cases, the only treatment necessary is drying to reduce the moisture content.

In addition to natural sources, CO_2 in relatively high concentrations is generated as a by-product from many different processes, such as hydrogen production, ammonia production, ethylene oxide manufacture, natural gas processing and ethanol fermentation. The raw CO_2 from these sources serves as feeds for industrial production. The nature and level of impurities vary depending on the source. The industrial production process is customized to remove different impurities that could be present in raw CO_2 from diverse sources and to meet differing product specifications for various end uses. The capacity of a typical industrial CO_2 production plant ranges from 100 to 1,000 tonnes/day. Due to high transportation costs, CO_2 plants are usually installed close to the demand centers.

In the US, 73% of the merchant CO_2 production capacity is owned by four major industrial gas companies: Praxair, Linde, Airgas and Air Liquide (Garvey, 2009). There are several companies with the expertise in supplying the turnkey industrial CO_2 production plants such as Salof (2010), Union Engineering (2010), Wittemann (2010) and Toromont Energy Systems (2010). They have experience in building plants according to the specifications provided by the industrial gas companies.

The CO_2 purification process proposed for the oxy-fuel flue gas is somewhat different from the industrial process for several reasons. The volumes of CO_2 that must be captured and sent to geological storage sites will be very large. As a result, it is expected that CO_2 will be transported through pipeline in a supercritical state. This is in contrast to a refrigerated liquid CO_2 product that is produced from the industrial plants and distributed to the end users, usually by trucks. The combination of trace impurities found in the flue gas is different from that typically found in raw CO_2 streams in the industrial CO_2 plants. Although the industrial plants have dealt with most of the trace impurities that will be present in the flue gas individually, none of the industrial CO_2 sources contain them together and at concentrations typically found in the flue gas. The purity requirements for sequestration may not be as stringent as those for industrial use. Thus, the needs for the oxy-fuel flue gas purification are distinctly different from those of industrial CO_2 production with respect to production capacity, nature of impurities, product quality specifications and final state of product. Despite these differences, the experience gained in removal of a wide range of trace contaminants to meet very strict food and beverage grade standards can be leveraged for purifying CO_2-rich flue gas generated in the oxy-fuel combustion process.

Section 11.2 describes the industrial production process for producing the merchant product. Subsequent sections are focused on the process for producing CO_2 for sequestration from the oxy-fuel flue gas.

11.2 Industrial carbon dioxide (CO_2) production process

Crude CO_2 from different industrial sources usually contains greater than 90% (by vol.) of CO_2 and various trace impurities depending on the source, as shown in Table 11.1 (CGA, 2004). The industrial CO_2 production process converts crude CO_2 into high purity liquid CO_2 product. High purity is achieved by several purification steps operating at pressures up to 23 bara (334 psia) and at temperatures ranging from 300°C to –57°C depending on the type of impurities present in the crude CO_2.

Table 11.1 Possible trace impurities by sources (CGA, 2004)

Component	Combustion	Wells/ geothermal	Fermentation	Hydrogen or ammonia	Ethylene oxide
Aldehydes	X	X	X	X	X
Amines	X			X	
Benzene	X	X	X	X	X
Carbon monoxide	X	X	X	X	X
Carbonyl sulfide		X	X	X	
Cyclialiphatic hydrocarbons	X	X		X	X
Dimethyl sulfide		X	X		
Ethanol	X	X	X	X	X
Ether		X	X	X	X
Ethyl acetate		X	X		X
Ethyl benzene		X		X	X
Ethylene oxide					X
Halocarbons	X				X
Hydrogen cyanide	X				
Hydrogen sulfide	X	X	X	X	X
Ketones	X	X	X	X	X
Mercaptans	X	X	X	X	X
Mercury	X				
Nitrogen oxide	X		X	X	X
Phosphine					
Radon		X			
Sulfur dioxide	X	X	X	X	
Toluene		X	X	X	X
Vinyl chloride	X				X
Volatile hydrocarbons	X	X	X	X	X
Xylene		X	X	X	X

11.2.1 CO_2 purity specifications

Different CO_2 purity specifications for beverage, medical, food processing and commercial applications are described by CGA (2004). The specifications for food processing, beverage and medical grade CO_2 are stringent with very low tolerances for various trace impurities. As an example, the beverage grade specification is shown in Table 11.2 (CGA, 2004). Although the specifications for various uses differ somewhat, the purification process is usually the same; only the tests necessary to validate the CO_2 quality are unique for different uses.

Table 11.2 Beverage grade CO_2 purity specification (CGA, 2004)

Component	Specification units in ppm (v/v) unless stated
CO_2	> 99.9% v/v
Acetaldehyde	< 0.2
Ammonia	< 2.5
Benzene	< 0.02
Carbon monoxide	< 10
Carbonyl sulfide	[a]
Hydrogen cyanide	Not detected[b]
Hydrogen sulfide	[c]
Methanol	< 10
Nitric oxide	< 2.5
Nitrogen dioxide	< 2.5
Oxygen	< 30
Sulfur dioxide	[a]
Total sulfur	< 0.1[a]
Total hydrocarbons (as methane)	< 50
Non-methane hydrocarbons	< 20
Non-volatile residue	< 10 ppm (w/w)
Oil/grease	< 5 ppm (w/w)
Phosphine	< 0.3
Total sulfur	< 0.1
H_2O	< 20
Dew point	–55.6°C

[a] If the total sulfur content exceeds 0.1 ppm (v/v) as sulfur, then the species shall be determined separately and the following limits apply:

Carbonyl sulfide	< 0.1 ppm (v/v)
Hydrogen sulfide	< 0.1 ppm (v/v)
Sulfur dioxide	< 1.0 ppm (v/v)

[b] Applies to CO_2 from combustion and coal gasification sources. Current detection level is 0.5 ppm

[c] The use of vapor sample is required for the United States Pharmacopeia

11.2.2 Process description

The industrial CO_2 production process includes compression, warm-end purification, liquefaction and final purification, and an ammonia refrigeration system. Figure 11.1 shows a schematic diagram of the industrial CO_2 production process.

Raw CO_2 compression

Raw CO_2 feed is generally obtained at near ambient pressure from the industrial source. Referring to Fig. 11.1, the raw CO_2 is first sent to a knock-out drum to separate any free liquid. The raw CO_2 is then compressed to 1.4–1.8 bara (20–26 psia) using a blower and then cooled to about 10°C (50°F) in two coolers using

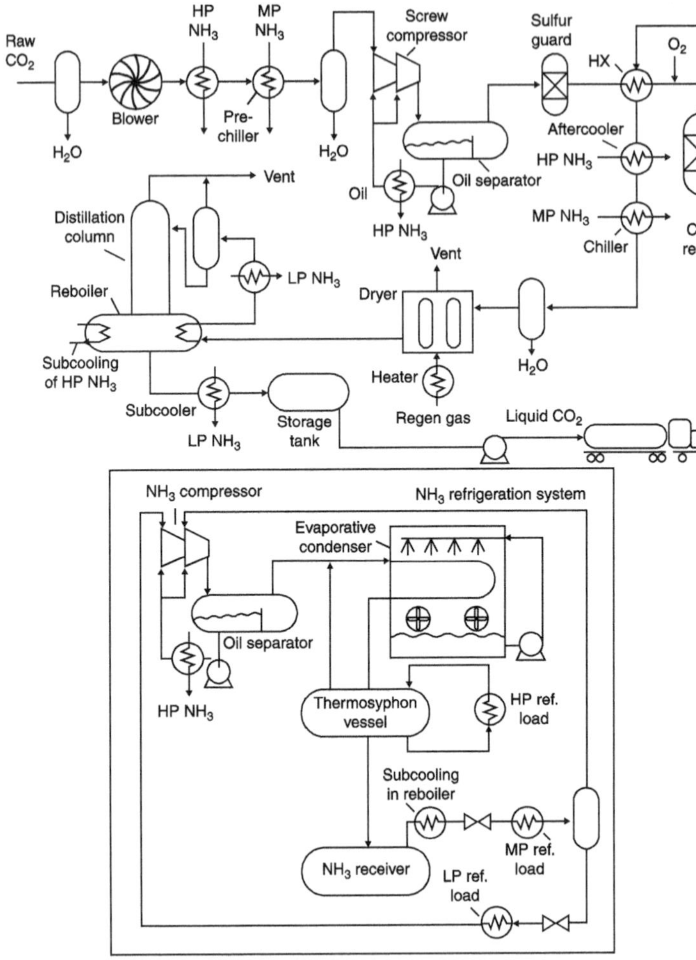

11.1 Process schematics for industrial CO_2 production.

high pressure and medium pressure ammonia streams. A phase separator is used to remove the condensed water. In some cases, instead of using the coolers and a phase separator, a direct contact scrubbing tower is used in which water wash is used to remove any water soluble impurities and particulate matter from the raw CO_2. The raw CO_2 is then compressed to about 23 bara (334 psia) using a two-staged oil flooded screw compressor. In old designs, the gas is cooled between stages to minimize power consumption and keep temperatures within compression equipment below specified limits. In newer designs, a compound machine is used without an intercooler. After the last stage of compression, oil is separated, cooled and recycled to the compressor. A coalescent filter (not shown) is used to reduce oil content in the gas stream to less than 1 ppm (w/w).

Warm-end purification

The compressed CO_2 is processed in different unit operations to remove different impurities. For removal of trace quantities (usually less than 100 ppm) of hydrocarbons and carbon monoxide, a catalytic oxidation (catox) reactor containing platinum catalyst is used. The compressed CO_2 is first passed through a sulfur guard bed to remove any traces of sulfur compounds to protect the catox catalyst. For removal of trace quantities of H_2S, ZnO or metal-impregnated ZnO is used. The non-regenerable ZnO bed is sized to last at least a month before it needs to be replaced. After sulfur removal, the compressed CO_2 is heated in a heat exchanger against the reactor effluent. An electric heater (not shown) is used to further heat the CO_2 to a temperature, typically 150–300°C (302–572°F), at which hydrocarbons and oxygen will react. Oxygen is added to the heated CO_2 and the mixture is introduced into the catox reactor, wherein hydrocarbons are oxidized into CO_2 and water vapors. The hot reactor effluent is first cooled in a heat exchanger against the feed to the catox reactor and then in an aftercooler using high pressure ammonia. The compressed CO_2 is further cooled to 10°C in a chiller using medium pressure ammonia to condense out as much water as possible before drying. A phase separator is used to separate water. The chilled CO_2 is then sent to a dryer.

When a catox reactor is not needed, the compressed CO_2 is cooled after oil removal to separate out any condensed water in a knock-out drum. In some plants, a high pressure water scrubber is used to cool the compressed gas and to reduce its water content, while at the same time removing any water soluble impurities from CO_2. Carbon beds are also employed in some plants when it is necessary to control odor and/or to remove trace quantities of volatile organic compounds.

Usually, drying is the last step in the warm-end purification process. Alumina is commonly used for drying the CO_2 stream. The dryer consists of two alternating vessels of fixed alumina beds with one bed on the feed step while the other is being regenerated. During the feed step, moisture in the gas stream is reduced to less than 10 ppm. For regeneration, a heated regeneration gas is used to drive out moisture from the alumina bed. The bed is cooled before it is used for the feed step by passing a cold regeneration gas through it. CO_2 vapors from the storage tank, vent gas from the distillation column (described later) or air is used as the regeneration gas. The dried compressed CO_2 is sent to the liquefaction and final purification section.

If the raw CO_2 contains larger concentrations of sulfur compounds or hydrocarbons, additional purification steps are necessary. When feed contains a large concentration of H_2S, a process such as Lo-CAT® is used (Merichem, 2010). A liquid redox system in this process removes H_2S from the gas stream in the absorber using a solution containing chelated iron and then the solution is sent to an oxidizer for regeneration and to produce an elemental sulfur. Alternatively, a Sulfatreat process can be used for bulk sulfur removal (Sulfatreat, 2010). The Sulfatreat process uses iron-oxide based granular material to remove hydrogen sulfide from wet gas streams.

If significant quantities of hydrocarbons are present, the sequence of purification steps described earlier will be different. The raw CO_2 from natural wells is usually available at pressure and it contains greater than 1% hydrocarbons. In this case, the high pressure raw CO_2 is first chilled to less than 10°C to knock out any hydrocarbon liquids. The raw CO_2 is then dried to remove moisture. The dried raw CO_2 containing hydrocarbons is then fed to a wash column from the bottom. In a condenser at the top of the column, low pressure ammonia provides refrigeration to condense a small fraction of feed to generate reflux in the column. As liquid descends in the column, it removes hydrocarbons from the rising vapor of raw CO_2. The reflux is controlled such that the liquid recovered from the bottom consists mainly of hydrocarbons. Most of the C2+ hydrocarbons are removed in the wash column, while methane remains in the CO_2 stream. For removal of methane and any other residual hydrocarbons in the raw CO_2 recovered from the top of the column, a catox reactor and associated heat exchangers are used as described earlier. Another dryer unit downstream of the catox reactor is required to remove moisture. The dried raw CO_2 is then sent to the liquefaction and final purification section.

Liquefaction and final purification

Referring to Fig. 11.1, the dried CO_2 stream is processed in this section to liquefy CO_2 and to reduce levels of non-CO_2 components to meet product specifications. An ammonia refrigeration system is used to provide the refrigeration necessary to liquefy CO_2. The dried CO_2 stream is first cooled to –14°C (7°F) against boiling CO_2 in the reboiler associated with the distillation column. It is further cooled to –26.7°C (–16°F) by low pressure ammonia to condense most of the CO_2 contained in the feed. The partially condensed feed stream is separated in a phase separator into liquid and vapor streams. The liquid stream is fed to the top of a distillation column. The column overhead vapor stream and vapor stream from the phase separator are combined to form a vent stream. This vent stream can be cooled using low pressure ammonia refrigeration to condense additional CO_2, which can be fed back to the column for increasing the CO_2 recovery. As the liquid CO_2 descends in the column, more volatile components are stripped off by rising vapors from the reboiler and the purified liquid CO_2 enters the reboiler at the bottom of the column. In the reboiler, the purified CO_2 is partially reboiled to generate vapors, which rise through the column. The heat for reboiling CO_2 is provided by the dried CO_2 stream and high pressure liquid ammonia, both of which get cooled in the process. The purified CO_2 is subcooled to less than –26°C using low pressure ammonia and then throttled to 15 to 17 bara (218 to 247 psia) in the storage tank. The overhead gas stream from the column contains the impurities and CO_2. The vent gas from the column can be used as a regeneration gas for the dryer.

To increase recovery of CO_2 from the process, the column overhead stream can be further processed according to a method described by Howard (1999). In this variation, the column overhead vapor is cooled to –45°C (–49°F) to condense additional CO_2. The condensed CO_2 is separated in a phase separator and then throttled to generate refrigeration. The CO_2-rich liquid is warmed against the column overhead vapor and then recycled to the second stage of the raw CO_2 compressor. The CO_2-lean stream from the phase separator is vented.

Ammonia refrigeration system

The ammonia refrigeration system provides the cooling duties in various heat exchangers in the process. Referring to Fig. 11.1, ammonia vaporized at low pressure is compressed in a two-stage oil flooded compressor to 13.5 bara (196 psia) and passed through a coalescent filter to reduce its oil content to less than 1 ppm. The compressed ammonia is cooled in an evaporative condenser to produce pressurized liquid ammonia. The liquid ammonia flows into a holding tank, from where it is circulated as a coolant in various raw CO_2 and oil coolers as indicated by 'HP NH_3' labels in Fig. 11.1. Some of the modern plants use a thermosyphon design for the holding tank so that the ammonia recirculation is induced naturally. Such design eliminates the need for an ammonia recirculation pump, which could be a potential source of leak. In some plants, instead of using high pressure ammonia as a coolant in various heat exchangers, cooling water or a closed loop glycol cooling system is used. The high pressure liquid ammonia from the thermosyphon vessel flows into an ammonia receiver. The liquid ammonia from the receiver is first subcooled in the column reboiler and then throttled to an intermediate pressure to cool it further. The cooled medium pressure ammonia is used in a pre-chiller and a chiller, as indicated by 'MP NH_3' labels in Fig. 11.1, for cooling raw CO_2 streams to about 10°C. The ammonia vapor generated at medium pressure is recycled to the ammonia compressor. The liquid ammonia from the chillers is throttled to just above ambient pressure to cool it to –29°C. The low pressure ammonia is used to provide refrigeration for condensing CO_2 in the feed and for subcooling the product CO_2 before it is sent to the storage tank, as indicated by labels 'LP NH_3' in Fig. 11.1. The vaporized low pressure ammonia is returned to the ammonia compressor.

The power consumption in the industrial plants will depend on many factors such as capacity of the plant, pressure at which the raw CO_2 is available, concentration of CO_2 in the feed and ambient conditions. A majority of raw CO_2 sources will contain greater than 98% of CO_2 by volume on a dry basis. With this raw CO_2 purity and for a plant with 300 tonnes/day capacity, the power consumption in the industrial plant will vary between 170 and 190 kWh/tonne CO_2.

11.3 Oxy-fuel flue gas CO_2 purification process

While the process for purifying the flue gas from oxy-fuel combustion shares many similarities with the industrial CO_2 production, there are also significant differences between the two processes. The flue gas from the oxy-fuel-fired power plant is expected to contain 75–85% CO_2 on a dry basis. Most likely, it will have to be purified to at least 95% CO_2 purity. A large fraction of the impurities in the flue gas comes from the atmospheric gases. The flue gas will also contain small amounts of SOx, NOx, carbon monoxide, particulate matter, mercury and other trace impurities. The extent of purification needed for oxy-fuel flue gas will depend on whether CO_2 is captured for EOR or for geological storage. The purified CO_2 will have to be compressed to a supercritical state and transported by pipeline to its eventual destination. In the following description, the plant designed to produce CO_2 from the oxy-fuel flue gas is referred to as the CO_2 processing unit (CPU); this term is intended to encompass cooling, compression and purification equipment needed to process the CO_2-rich flue gas exiting the boiler into the purified CO_2 stream ready for pipeline transport.

11.3.1 Requirements for a utility-scale oxy-fuel power plant

For a given capacity of a utility-scale oxy-fuel power plant, the volume of flue gas will depend on the efficiency of the power plant. Higher efficiency plants will consume less fuel and generate less CO_2. Therefore, it will be prudent to design the plant with as high efficiency as possible to minimize the cost of CO_2 capture. Generally speaking, a utility-scale power plant with 500 MW gross output with ultra supercritical steam cycle will generate greater than 8400 tonnes CO_2/day at a full operating rate. A representative oxy-fuel flue gas stream (temperature, pressure and composition) is shown in Table 11.3. (The composition shown in Table 11.3 was estimated with the assumptions listed in Table 11.7.) In addition, it was assumed that the amount of NOx generated is significantly reduced in the oxy-fuel combustion process and the FGD (flue gas desulfurization) unit is used

Table 11.3 A representative oxy-fuel flue gas stream

Pressure	1 atm
Temperature	150°C
Composition by volume:	
CO_2	54.89%
N_2	9.0%
Ar	2.5%
O_2	3.5%
H_2O	30.0%
HCl	100 ppm
NO	200 ppm
NO_2	20 ppm
SO_2	500 ppm
SO_3	5 ppm
Hg	15 ppb
CO	275 ppm
Particulate matter	1.5 mg/Nm^3

for SOx control for a portion of the flue gas recycle stream that is used as the primary air. The composition of flue gas could vary from case to case depending on the type of coal used, oxygen purity, air ingress into the boiler and any provisions for the SOx/NOx control.

11.3.2 CO_2 purity specifications

The CO_2 quality specifications for EOR are defined mainly based on the current practice in the US. Kinder Morgan is a major supplier of CO_2 to EOR operations. Their pipeline CO_2 specifications are listed in Table 11.4 (Hirl, 2009).

Currently, there are no established specifications for CO_2 to be sequestered in geological formations. Any specifications that are developed should take into account the impact of impurities on various components of the CCS (CO_2 capture and storage) chain. Technical and economical feasibility of producing CO_2 with the desired purity, compatibility with the pipeline materials, public acceptance and safety issues for pipeline transportation and the impact on the effectiveness of geological storage must be considered for developing the specifications (SNC-Lavalin Inc., 2004). If atmospheric gases at significant concentrations are left in CO_2, it may not be practical to form supercritical CO_2 (World Resources Institute, 2008). If oxy-fuel flue gas with about 80% CO_2 (by volume) were to be sequestered, the storage capacity requirements would increase significantly, by a factor of two to four according to preliminary estimates by Sass *et al.* (2009). The presence of about 20% non-CO_2 components will also increase the cost of transportation and injection by 10–25% (Sass *et al.*, 2009). It is therefore preferable to have a minimum purity of 95% (by vol.) CO_2. To avoid corrosion in the pipeline, it is desirable to reduce moisture content below the dew point temperature expected in the pipeline. The specifications for CO_2 supplied via pipelines in the US requires moisture to be less than 632 ppm (or 30 lb/MMscf of CO_2 as listed in Table 11.4). However, moisture would not be an issue for the oxy-combustion technology. The purification of CO_2 requires cooling the flue gas close to the

Table 11.4 Kinder Morgan pipeline CO_2 specification (Hirl, 2009)

Component	Specification
CO_2	> 95 mole%
H_2O	< 30 lb/MMscf product
H_2S	< 20 ppmw
Total sulfur	< 35 ppmw
O_2	< 10 ppmw
N_2	< 4 mole%
Hydrocarbons	< 5 mole%

triple point of CO_2 (–57°C) in a cold box, and to avoid freezing at this temperature, moisture would be removed in a dryer unit upstream of the cold box to less than 1 ppm. The specification limits of impurities such as SOx, NOx and mercury must consider public safety in the event of a release of CO_2 from pipeline. Dispersion modeling must be carried out to ensure that their concentrations in the atmosphere near pipelines do not exceed safe human exposure limits. The impact of trace impurities on the injection and storage systems has not been studied rigorously to provide definitive input for the CO_2 purity specifications. A study by Loizzo (2009) indicates that the presence of SO_2 in the saline aquifer can result in the precipitation of anhydrite ($CaSO_4$) and barite ($BaSO_4$), which can lead to reservoir plugging and injectivity reduction. Sulfate could be a problem in carbonate-rich formation, but not in pure sandstones of feldspar-rich formations (Sass *et al.*, 2009). None of these studies define specific limitations on any trace impurities.

In the absence of any established standards for sequestration, many different CO_2 quality specifications have been proposed. Based on the scenarios presented by Kluger and Marion (2008) and Kathar (2009), three CO_2 purity specifications are proposed here in Table 11.5: medium purity, high purity and very high purity. The important constraint in the medium purity specification is related to CO_2 purity (minimum 95%). The high purity specification includes stringent constraints for atmospheric gases (nitrogen, oxygen and argon) and carbon monoxide. In the very high purity specification, more stringent SOx and NOx constraints are also added. It will be shown later how the CO_2 purification process for oxy-fuel flue gas can be modified to achieve these different purities.

Table 11.5 Proposed CO_2 purity specifications for sequestration

	Medium purity	High purity	Very high purity
Composition by volume:			
CO_2	> 95%	> 99.5%	> 99.5%
$N_2 + O_2 + Ar$	< 5%	< 10 ppm	< 10 ppm
H_2O	< 10 ppm	< 10 ppm	< 10 ppm
NOx	< 1500 ppm	< 1500 ppm	< 100 ppm
SOx	< 1500 ppm	< 1500 ppm	< 100 ppm
CO	< 100 ppm	< 10 ppm	< 10 ppm

11.3.3 Process description

The overall processing approach would be the same for producing CO_2 to meet either of the purity specifications listed in Table 11.5. Additional purification equipment will be included to meet the more stringent specifications. The CO_2 compression and purification process encompasses a combination of unit operations and equipment technologies from other industrial gas technologies.

The process leverages experience in producing purified liquid CO_2 from a variety of industrial and natural sources as well as experience in cryogenic industrial gas processes. The warm-end clean-up system of the process is similar to the industrial CO_2 plants in process design and operation, while the sub-ambient temperature purification step uses the cold box equipment that is similar to that used in the cryogenic separation and liquefaction processes such as air separation, hydrogen carbon monoxide separation and liquefied natural gas (LNG) production. An auto-refrigeration based process is used to produce gaseous CO_2 from the cold box. This process does not require the ammonia refrigeration system and thus makes the process simpler while eliminating the need for on-site ammonia storage and avoiding environmental and safety issues associated with ammonia. Elimination of the ammonia refrigeration system allows the process to be operated at much lower temperature (down to –57°C), which is thermodynamically advantageous for CO_2 purification, improving the CO_2 recovery possible in the cold box to approximately 90%. Use of a multi-stream brazed aluminum heat exchanger (BAHX) in the cold box eliminates the need for a series of shell and tube heat exchangers resulting in simpler equipment and compact layout.

Figure 11.2 shows a schematic diagram of the CPU. The major steps in the process include flue gas cooling/condensation, raw CO_2 compression and warm-end purification, cold box purification, product compression and cold box vent stream processing.

Flue gas cooling/condensation

The CO_2-rich flue gas from the boiler at about 150°C is cooled first in the flue gas cooler/condenser. A series of heat exchanger coils are placed in a large duct for cooling the flue gas. In the initial set of coils, low grade heat from the flue gas is

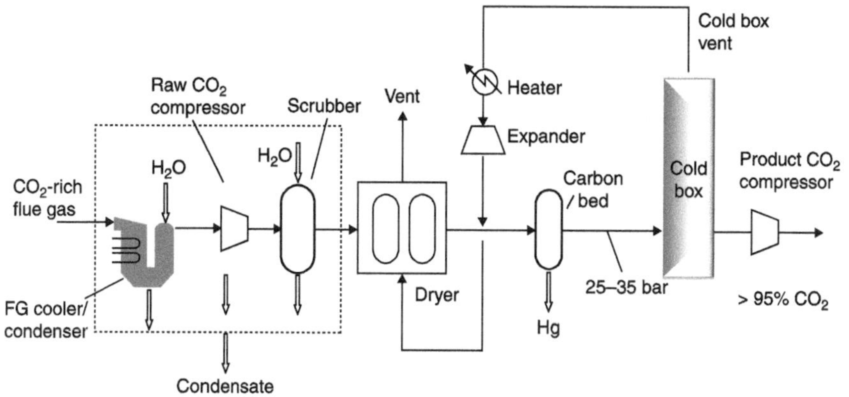

11.2 CO_2 processing unit schematics.

recovered by preheating the boiler feed water. In the final set of coils, cooling water is used for further cooling the flue gas to near ambient temperature. Finally, the flue gas is contacted directly with water in a spray tower to remove any water soluble impurities such as HCl, HF, NH_3, SO_3 and NO_2. It is expected that some of the particulates will also be removed.

Raw CO_2 compression and warm-end purification

The flue gas is then sent to a raw CO_2 compression system where it will be compressed from ambient pressure to 25–35 bara (362.5–507.5 psia). The pressure for purification is selected to minimize the overall power consumption of the process. The pressure required to purify CO_2 increases with the decrease in CO_2 concentration (on a dry basis) in the flue gas. The higher product purity (> 99.5%) requirement also requires higher pressure. For the flue gas stream of Table 11.3, the flue gas will be compressed to about 25 bara. For large volumes of flue gas, a multi-stage centrifugal compressor is used to minimize capital and operating costs. The compression train includes water-cooled shell and tube heat exchangers and condensate knock-out drums after each compression stage. Packaged compression equipment including compressors, intercoolers and knock-out drums is usually supplied by suppliers such as Dresser-Rand (2010), Man Turbo (2010), Atlas Copco (2010) and Mitsubishi (2010).

Within the compression train, some of the trace impurities such as SOx, NOx and Hg will drop out in the condensate collected in the knock-out drums. Within the compression train, some of SO_2 and NO will be oxidized to SO_3 and NO_2. These compounds readily react with water to form sulfuric and nitric acids, which will also be recovered in the condensate. Similarly, any oxidized mercury present in the flue gas will dissolve in the condensate. The presence of water and acidic impurities in the raw gas compression equipment requires stainless steel construction for all parts in contact with the wet flue gas. The compressed CO_2 is cooled in an aftercooler and then fed to another water scrubber to remove any residual water soluble impurities.

The compressed flue gas from the scrubber is fed to a dryer unit to reduce moisture content to less than 1 ppm. At least two beds are used such that one bed is used for drying the feed while the other bed is being regenerated. The regeneration step involves first heating the bed to desorb moisture by passing the heated regeneration gas through it followed by cooling the bed to the feed temperature by passing the cold regeneration gas through. The dried compressed CO_2 is passed through the carbon beds to remove any mercury that could be present in the flue gas. Two beds in a lead–lag arrangement are used. When mercury breaks through from a leading bed, the bed is taken out of service to replace spent carbon with fresh carbon. It is then put back in the service as a lagging bed. Generally, the carbon beds are sized such that the change-out of carbon is no more frequent than once every three months.

Cold box purification

Final purification of the CO_2 takes place in a cold box. The cold box uses a BAHX technology that allows heat exchange between multiple streams and enables very close approach temperatures (1°C). The use of BAHX results in very compact layout as one BAHX can replace several shell and tube heat exchangers. The cold box could include several BAHXs, distillation columns and phase separators. The impurities in the feed to the cold box include atmospheric gases (O_2, N_2 and Ar) at percentage levels and trace impurities (SO_2, NO and CO) at ppm levels. Two different cold box configurations are used, which one depending on the CO_2 purity required.

A schematic of the cold box for producing CO_2 at greater than 95% purity is shown in Fig. 11.3. The design uses a partial condensation process in which the compressed, dry CO_2 stream is cooled in a primary heat exchanger (PHX) to a temperature of –45 to –50°C, at which a majority of the CO_2 condenses. The partially condensed CO_2 stream is fed to a phase separator to separate it into liquid and vapor streams. More volatile gases such as oxygen, nitrogen, argon and carbon monoxide are mostly recovered in the vapor stream. The liquid stream contains greater than 95% by volume CO_2 with the balance mainly comprising atmospheric gases. The purified liquid CO_2 is then throttled to provide refrigeration in the cold box. The throttling is controlled to ensure that the lowest temperature in the cold box does not drop below triple point of CO_2 (–57°C) to prevent solid CO_2 formation. Both the purified CO_2 stream and the vent stream are warmed against the incoming feed stream. The purified CO_2 is recovered at 12 to 13 bara.

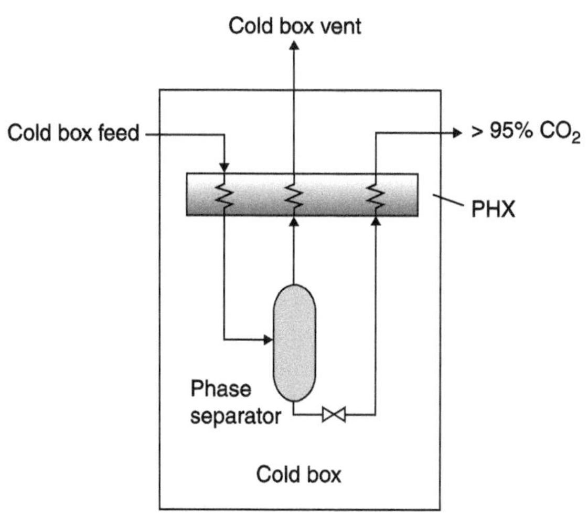

11.3 Cold box for partial condensation process.

Other variations of the configuration shown in Fig. 11.3 are possible for the partial condensation based CO_2 purification. In one variation presented by Shah (2005), a two-stage partial condensation process is used to improve the CO_2 recovery and to reduce the power consumption. In this process, the cold box feed is first cooled to about –30°C to partially condense and separate the purified liquid CO_2 stream. The vapor from the first phase separator is further cooled to –45 to –50°C for a second stage of separation. Two product CO_2 streams can be throttled to two different pressures before warming them in the PHX. Another variation is possible by splitting the purified CO_2 from the single-stage process of Fig. 11.3 into two separate streams and then expanding them to two different pressures. In both of these variations, at least a portion of product CO_2 is obtained at higher pressure (up to 21 bara), thus lowering the power required for product compression. This variation is utilized in the cold box of the high purity case described later.

Greater than 98% of the SO_2 contained in the feed to the cold box remains in the purified CO_2 due to its lower volatility compared with CO_2. The concentration of SO_2 in the cold box vent stream will usually be less than 20 ppm. Another trace impurity, NO, reacts with O_2 in the cold box feed to produce NO_2 and this reaction speeds up as the feed stream is cooled in the cold box. Like SO_2, NO_2 is also less volatile than CO_2 and therefore mostly remains in the purified CO_2 stream. Any NO in the feed that is not oxidized is primarily recovered in the cold box vent stream. A majority of carbon monoxide contained in the cold box feed will also be recovered in the vent stream. The concentration of CO_2 in the cold box vent stream will range between 30% and 40% (by volume) depending on the operating temperature and pressure of the phase separator. The recovery of CO_2 from the cold box will increase with the increase in concentration of CO_2 in the cold box feed.

If greater than 99.5% CO_2 (Table 11.5) is desired, the cold box configuration shown in Fig. 11.4 is used (Shah and Howard, 2010). A distillation column is used in the cold box to achieve more stringent specifications for the atmospheric gases and carbon monoxide. In this process, the compressed and dried feed stream is partially cooled in the main heat exchanger and then it is used to reboil the liquid at the bottom of the stripping column. It is then further cooled in the main heat exchanger until a majority of CO_2 in the feed stream is liquefied. The partially liquefied feed is introduced in the stripping column to produce the desired purity CO_2 from the bottom of the column. Using the column in the cold box, it is possible to achieve the oxygen specification of less than 10 ppm that is needed for EOR. To minimize the power consumption for the product CO_2 compression, the liquid CO_2 product from the column is split into two streams and those streams are expanded to two separate pressures. Both product liquid streams are vaporized and warmed against the feed stream in the heat exchanger and recovered as gaseous purified CO_2 streams at two pressures. The column overhead vapor stream is also warmed against the feed stream in the same heat exchanger and recovered as the cold box vent stream.

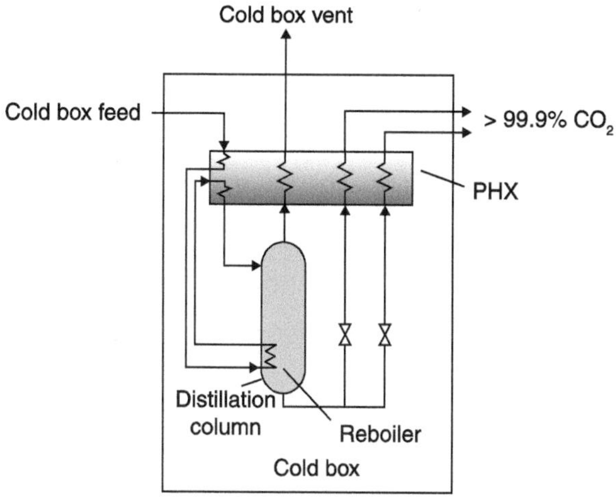

11.4 Cold box for distillation process.

Product CO_2 compression

The purified CO_2 stream from the cold box is compressed in a multi-stage, inter-cooled centrifugal compressor to a pressure of 103.4 bara (1500 psia) or higher. An aftercooler cools the compressed CO_2 stream to near ambient temperature. Since, the purified CO_2 is free of moisture, carbon steel is used as a material of construction for all the equipment and piping downstream of the cold box. When CO_2 is recovered at two pressures, as in the case of the cold box configuration of Fig. 11.4, the lower pressure CO_2 is first compressed to a pressure of the higher pressure CO_2 stream. Both CO_2 streams are then combined and compressed to a final desired pressure for the pipeline transport.

Cold box vent stream processing

The cold box vent stream is obtained at about 24 bara. To recover power, the vent stream is heated and then expanded in an expander. The expanded vent stream is used as a regeneration gas for the dryer unit. The moisture laden vent stream is vented to the atmosphere as the CPU vent stream. The volume of the flue gas from the oxy-fuel power plant is usually one-fourth of the volume of an air-fired power plant. After processing this much smaller volume of the flue gas in the CPU, the volume of the CPU vent stream will be less than 3% of the flue gas emitted from the air-fired plant.

Achieving different CO_2 purities

Table 11.6 shows CO_2 purities achieved by processing the flue gas shown in Table 11.3 in different purification systems. With the CPU process shown in Fig. 11.2,

Table 11.6 Product CO_2 compositions from different purification processes

	Medium purity	High purity	Very high purity
Composition by volume:			
CO_2	96.82%	99.935%	99.994%
$N_2 + O_2 + Ar$	3.11%	9 ppm	9 ppm
H_2O	1 ppm	1 ppm	1 ppm
NOx	170 ppm	165 ppm	29 ppm
SOx	475 ppm	475 ppm	21 ppm
CO	60 ppm	< 1 ppm	< 1 ppm

the medium purity CO_2 is obtained when the cold box shown in Fig. 11.3 is used and the high purity CO_2 is obtained when the cold box of Fig. 11.4 is employed. In estimating these purities, it was assumed that about half of SOx and NOx are removed in various condensate streams collected in the warm end of the process and about 80% of NO is oxidized to NO_2 in the cold box.

To achieve the very high purity CO_2 specified in Table 11.5, two different approaches can be used. In a conventional approach, an SCR (selective catalytic reduction) unit can be installed in the boiler island of the power plant to reduce the NOx in the CPU feed to a level that will result in less than 100 ppm in the final product. Alternatively, an extra distillation column can be used in the cold box to separate NO_2 from CO_2. Similarly, for reducing SOx level in purified CO_2, either an FGD unit can be used after flue gas from the boiler is cooled to about 60°C or a distillation column can be added in the cold box for SO_2–CO_2 separation. There are several new methods proposed for removing SOx and NOx from the compressed flue gas before it is dried. The very high purity CO_2 shown in Table 11.6 is estimated for one of these new methods discussed later in the chapter.

11.3.4 Cost and performance

The cost and performance for a 600 MW power plant with CO_2 capture are presented here. The sensitivity of the CO_2 capture cost to air ingress and CO_2 purity is also presented. Key assumptions of the technoeconomic analysis are adapted from Shah (2007) and listed in Table 11.7. The power plant is based on a supercritical steam cycle with 39% (HHV) efficiency without CO_2 capture. The oxygen purity is assumed to be 95% (by volume). With the assumptions listed in Table 11.7, the flue gas composition shown in Table 11.3 is obtained. The flue gas is purified to medium and high CO_2 purities (shown in Table 11.6) using the cold box configurations shown in Fig. 11.3 and 11.4, respectively. With a partial condensation based cold box (Fig. 11.3), the CO_2 product with 96.8% purity is obtained. For the cold box with a distillation column (Fig. 11.4), the CO_2 product with 99.9% purity is obtained. The purified CO_2 is compressed to 103.45 bara (1500 psia) at the plant battery limit.

Table 11.7 Technoeconomic analysis assumptions

Base power plant capital cost	US$1600/kW net
Net output	600 MW
Base plant efficiency	39% HHV
Operating rate	90%
Coal	PRB (Powder River Basin)
Coal price	$1.5/MMBtu (HHV)
Air leak	3% of wet flue gas
O_2 purity	95%
O_2 in flue gas exiting boiler	3%
Product CO_2 pressure	103.5 bara (1500 psia)
FGD	Partially used for oxy-fuel case
SCR	Shut down for oxy-fuel case
CO_2 pipeline and injection	Included

Table 11.8 shows the comparison between medium and high CO_2 purity processes. Increasing the CO_2 purity from 96.8% to 99.9% has a minimal impact on CO_2 recovery and capture cost. The CO_2 recovery decreases by only about 1.6% while the cost of CO_2 capture goes up by only about 4%. The costs of compressors account for greater than 75% of the CPU cost, while the cold box cost accounts for less than 5% of the CPU cost. For the higher purity case, the pressure of the purified CO_2 from the cold box decreases. As a result, the capital and operating costs of the product compressor increase slightly. The incremental capital cost for the cold box of the high purity case is insignificant. The net impact of the higher capital cost of CPU on the CO_2 capture cost is US$0.9/tonne. The specific power in the CPU increases from 148 kWh/tonne for the medium purity case to 164 kWh/tonne for the high purity case. The increased power consumption for the high purity case increases the CO_2 capture cost by about US$0.8/tonne. Thus, the total cost of CO_2 capture goes up by US$1.7/tonne.

Table 11.8 Comparison between medium and high CO_2 purity processes

Purification process	Medium purity	High purity
CO_2 purity, %	96.8	99.94
CO_2 recovery, %	89.6	88.2
Power consumed in CPU, kWh/tonne CO_2	148	164
CO_2 capture cost, US$/tonne	41.2	42.9

Note: 1 tonne = 1000 kg.

11.3.5 Effect of air ingress

The premise of oxy-fuel combustion is to eliminate inerts from the process and obtain flue gas that contains only CO_2 and water vapor so that pure CO_2 can be obtained by condensing and knocking out the water. Practically speaking though, excess oxygen is required for combustion and air ingress into the boiler cannot be avoided, so the flue gas will contain a significant amount of inerts. A majority of PC (pulverized coal) boilers in the US operate with balanced draft with a slight negative pressure. This causes ingress of air into the flue gas at various locations in the boiler and along the convective path of flue gas. Air ingress is defined as percentage based on the volume of the flue gas exiting the boiler. It can vary from 2% in modern boilers to as high as 15% in the older boilers. The effect of air ingress is shown in Fig. 11.5. The CO_2 recovery decreases significantly with the increase in the air ingress, which results in high CO_2 avoidance costs. At 10% air ingress, the cost of CO_2 avoided is estimated to be US$91/tonne. This is certainly higher than the cost of CO_2 avoided by alternative capture methods such as post-combustion and pre-combustion. For oxy-fuel technology to be a viable retrofit solution for CO_2 capture, it will be important to take steps to reduce air ingress rates in older boilers. In modern boilers designed for oxy-fuel technology, it should be possible to keep the air ingress to a minimum or even consider maintaining slight positive pressure in the boiler to keep air out.

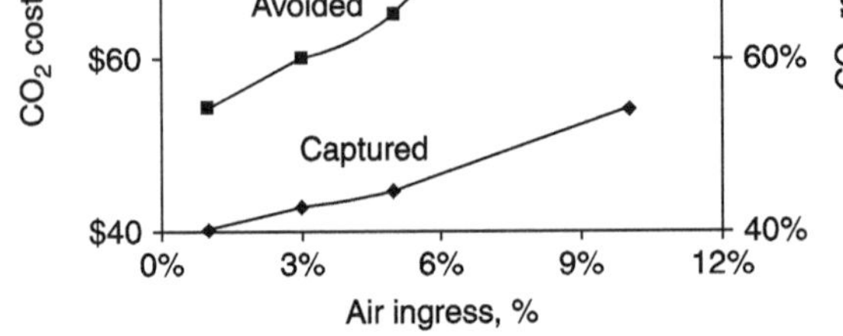

11.5 Effect of air ingress on CO_2 recovery and CO_2 capture costs.

11.4 Recent advances in the oxy-fuel flue gas CO_2 purification technology

The recent advances have focused on development of near-zero emissions CO_2 purification technologies. The emissions of CO_2, SOx, NOx and mercury can be reduced by 99% using these methods. In addition, advances in the CO_2 compressor technology promise to reduce the overall cost of CO_2 capture.

11.4.1 Near zero emissions CO_2 processing unit

Shah *et al.* (2009) have proposed a near zero emissions oxy-combustion CO_2 purification technology as shown in Fig. 11.6. This technology adds two process units to the process shown in Fig. 11.2: one for SOx/NOx removal and one for recovering CO_2 from the cold box vent stream. Two alternative SOx and NOx removal approaches are proposed. The CO_2-lean stream from the VPSA unit is heated and then passed through a catox reactor for converting carbon monoxide into CO_2. The effluent from the catox reactor is expanded to recover power. The expanded CO_2-lean stream is used as a regeneration in the dryer and eventually vented to atmosphere. The vent stream from the near zero emissions CPU of Fig. 11.6 will contain very low levels of CO_2 and criteria pollutants (SOx, NOx, CO, Hg and PM).

This technology can reduce the emissions of CO_2, SOx, NOx, mercury, carbon monoxide and particulate matter (PM) by 99% compared with the emissions from air-fired power plant and produce high purity CO_2 relatively free of trace impurities. This high environmental performance can be achieved with a potentially lower investment for SOx/NOx control compared with that required in the conventional plants. The details of two SOx/NOx removal methods and CO_2 recovery from the cold box vent are described below.

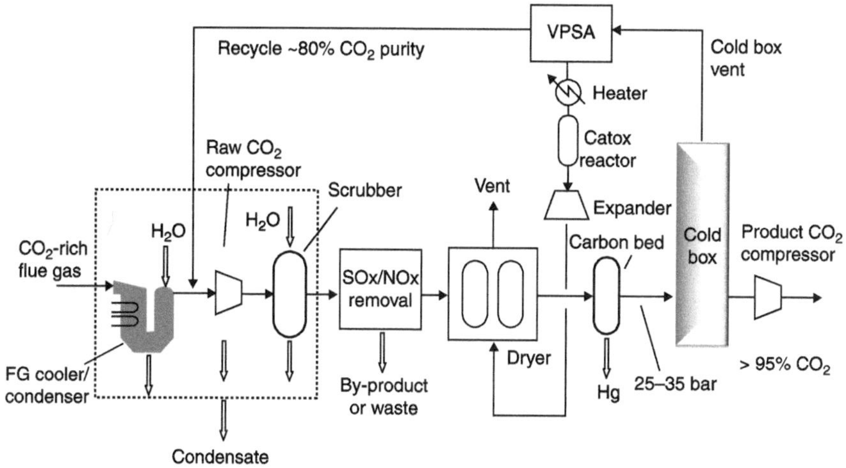

11.6 Process schematics for near zero emissions CO_2 processing unit (Shah *et al.*, 2009). VPSA, vacuum pressure swing adsorption.

11.4.2 Integrated pollutant removal

In a conventional power plant, SOx (sulfur oxides such as SO_2 and SO_3) are removed by the FGD unit. In the FGD unit, CaO or $CaCO_3$ is reacted with SO_2 to form calcium sulfate and/or calcium sulfite. Both capital and operating costs of the FGD unit are high. For the NOx (nitrogen oxides such as NO and NO_2) removal, SCR is used. The SCR unit is also capital intensive. In the oxy-fuel process, since flue gas has to be compressed, it is possible to remove SOx and NOx at high pressure and reduce capital and operating costs.

Two processes described below utilize reactions from the historic lead chamber process (Fairlie, 1936) and nitric acid manufacture process (EPA, 2010). In the lead chamber process, the important reactions are:

$$NO + \tfrac{1}{2}O_2 \rightarrow NO_2 \qquad [11.1]$$

$$NO_2 + SO_2 + H_2O \rightarrow NO + H_2SO_4 \qquad [11.2]$$

$$NOx + H_2SO_4 \leftrightarrow H_2SO_4.NOx \qquad [11.3]$$

For nitric acid production, Reaction [11.4] and Reaction [11.1] are utilized.

$$3NO_2 + H_2O \rightarrow 2HNO_3 + NO \qquad [11.4]$$

In the process described by White *et al.* (2006), the flue gas is compressed to about 15 bara, at which oxidation of NO to NO_2 by Reaction [11.1] will be very favorable. Reaction [11.2] will convert SO_2 into SO_3. The liberated NO from Reaction [11.2] is reused in Reaction [11.1]. Sufficient residence time is provided in a contact tower with a recirculating liquid to allow complete removal of SO_2 from the flue gas according to Reactions [11.1] and [11.2] (White *et al.*, 2006). After all of the SO_2 is converted to H_2SO_4, flue gas will be compressed to about 30 bara, at which NOx will be converted by Reactions [11.4] and [11.1] into nitric acid (White *et al.*, 2006). Another contact tower with recirculating liquid is used to provide sufficient residence time for NOx removal from the flue gas.

The SOx/NOx removal process proposed by Shah *et al.* (2009) is adapted from the catalytic chamber process developed by Tyco Laboratories (Gruber and Walitt, 1970). This process (Shah *et al.*, 2009) also utilizes the reactions from the lead chamber process and nitric acid manufacture for SOx and NOx removal. In addition, the process also incorporates a mercury removal step. The CO_2-rich flue gas is compressed to 25–35 bara and then passed through a series of contact towers. The first step involves mercury removal by sulfuric acid according to the Outokumpu process (Louie, 2005). After mercury removal, the flue gas is fed to a stripper, where it is contacted with the NO_2-rich concentrated sulfuric acid recycled from the downstream unit operations. The flue gas strips off NO_2 from the acid. The NO_2-rich flue gas is then passed on to the SO_2 reactor where Reaction [11.2] will convert SO_2 into sulfuric acid. Greater than 99% of SO_2 is expected to be removed from the flue gas in this reactor. Due to NO_2 recycle, this process does not heavily rely on Reaction [11.1] for high SOx removal efficiency. The flue gas exiting the reactor will still contain high concentrations of NO and NO_2. The NOx compounds will be absorbed by sulfuric acid in the absorber according to Reaction [11.3]. The flue gas that is relatively free of SOx, NOx and mercury is dried, if needed, and then sent to a cold box for producing a purified CO_2 stream.

Sulfuric acid containing NOx (H_2SO_4.NOx) from the absorber and sulfuric acid generated in the SO_2 reactor are sent to the stripper. The NO_2-depleted sulfuric acid from the stripper is treated to remove residual NOx and to produce commercial grade sulfuric acid. The evolved NOx from this step is directed to another reactor for the production of nitric acid according to Reaction [11.4] and the NO generated from this reaction is recycled and mixed with the flue gas entering the stripper.

Another SOx/NOx process described by Shah *et al.* (2009) uses activated carbon for direct oxidation of SO_2 and NO according to Reactions [11.5] and [11.6]. In this process, the flue gas is compressed to 25–35 bara and then sent to an activated carbon bed system comprising at least two beds. One bed processes the flue gas stream while the other bed is being regenerated by washing with water followed by partial drying. The washing step will produce a dilute acid stream containing sulfuric and nitric acids. After SOx and NOx are removed, the flue gas is processed in the dryer, carbon beds and cold box. By using the cold box configuration shown in Fig. 11.4, this process is projected to produce very high purity CO_2 as shown in the last column of Table 11.6.

$$SO_2 + \tfrac{1}{2}O_2 \rightarrow SO_3 \qquad [11.5]$$

$$NO + \tfrac{1}{2}O_2 \rightarrow NO_2 \qquad [11.6]$$

The main benefits of the processes described above are lower capital investment required for SOx/NOx removal and production of higher purity CO_2 that is relatively free of trace impurities. For the process that produces commercial grade sulfuric acid, the costs associated with the disposal of the acid stream are avoided while generating revenue from the sale of the acid.

11.4.3 CO_2 recovery from the cold box vent

The recovery of CO_2 from the cold box decreases with the decrease in concentration of CO_2 in the feed to the cold box. For a boiler with air ingress of 3%, the flue gas is expected to have about 78% CO_2 on a dry basis. With this feed, the recovery of CO_2 is expected to be about 90%. The unrecovered CO_2 escapes in the vent stream from the cold box. The concentration of CO_2 in the vent stream will be between 30% and 40% and the pressure of the vent stream will be 24–34 bara. To improve recovery of CO_2 from the process, an adsorption based technology is proposed (Shah *et al.*, 2009). Referring to Fig. 11.6, the high pressure cold box vent stream is passed through a VPSA (vacuum pressure swing adsorption) unit.

The VPSA unit recovers the CO_2-rich stream at about 80% purity and at near ambient pressure (Kumar, 2009). This stream is recycled and mixed with the raw CO_2 just upstream of the raw CO_2 compressor. Overall CO_2 recovery close to 99% is achieved. The higher recovery increases capital and operating costs of the CPU; however, the capital and operating costs of the ASU (air separation unit) are unchanged. The combined costs of the ASU and the CPU increase by about 5% for the higher recovery case. Since the CO_2 recovery increases by almost 10%, the overall cost of CO_2 capture actually decreases (Shah, 2009). The use of the VPSA in combination with the cold box makes it possible to maintain high CO_2 recovery even when the air ingress in the boiler is high and the CO_2 concentration in the flue gas is low. For example, when air ingress into boiler reaches 10%, the flue gas CO_2 concentration will be 61% and the CO_2 recovery from the cold box alone would be 71%. By employing VPSA along with the cold box, the CO_2 recovery can be increased to 97%. Figure 11.7 shows the benefit of combining VPSA with the cold box for achieving high CO_2 recovery even when air ingress is high.

Another way to recover CO_2 from the cold box vent is to use membranes (White, 2008). The high pressure cold box vent is fed to membranes, which allow CO_2 and oxygen to permeate on the other side. The permeate rich in CO_2 and oxygen is recycled back to the boiler. This approach increases the overall CO_2 recovery to greater than 97% and reduces oxygen requirements from the ASU by about 5% (White, 2008).

11.4.4 Advanced compression and heat integration

The CO_2 compressors will consume 7 to 10% of the gross power generated in the power plant depending on the steam cycle efficiency. In addition, the cost of the

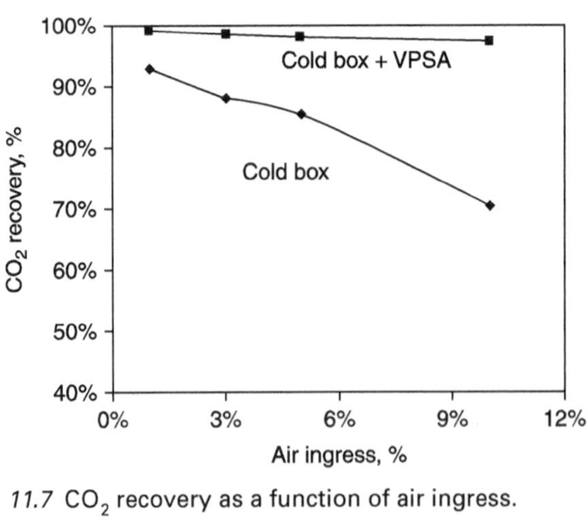

11.7 CO_2 recovery as a function of air ingress.

CO_2 compressors will account for greater than 75% of the equipment costs in the CPU. Improving the compression technology is therefore essential in reducing the cost of CO_2 capture. Ramgen Power Systems is developing a shock wave compressor technology that has a potential to save both capital and operating costs (Ramgen, 2010). Ramgen's compressor technology is based on proven supersonic aircraft technology. The benefits of supersonic compression technology are high efficiency, high single-stage pressure ratios, lower footprint and lower capital costs. By achieving a high pressure ratio per stage, the number of compression stages required is reduced, which results in a less complex system with lower footprint. High pressure ratio also results in discharge temperatures from each stage that are high enough (180–200°C) to make it worthwhile to recover the heat. In a conventional multi-stage centrifugal compressor, the heat of compression is rejected to atmosphere as the discharge temperatures are too low (80–90°C) for recovering the heat for any meaningful purpose in the process. Although the high pressure ratio per stage results in higher power consumption, the recovery of heat of compression for the steam cycle can overcome this deficit and improve the net efficiency of the power plant, if the efficiency of this novel compressor is sufficiently higher than that of the conventional compressor. It is this author's estimate that the adiabatic efficiency for each stage of the novel compressor needs to be at least 5% higher than that of the conventional compressor for it to improve the efficiency of the power plant.

11.5 Environmental performance of oxy-fuel power plant

Since the entire flue gas stream is processed in the CPU to produce a purified CO_2 stream, the vent stream from the CPU to atmosphere is very small in volume and most of the criteria pollutants are removed in the process as liquid or solid waste resulting in near zero atmospheric emissions. Table 11.9 shows compositions of

Table 11.9 Vent streams from different CO_2 purification processes

Purification process	Medium purity	High purity	Very high purity
Composition by volume: (on a dry basis)			
CO_2	36.8%	30.0%	4.3%
N_2	38.2%	42.0%	57.6%
O_2	14.3%	16.3%	22.1%
Ar	10.5%	11.6%	16.0%
NOx	76 ppm	85 ppm	25 ppm
SOx	15 ppm	10 ppm	1 ppm
CO	1152 ppm	1282 ppm	2 ppm
Hg	0	0	0

the CPU vent streams obtained from different processes. The medium and high purity cases correspond to the overall process shown in Fig. 11.2 and the cold box configurations shown in Fig. 11.3 and 11.4, respectively. The process for very high purity CO_2 corresponds to the overall process of Fig. 11.6, activated carbon process for SOx/NOx removal (described earlier) and the cold box of Fig. 11.4. Compared with air-fired operation, the emissions of CO_2, SOx, NOx and Hg are reduced by 99% or more for the very high purity case. Thus, the oxy-fuel CO_2 purification process promises to be a truly near zero emissions process.

11.6 Future trends

Although the scale of existing industrial plants is relatively small, much of the experience gained from building and operating these CO_2 plants will be applicable for designing the much larger scale CPUs required for oxy-fuel power plants. Experience in construction materials selection, trace contaminant removal and analysis, adsorbent/impurity interactions, thermodynamic behavior of CO_2 mixtures and plant control philosophy can be applied to CPUs. The experience gained from designing a wide range of complex cold boxes for cryogenic industrial gas processes will also be applicable for designing the cold boxes needed in the CPU. However, before a larger scale oxy-fuel power plant is built, technology demonstration is required at a scale sufficient to provide knowledge specific to the purification of oxy-fuel flue gas and generate parameters for design and engineering of the larger scale plants. The demonstration of operability of an integrated plant comprising air separation unit, oxy-fuel fired boiler and CPU is also needed.

There are several ongoing or planned oxy-fuel demonstration projects in the world. Vattenfall is operating a 30 MW_{th} pilot oxy-fuel demonstration project that includes the CPU (Thebault *et al.*, 2009). The CO_2 purification plant design is similar to that of an industrial CO_2 plant. The test results show significant removal of SOx and NOx in the condensate collected from the raw CO_2 compression train. At the Vattenfall site, another smaller scale (slip stream equivalent to 1 MW_{th}) demonstration of a CO_2 purification process designed for the oxy-fuel plant is planned (White and Fogash, 2009).

The Callide oxy-fuel demonstration project in Australia is planning to install a 75 tonnes/day CO_2 plant for purifying flue gas with specific plans to better understand the fate of mercury and NOx (Spero, 2009).

The CIUDEN project in Spain is planning to install a CPU that is a scaled down version of the design necessary for a utility-scale power plant (Lupion *et al.*, 2009). The flue gas in the CIUDEN project will be obtained from either a 20 MW_{th} PC boiler or a 30 MW_{th} circulating fluidized bed (CFB) boiler (Lupion *et al.*, 2009).

Much larger scale demonstration projects are planned in Europe. Vattenfall has proposed a 250 MW_e oxy-fuel project at their Jänschwalde station in Germany to be operational by 2015 (Stromberg, 2009). Endesa and CIUDEN are planning on a 300 MW_e oxy-CFB project in Compostilla, Spain to be operational by 2015 (Lupion *et al.*, 2009). As the demonstration projects get larger, the CO_2 purification technology used in these projects is likely to be based on an auto-refrigeration process that does not use the ammonia refrigeration system used in the smaller industrial CO_2 plants. The technology will also evolve towards achieving near zero emissions. Successful demonstration of an integrated plant at 250–300 MW_e scale between 2015 and 2020 will lead to commercial deployment of a utility-scale (> 500 MW_e) plant after 2020.

11.7 Conclusions

The purification of CO_2-rich flue gas from the oxy-combustion process combines the experience gained in the purification of CO_2 from industrial sources and the cryogenic industrial gas separation and liquefaction processes. Recent advances in the purification technology hold promise to attain near zero emissions from the oxy-fuel power plant while producing very high purity CO_2 relatively free of trace impurities. With these developments, it will be possible to recover up to 99% of CO_2 and reduce emissions of SOx, NOx, particulates and mercury through the vent stack by 99% or more in comparison with air-fired operation. The demonstration of the CO_2 purification unit and its integrated operation with the oxy-fuel boiler and air separation unit is essential to ensuring that the oxy-fuel technology will be ready for commercial deployment by 2020.

11.8 Acknowledgements

The author would like to thank colleagues Abbey Bacak and Neil Prosser, both of Praxair, Inc., for providing assistance in preparation of this chapter.

11.9 References

Atlas Copco (2010), available from: http://www.atlascopco-gap.com/ [accessed 20 April, 2010].

CGA (2004), '*Commodity Specification for Carbon Dioxide*', CGA G-6.2-2004, 5th edition, Compressed Gas Association.

Dresser-Rand (2010), 'Datum® Centrifugal Compressors', Available from: http://www.dresser-rand.com/literature/turbo/85188-09-DATUMsm.pdf [accessed 20 April, 2010].

EPA (2010), 'Nitric Acid', available from: http://www.epa.gov/ttn/chief/ap42/ch08/final/c08s08.pdf [accessed 15 April, 2010].

Fairlie, A M (1936), *Sulfuric Acid Manufacture*, New York, Reinhold Publishing Corporation.

Garvey, M D (2009), 'Carbon Dioxide – A Market on the Move', *Cryogas International*, May 2009, 28–31.

Gruber, A and Walitt, A (1970), '*Development of the Catalytic Chamber Process*', Final Report, Contract No. CPA 70-59, US Environmental Protection Agency.

Hirl, M J (2009), 'Project Lincoln – An Assessment of CO_2-EOR Development Opportunities in the State of Illinois', An Illinois Department of Commerce and Economic Opportunity Funded Project, Grant No. 08-483004, June 2009.

Howard, H E (1999), 'Carbon Dioxide Production System with Integral Vent Gas Condenser', US Patent 5,927,103. 27 July, 1999.

Kathar, A (2009), 'CO_2 Quality and Other Relevant Issues', paper presented at 2nd Working Group Meeting on CO_2 Quality, Cottbus, Germany, 7 September, IEA GHG.

Kluger, F and Marion, J (2008), 'Alstom Development of Oxy-fuel PC and CFB Power Plants', paper presented at 3rd Workshop of the International Oxyfuel Combustion Research Network, Yokohama, Japan, 5–6 March, 2008, IEA GHG.

Kumar, R (2009), 'Process and Apparatus to Recover Medium Purity Carbon Dioxide', US Patent 7,618,478 B2. 17 November, 2009.

Loizzo, M (2009), 'Oxyfuel Flue Gas, Steel and Rock Implications for CO_2 Geologic Storage', paper presented at 2nd Working Group Meeting on CO_2 Quality, Cottbus, Germany, 7 September, 2009, IEA GHG.

Louie, D K (2005), *Handbook of Sulphuric Acid Manufacture*, 1st edition, Ontario, Canada, DKL Engineering Inc.

Lupion, M, Navarette, B, Otero, P and Cortes, V (2009), 'CIUDEN CCS Technological Development Plant on Oxy-combustion in Coal Power Generation', paper presented at the 1st IEA Oxyfuel Combustion Conference, Cottbus, Germany, 8–11 September, 2009, IEA GHG.

Man Turbo (2010), 'RG – Integrally Geared Compressors', available from: http://www.mandieselturbo.com/files/news/filesof12079/MTM_Produktblatt_RG_1018_e.pdf [accessed 20 April, 2010].

Merichem (2010), 'Lo CAT® Applications', available from: http://www.gtp-merichem.com/products/lo-cat/applications/index.php [accessed 20 March, 2010].

Mitsubishi (2010), available from: http://www.mhi.co.jp/en/products/detail/compressor.html [accessed 20 April, 2010].

Ramgen (2010), 'Ramgen's Low-Cost, High-Efficiency CO_2 Compressor Technology', available from: http://www.ramgen.com/apps_comp_unique.html [accessed 16 April, 2010].

Salof (2010), available from: http://www.salofrefrigeration.com/ [accessed 20 March, 2010].

Sass, B M, Farzan, H, Prabhakar, R, Gerst, J, Sminchak, J, Bhargava, M, Nestleroth, B and Figueroa, J (2009), 'Considerations for Treating Impurities in Oxy-Combustion Flue Gas Prior to Sequestration', *Energy Procedia*, 1, 535–542.

Shah, M M (2005), 'Capturing CO_2 from Oxy-Fuel Combustion Flue Gas', paper presented at the Inaugural Workshop of Oxy-Fuel Combustion Network, Cottbus, Germany, 29–30 November, 2005, IEA GHG.

Shah, M M (2007), 'Oxy-Fuel Combustion for CO_2 Capture from New and Existing PC Boilers', paper presented at Electric Power 2007 Conference, Chicago, IL, USA, 1–3 May, 2007, Electric Power.

Shah, M M (2009), 'CO_2 Processing Unit (CPU) for Oxyfuel Flue Gas', paper presented at 2nd Working Group Meeting on CO_2 Quality, Cottbus, Germany, 7 September, 2009, IEA GHG.

Shah, M M and Howard, H E (2010), 'Carbon Dioxide Purification Method', US Patent 7,666,251 B2. 23 February, 2010.

Shah, M, Kumar, R, Degenstein, N and Zanfir, M (2009), 'Near-Zero Emissions Oxy-Combustion Flue Gas Purification', paper presented at the Annual NETL CO_2 Capture

Technology for Existing Plants R&D Meeting, Pittsburgh, PA, USA, 24–26 March, 2009, NETL.

SNC-Lavalin Inc. (2004), '*Impact of Impurities on CO_2 Capture, Transport and Storage*', Report Number PH4/32, IEA GHG R&D Programme, UK.

Spero, C (2009), 'Callide Oxyfuel Project – Status and Development', paper presented at the 1st Oxyfuel Combustion Conference, Cottbus, Germany, 8–11 September, 2009, IEA GHG.

Stromberg, L (2009), 'Oxyfuel – The Way Forward and the Drivers', paper presented at the 1st Oxyfuel Combustion Conference, Cottbus, Germany, 8–11 September, 2009, IEA GHG.

Sulfatreat (2010), available from: http://www.sulfatreat.com [accessed 23 March, 2010].

Thebault, C, Yan, J, Jacoby, J and Anheden, M (2009), 'Behaviors of NOx and SOx in CO_2 Compression and Purification Processes', paper presented at the 1st Oxyfuel Combustion Conference, Cottbus, Germany, 8–11 September, 2009, IEA GHG.

Toromont Energy Systems (2010), available from: http://www.toromontsystems.com/ [accessed 20 March, 2010].

Union Engineering (2010), available from: http://www.union.dk/ [accessed 20 March, 2010].

White, V (2008), 'Purification of Oxyfuel Derived CO_2', paper presented at 3rd Workshop of the International Oxyfuel Combustion Research Network, Yokohama, Japan, 5–6 March, 2008, IEA GHG.

White, V and Fogash, K (2009), 'Purification of Oxyfuel Derived CO_2: Current Developments and Future Plans', paper presented at the 1st Oxyfuel Combustion Conference, Cottbus, Germany, 8–11 September, 2009, IEA GHG.

White, V, Allam, R and Miller, E (2006), 'Purification of Oxyfuel-Derived CO_2 for Sequestration or EOR', paper presented at the 8th International Conference on Greenhouse Gas Control Technologies, Trondheim, Norway, 19–22 June, 2006.

Wittemann (2010), available from: http://www.pureco2nfidence.com/launch/ [accessed 20 March, 2010].

World Resources Institute (2008), '*CCS Guidelines: Guidelines for Carbon Dioxide Capture, Transportation and Storage*', World Resources Institute, Washington, DC.

Part III

Advanced oxy-fuel combustion concepts and developments

12
Direct oxy-coal combustion with minimum or no flue gas recycle

H. KOBAYASHI and L. E. BOOL, Praxair, Inc., USA

Abstract: Oxy-coal combustion without flue gas recycle can potentially simplify the boiler design and improve the efficiency. In this chapter prior work on oxy-fuel combustion without flue gas recycle is reviewed first. Key technical issues and design considerations are discussed next. The chapter then presents several boiler design concepts to address technical issues.

Key words: oxy-fuel, oxy-coal, combustion, boiler, flue gas recycle, zero recycle, CO_2.

12.1 Introduction

Oxy-fuel combustion produces much higher adiabatic flame temperatures than air-fuel combustion. In order to convert an existing air-fired boiler to oxy-coal firing flue gas recirculation (FGR) is required, as discussed in other chapters. In fact all of the proposed oxy-coal boiler conversion projects are based on FGR to control the flame temperature and to maintain proper balancing of heat transfer to different parts of the steam cycle. High adiabatic flame temperature and the resulting high heat flux are often erroneously cited as the reason for the FGR requirement. As discussed in the following sections, the actual reason is the configuration of the existing air-fired boiler design, i.e., the radiant furnace section followed by a convective section, which has been optimized in the last century in order to maximize the steam cycle efficiency for air firing. For the construction of a new oxy-coal fired boiler, there are no fundamental heat transfer issues requiring FGR in the boiler design. In fact, there are over 1000 industrial furnaces that have been successfully converted from air-fuel firing to full oxy-fuel firing without FGR.[1] They are typically high temperature process furnaces such as steel reheating, glass melting, aluminum melting, copper melting and hazardous waste incineration furnaces. Precise control of furnace temperature and heat flux profiles in these process furnaces is often more critical than for utility boilers. Both high flame temperature oxy-fuel burners and advanced oxy-fuel burners utilizing internal FGR were successfully applied. The heat flux profile was often improved after the oxy-fuel conversion by selecting a proper burner design and optimizing the burner placement and the firing rate distribution without any changes in the furnace geometry. These commercial examples have clearly demonstrated that oxy-fuel combustion can reproduce the desired temperature and heat flux profiles in any air-fuel fired high temperature furnaces.

An oxy-coal firing system that does not require flue gas is often called 'zero recycle oxy-coal combustion'. However, safety and operational concerns may require some recirculated flue gas to transport the pulverized coal into the boiler. Therefore a more accurate expression is 'near zero recycle oxy-coal combustion'. In order to be viable, the boiler and burner designs must mitigate the potential drawbacks associated with high flame temperatures and high pollutant concentrations. For example the design must manage the heat flux profile to avoid extreme material temperatures and slagging and corrosion problems. The design would also need to be able to handle high concentrations of pollutants such as SOx and particulates that could potentially lead to high corrosion and slagging/fouling rates. Finally, the significant equipment and design modifications required to mitigate these potential problems make it likely that near zero recycle boilers will be new build rather than retrofits.

12.2 Prior work on near zero flue gas recycle oxy-fuel fired boilers

There are only a few examples of design studies and pilot scale testing in the literature for oxy-fuel firing of boilers without FGR. An industrial scale oxy-heavy oil fired boiler demonstration project was conducted by a Japanese consortium in 1993–2000.[2,3] The goal was to develop a high performance boiler by combining oxy-fuel firing and a condensing heat exchanger to achieve a boiler efficiency of 105% LHV (98.9% HHV) and low NOx emissions. A pilot scale boiler was built without FGR and successfully tested. Based on the experience, a 17 tpd steam industrial scale boiler demonstration plant, consisting of a saturated steam boiler (38,000 lb/h at 355 psig) without FGR, an integrated PSA oxygen generation system driven by a back pressure steam turbine, and a condensing heat exchanger, was built and operated. The new boiler was more compact and the furnace volume was reduced to about 70% of the conventional air-fired boiler by taking advantage of the high temperature oxy-fuel flame and the consequent high heat fluxes available. The main boiler design modification required was to accommodate the change of the heat absorption ratio between the furnace and the convective banks. Since most of the heat available in the oxy-fuel combustion is in the high temperature furnace zone, the furnace heat absorption and the convective bank heat absorption were designed to be about 85% (vs. ~60% for a typical air-fired boiler) and 10% respectively. An additional 10% heat absorption was designed for the downstream economizer and condensing heat exchanger. The boiler was fired by a single 40 MMBtu/h oxy-heavy oil burner at a volumetric heat release rate of about 63,000 Btu/h/ft^3. Although the burner was designed to produce a typical high temperature oxy-fuel flame, the heat flux distribution over the water walls was well controlled. At full load the average and the maximum local heat fluxes measured were about 120,000 and 165,000 Btu/h/ft^2 respectively. Although higher heat fluxes resulted in somewhat higher tube surface temperatures, the increases were well within the temperature limit of the conventional tube material used for this industrial boiler design.

This project demonstrated the feasibility of oxy-heavy oil firing without FGR for a saturated steam industrial boiler. However, the result cannot be simply applied for large utility boilers with high superheat temperatures such as a supercritical steam boiler. Since the normal steam temperature was about 436°F, boiler tubes were able to take a very large heat flux without exceeding the tube temperature limit. Industrial boilers are also much smaller than modern utility boilers, which makes the ratio of the heat transfer surface area to the furnace volume larger. Consequently the average heat flux to the boiler walls is reduced for a given firing density. The small flame size in the middle of the furnace facilitated uniform radiation heat fluxes to all boiler walls without local overheating. The furnace heat release rate, which is defined as the total energy input (i.e., the fuel input plus preheated air energy) divided by the total available flat heat sink area, of a typical dry bottom coal fired utility boiler is in a range of 60,000 to 100,000 Btu/h/ft^2 and the furnace exit gas temperature (FEGT) is in a range of 1900 to 2300°F.

In another study the conversion of a natural gas fired industrial boiler to oxy-fuel firing was analyzed with and without FGR for potential CO_2 capture applications for petroleum refineries.[4] The aim of the study was to establish the value of the boiler design without FGR such as boiler compactness and additional efficiency gains. The base boiler selected for the study was rated at 213 MWt to produce superheated steam at 1350 psig and 950°F. The conceptual design of a new oxy-fuel boiler without FGR was based on the steam generator arrangement similar to the conventional boiler, but the boiler volume was reduced and the cross-sectional area of the flue gas passage for the convective section was reduced to account for the reduced flue gas volume. With such a design heat fluxes to the furnace waterwalls in excess of 325,000 Btu/ft^2/h appeared possible. High heat fluxes reduced the required heat transfer area, but necessitated the tube material to be upgraded to a more expensive ferric alloy material used in large supercritical utility boilers. A tube metal temperature rise over the steam temperature in excess of 200°F was estimated under the condition described.

Although a cursory analysis indicated that the boiler would operate in the nucleate boiling regime, a more detailed analysis of the circulation system was recommended to confirm the finding. The small available flow of the combustion products required the installation of superheater surface in a much higher temperature zone. The superheater was designed with gas inlet temperature of approximately 3000°F, which necessitated selection of a higher strength alloy for the high temperature superheater tubes. The net power output of the plant without FGR improved about 3% over the FGR plant mainly due to improved heat recovery in the condensing heat exchanger. Although higher tube material costs of the no-FGR boiler partly offset the cost savings from the smaller boiler size, the study showed somewhat lower overall capital costs, electricity costs and CO_2 capture cost for the zero recycle concept.

More recently oxy-natural gas and oxy-coal firing were tested in an existing 15 MWt industrial packaged boiler with and without flue gas recirculation.[5,6] The boiler was designed to produce up to 60,000 lb/h with mild superheat (135 pisg, 454°F). Local heat flux to a boiler tube in the side wall about 30 inches from the burner wall was estimated using measurements by thermocouples implanted to the tube under different firing conditions. Highest heat fluxes, estimated by the thermocouple measurement at the specific location, for the high flame temperature oxy-natural gas burner without FGR were in a range of 30,000 to 40,000 Btu/h/ft^2 as compared with 10,000–15,000 Btu/h/ft^2 measured for air-natural gas firing at about 30 MMBtu/h. (Note: the furnace average heat fluxes per flat projected waterwall surface area were not reported.) The boiler efficiency was increased from 88.4% for air-firing to 94.8% for oxy-natural gas firing without FGR. No damages to boiler refractory materials and tubes were observed. Preliminary screening of boiler samples showed no adverse effect in terms of slagging and fouling after oxy-coal firing. This project also demonstrated the feasibility of high temperature oxy-fuel firing without FGR. As discussed previously for the Japanese project, however, the results from industrial boilers should not be considered as a demonstration of technical feasibility for large utility boilers.

12.3 Design considerations for near zero flue gas recycle

Over a hundred years of power plant development has resulted in highly complex and optimized boiler designs to maximize the steam power cycle efficiency. These designs require careful attention to the heat absorption pattern and the heat flux distributions. This is particularly true in coal-fired boilers where mineral matter in the coal can lead to extensive slagging, fouling and submicron ash formation depending on the flame temperature. In order to design a near zero recycle oxy-coal boiler it is critical to understand how oxy-coal combustion impacts these critical design criteria.

One of the most important design parameters is the heat absorption pattern in the boiler. The proportion of heat absorbed to preheat the boiler feedwater to the boiling temperature, evaporate the water, and superheat the steam must be carefully controlled to achieve the required steam characteristics. This heat absorption pattern, coupled with temperature limitations of tube materials, tends to define conventional boiler designs. For example, high radiative heat flux from the flame zone is used to evaporate boiler feedwater preheated in the economizer. The relatively low temperature of the water and high heat transfer in boiling tubes keeps the tube temperature within allowable limits. The FEGT is then cool enough that heat flux to the superheat tubes is reduced. The lower heat flux in the convective section allows tube temperature limits to be avoided, even with high steam temperatures and relatively low steam side heat transfer coefficients.

Given the importance of matching the heat flux to the local water/steam conditions it is critical to understand how near zero recycle oxy-coal combustion impacts heat transfer. It is well known that radiative heat transfer from pure oxy-coal flames is much higher than equivalent air-coal flames. This increased heat transfer is due to a combination of higher flame temperature and increased gas emissivity by the high concentrations of emissive gases (CO_2 and H_2O) in the flue gas. In pure oxy-fuel combustion the nitrogen or recirculated flue gas is eliminated and the gas does not need to be heated to furnace temperature. This leads to substantially higher flame temperatures. Adiabatic flame temperature of a typical bituminous coal at different oxygen enrichment levels in air is shown in Fig. 12.1. With ambient temperature air the adiabatic flame temperature is about 3300°F, which is increased to about 3600°F with a typical air preheat temperature of 600°F. The top line represents the adiabatic flame temperature that would be calculated using conventional heat capacity correlations. These extreme temperatures are not seen in practice due to dissociation of flue gas components in the flame. This dissociation binds potential thermal energy (sensible heat) as chemical energy. As the oxygen concentration increases more energy goes into dissociation, limiting the maximum available flame temperature at about 5000°F with pure oxygen.

High oxy-fuel flame temperature raises special concerns for coal combustion in boilers due to increased melting and evaporation of ash components, which would increase slagging and fouling of boiler tubes. For example, kaolinite, a common mineral species in coal, melts at 3245°F. Air-coal flames can exceed this temperature under specific conditions, but oxy-coal flames can routinely exceed

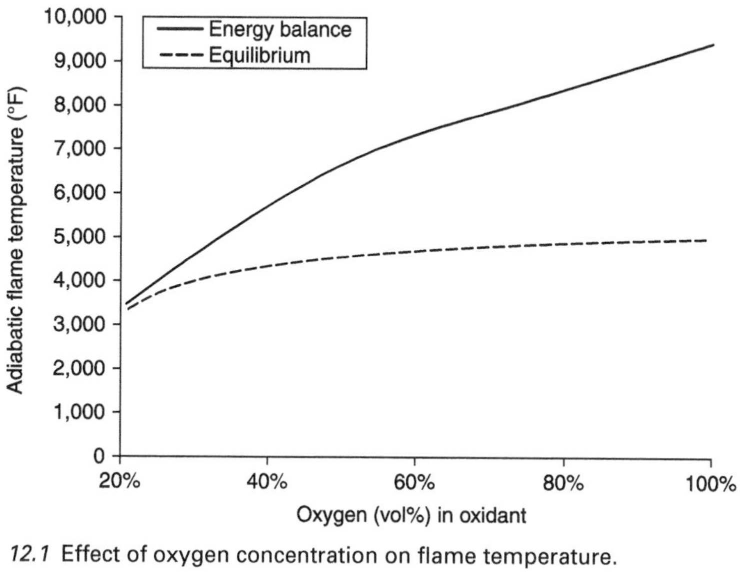

12.1 Effect of oxygen concentration on flame temperature.

this temperature. In addition, the elimination of nitrogen under near zero recycle oxy-coal conditions increases the concentration of fly ash in the gas by approximately three-fold. This increase in concentration, coupled with reduced furnace volume, leads to a potential increase in deposition rates. An equilibrium study was performed using the NASA CET86 equilibrium code to evaluate ash vaporization from a US bituminous coal under air-coal and near zero oxy-coal combustion conditions. This study suggested the amount of ash vaporized under these oxy-coal conditions was more than seven times that under air-coal conditions. As vaporized ash species condense they often form submicron fume, which may be difficult to capture in the air pollution control equipment. Finally, the conversion to oxy-fuel will increase the concentration of corrosive species, such as H_2S, which forms under fuel rich conditions. Waterwall corrosion has been shown to increase with H_2S concentration and tube temperature.[7] Therefore extra care needs to be taken to mitigate corrosion under oxy-fuel conditions.

For reliable commercial operation it is highly desirable to keep the boiler thermal conditions for near zero recycle oxy-coal combustion within the range of the well proven boiler designs for air firing. The key design parameters include the peak boiler tube surface temperature, the peak heat flux, and the peak flame and gas temperatures. FEGT should also be kept similar to those in conventional air-fired boilers to control the thermal conditions of the convective pass. As mentioned previously, many industrial furnaces have been converted to full oxy-fuel firing without FGR. Furnace temperature and heat flux profiles as well as FEGT were kept the same without changing the furnace geometry. The basic design principles developed for these conversions are reviewed briefly below as they are also applicable to the furnace section of the near zero recycle oxy-coal fired boiler.

Figure. 12.2 compares the heat available or absorbed in a furnace for hot air-coal and oxy-coal firing as a function of FEGT. At FEGT=2000°F, about 58% of

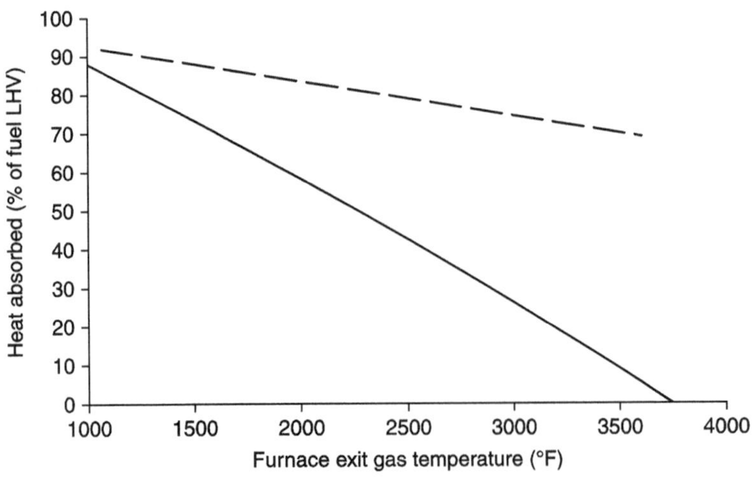

12.2 Heat absorption and furnace exit temperature.

heat input is absorbed in the furnace under air-firing and about 84% is absorbed under oxy-firing. This difference is due to the sensible heat contained in the nitrogen under air firing, which is eliminated under oxy-firing. To maintain the same heat absorption and the same FEGT in the furnace the fuel input under oxy-firing needs to be reduced to 69% (=58/84) of air-firing from overall heat balance. This fuel reduction is the main benefit and the motivation for many oxy-fuel conversions of industrial furnaces.

In order to reproduce the temperature and heat flux distribution of the original air-fired furnace two different design approaches have been developed. The first approach was to design an oxy-fuel burner that exhibits the same flame characteristics as the air burner replaced. Advanced low flame temperature oxy-fuel burners utilizing high velocity oxygen jets and internal flue gas recirculation were developed in the early 1980s and their flame characteristics were demonstrated to match those of equivalent air burners.[8] With these burners it became possible to replace existing air burners with new oxy-fuel burners to reproduce the same heat flux pattern without changing the furnace design. The second approach was to use high flame temperature oxy-fuel burners and control the furnace heat flux distribution by proper selection of oxy-fuel burners, burner placement and the firing rate distribution. Computational fluid dynamics (CFD) simulation was often used to optimize the burner configurations in order to improve the heat flux distribution over the existing air fired condition. Over 300 glass melting furnaces have been successfully converted to oxy-fuel firing without FGR in this fashion.[9]

In a boiler, however, only a portion of the heat is transferred in the furnace, and the balance is transferred in the convective pass. The residual heat in the oxy-coal flue gas at 2000°F is not sufficient to satisfy the 'convective' duty in a conventional air-fired boiler. A design option to satisfy the overall boiler heat requirement is to add a second oxy-coal fired furnace specifically designed to provide heat for superheaters and reheaters. Another design option is to expand the furnace to provide additional furnace zones for radiatively heated superheaters and reheaters. The oxy-coal firing rate for the expanded furnace has to be increased to satisfy the additional heat loads in this case. Since both options will increase the overall boiler size and the cost, design innovations are needed to reduce the overall boiler volume. The two design options and potential design concepts to reduce the furnace size are discussed in the following sections.

12.4 Separate fired chambers for different steam circuits

One way to balance the heat transfer distribution between boiling and superheating/reheating with near zero oxy-coal combustion is to separately fire different steam circuits in two or more combustion chambers. For example, a concept with three parallel chambers, each with a small convective economizer or feedwater heater section, for boiling, superheating and reheating was proposed.[5] This method guarantees that the appropriate amount of heat is transferred for each part of the steam circuit. The main chamber would be designed like a traditional furnace to take water to a specific temperature and pressure. For a subcritical boiler this would be the evaporative section where liquid water is evaporated to form saturated steam at design pressure. In a supercritical or ultrasupercritical boiler this would be a lower temperature supercritical water section. As discussed further on, in section 12.6, the average heat flux could be increased significantly by more uniformly distributing heat flux in the main chamber without overheating the tubes. The ability to handle a higher average heat flux without exceeding the peak flux limits would allow the furnace size to be minimized. The superheater and reheater chambers would also be designed like the traditional radiant furnace with steam tubes placed on all walls. The heat release rate has to be reduced to control the heat flux to superheater or reheater tubes within the traditional limits. As needed, internal recirculation in the furnace driven by the burner design could be used to control the peak heat flux. An advantage of the parallel chamber configuration is a reduced fan power requirement due to the low flue gas side pressure loss through the small parallel convective sections. A drawback of this concept, however, is the large superheater and reheater chamber volumes required to reduce the heat flux.

An alternative design concept is the separate fired combustion chambers connected in series. In a two-chamber configuration flue gas from the main 'boiling' chamber, after being cooled in an optional convective section, goes to the second superheater/reheater chamber where additional burners are used to transfer heat to the steam. The cooled flue gas from the main chamber plays the same role as the externally recirculated flue gas and the peak flame temperature is reduced. In fact, model calculations suggest that the superheater/reheater chamber would have flame temperatures closer to the conventional air-fired boiler. These lower peak flame temperatures, coupled with careful furnace design, would enable proper control of heat flux to superheater/reheater tubes. Finally the flue gas from the superheater/reheater chamber passes through a convective bank to preheat boiler feedwater and cool the flue gas to a desired temperature. An advantage of this design is a smaller superheat/reheat chamber volume as compared with the parallel configuration. The fan power requirement, however, will increase somewhat due to the increased flue gas volume in the superheater/reheater chamber.

12.5 Furnace with controlled radiant heating of superheaters and reheaters

Radiation heated pendant type superheaters and reheaters are commonly used in the conventional air-fired boiler design. They are placed near the furnace exit above the nose section of the furnace in order to limit the heat flux to steam tubes. Although the same arrangement can be adopted in the oxy-coal fired boiler, the space available above the nose section in the traditional boiler design is too small to accommodate for the large number of superheater and reheater tubes. Two furnace design modification concepts are proposed below without significantly increasing the size of the overall furnace volume. The first concept is to locate superheater and reheater tubes in the high heat flux zones of the furnace and to use 'screen tubes' to partially shade the tubes from the pure oxy-fuel flame. This concept is shown schematically in Fig. 12.3. The first row of tubes consists of traditional water-boiling tubes that can withstand high heat fluxes. Superheater and reheater tubes are placed behind the screen tubes in the second row. The fraction of the flame radiation that passes through the screen tubes can be well controlled by adjusting the spacing of the screen tubes and the distance between the first row and the second row.[10] The ratio of the heat flux to superheater/reheater tubes to screen tubes is proportionally turned down upon firing rate changes, which is an important benefit of this design. Thus, careful design of the screen tubes can control distribution of heat absorbed by superheater and reheater tubes located in the high heat flux zones of the furnace and keep the steam tube temperature within the allowable limit.

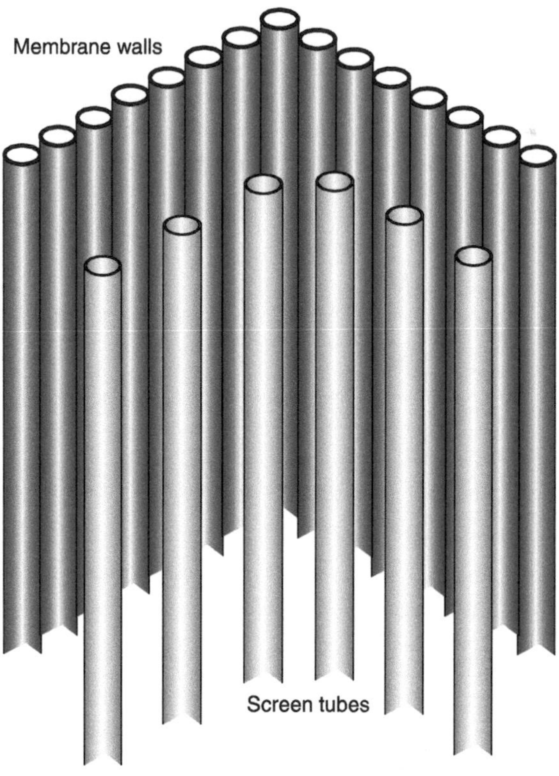

12.3 Schematic of screen tube boiler concept.

The second concept is to place some superheater/reheater tubes below the first row of burners. In a conventional boiler the heat flux to the tubes below the first row of burners, such as the hopper section, is often much lower than the heat flux in the burner region. By redesigning the bottom section of the furnace a large number of superheater and reheater tubes could be placed in this section with controlled heat fluxes from the combustion zone above. The overall heat fluxes to the top and bottom zones of a boiler furnace can be further controlled by changing the aspect ratio of the furnace. For example a tall furnace with a narrow width would reduce the radiant heat flux from the middle burner zone by reducing the view factor.

12.6 Furnace with distributed firing

In Fig. 12.4 gas temperature profiles of air-coal and oxy-coal combustion are compared at the same firing rate to illustrate the effects of installing near zero recycle oxy-coal combustion in a conventional boiler at the same firing rate. The predicted temperatures are approximate as a simplified radiation zone model was used to calculate the furnace thermal condition in the radiative section of a conventional 300 MW opposed wall fired boiler. The peak zone gas temperatures for air and oxygen firing are about 3300°F and 3800°F respectively in this example. The corresponding heat flux absorption profiles are shown in Fig. 12.5 together with a hypothetical distributed firing case. The calculated peak heat flux for oxy-firing, about 290,000 Btu/h/ft^2, is more than triple that for air-firing of about 80,000 Btu/h/ft^2. In the oxy-firing case the high heat flux and the small

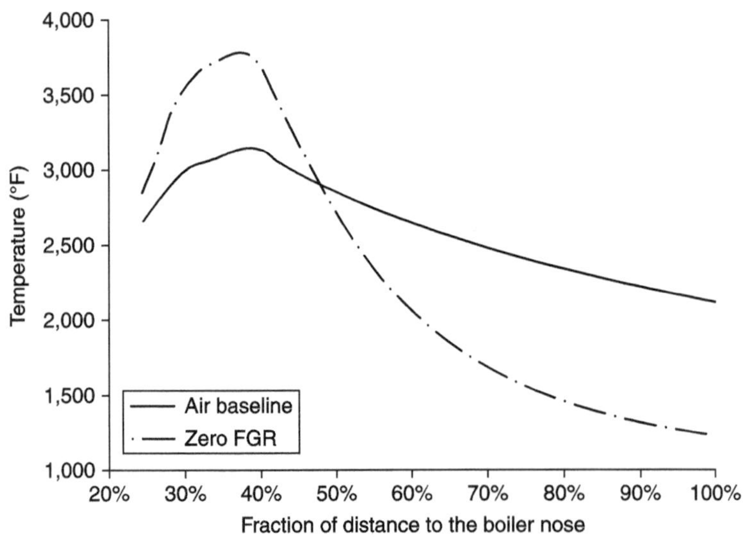

12.4 Estimated flame temperature in a conventional boiler for air and near zero combustion.

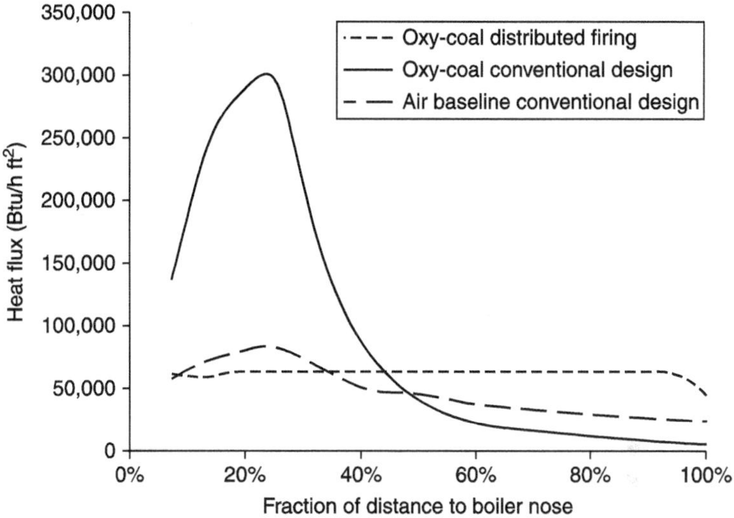

12.5 Estimated heat flux distribution – distributed firing concept.

flue gas volume leads to the flue gas rapidly cooling and the FEGT becomes substantially below that of the air-firing case. Clearly the thermal condition of this illustrative oxy-coal firing design is not practicable in a real boiler.

One of the most effective ways to minimize the furnace size is to make the heat flux distribution very uniform throughout the furnace. In many furnaces the materials selection and furnace arrangement are based on the peak heat flux that creates the highest tube surface temperature. In the air baseline heat flux distribution shown in Fig. 12.5 the furnace zones above and below the burner zones have much lower heat fluxes. The peak heat flux is about 30% higher than the average heat flux. Only a small portion of the furnace actually sees that elevated heat flux due to cooling of the furnace gasses as they rise in the furnace. Therefore the surface area of much of the furnace is essentially being underutilized. To illustrate the benefit of the distributed firing concept a hypothetical example of installing burners along the entire furnace length was modeled. The firing rate was reduced to 70% of the original oxy-fuel case to match the overall heat absorption to that of the baseline air case. The resulting heat flux distribution, as shown in Fig. 12.5, is very uniform and about 23% below the peak heat flux of the baseline air case. Hence the overall firing rate for the distributed firing case can be increased by the same amount without exceeding the peak heat flux observed in the baseline air case. Conversely a significant reduction in the furnace size is feasible if the same firing rate is maintained.

In the conventional wall-fired boiler the location of the top row of burners is set by the requirements to provide a sufficient gas residence time for char burnout and to provide the space for the introduction of overfire air for NOx control. These requirements could be relaxed substantially for direct oxy-coal firing. The small physical size of oxy-coal burners makes it easier to place them in desired locations, including in the walls of the bottom hopper section of the furnace. Since the char burnout rate becomes faster and the gas residence time is increased three-to-four-fold under oxy-coal combustion, the top row oxy-fuel burners could be placed at a much higher elevation without increasing unburned carbon in char.

Beside the additional burners and their placements, the burner design and firing strategies could also be used to generate a more uniform heat flux. One such strategy is to drive internal flue gas recirculation by high momentum oxy-fuel burners.[11] Furnace gasses in cooler zones are pulled into high velocity oxygen jets by the turbulent jet entrainment and reduce the flame temperature, and therefore the local heat flux. High furnace gas recirculation generates a more uniform temperature and heat flux distribution in the furnace. Another approach is to extend the heat release along the length of the furnace by deeply staging the burners in the burner zone. Under these fuel rich conditions only a portion of the heat is released in this zone. As supplemental overfire oxygen is fed at locations along the furnace more heat is released. This method can also control NOx formation from fuel-bound nitrogen, but may increase the slagging and corrosion potential in the reducing atmosphere zone of the furnace.

12.7 Furnace with multiple partition walls

Figure 12.6 depicts a furnace concept with multiple partition walls to increase the heat sink areas. In this example a large furnace is divided into four compartments using three vertical partition walls. The total waterwall area of the furnace with partition walls is increased to more than twice the original furnace while maintaining the same overall furnace volume. As a consequence, the average wall heat flux of the furnace with the partition walls can be reduced to less than half of the original furnace at the same total firing rate if the FEGT is kept constant. It is a powerful design option to control the average heat flux to walls. This design is in essence a modular furnace design concept where four small furnaces are combined as a single furnace. The modular design enables the demonstration of an oxy-coal fired furnace in a smaller scale and makes the scale-up to a large boiler easier.

There are other advantages. In this example each compartment has a larger height to width ratio, which reduces the radiant heat fluxes to the top and bottom zones and facilitates the placement of pendant type superheater and reheater tubes. Each flame is placed in the same thermal environment surrounded by furnace side walls, which would reduce the peak flame temperature. Once a flame has been optimized for uniform heat flux, high carbon burnout and low NOx emission, the same burner setting can be used for all other burners. In a conventional wall-fired furnace tuning of individual burners is very difficult as

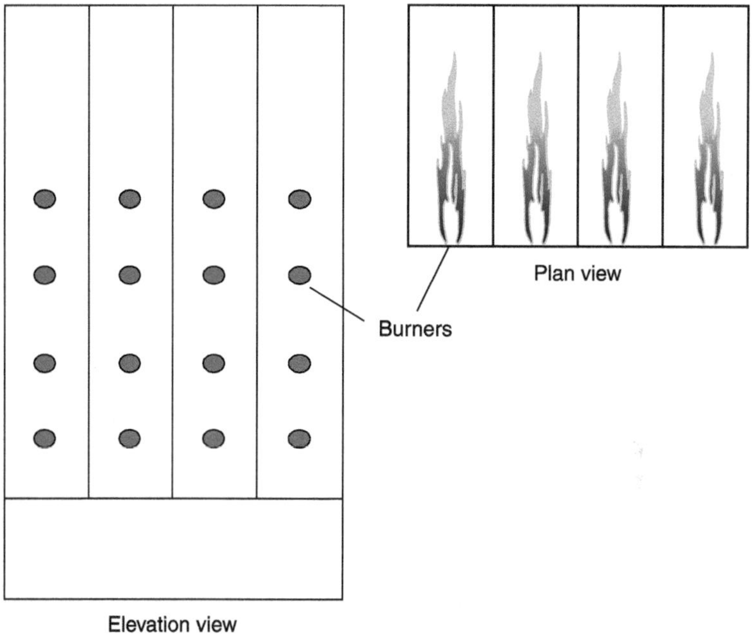

12.6 Schematic of partition wall design.

middle flames are surrounded by other flames and operate at a higher flame temperature than the flame adjacent to a side wall.

12.8 Conclusion

Oxy-fuel combustion with zero flue gas recycle has been demonstrated in many industrial furnaces, including two industrial boilers, without requiring major furnace design changes. Applications for utility boilers with high steam superheating, however, require modifications of the current boiler design. Due to the small flue gas volume the convective section has to be made much smaller and most of the superheater and reheater steam circuits have to be relocated and heated by radiation either in a separate furnace or within the expanded main furnace. Several furnace design concepts were presented to heat superheater and reheater tubes by radiation without exceeding the conventional heat flux limits while taking advantage of the high heat flux available from oxy-coal combustion. These concepts suggest that the new oxy-coal fired boiler could be made significantly smaller than the conventional boiler. The concentrations of ash particles and all gaseous species, however, will increase three-to-four-fold, which may significantly affect slagging, fouling and corrosion behaviors in the furnace. They require careful attention and further studies. As is the case with any new development, more detailed analyses, testing and optimization are required to develop a commercially reliable oxy-coal fired boiler with near zero flue gas recirculation.

12.9 References

1 Kobayashi, H. and Prasad, R., 'A Review of Oxygen Combustion and Oxygen Production Systems', Proceedings of Forum on High Performance Industrial Furnace and Boiler, 8–9 March, 1999, Tokyo, Japan, pp.74–1 to 74–8.

2 Taniguchi, H., 'Research and Development of Very High Performance Boiler with Oxygen Combustion and Latent Heat Recovery from Exhaust Gas', ASME International Joint Power Conference, New Orleans, 4–7 June, 2001.

3 *R&D Accomplishments from High Performance Boiler Development Project*, Final Report (in Japanese), New Energy and Industrial Technology Development Organization (NEDO) and The Japan Society of Industrial Machinery Manufacturers (JSIM), March 2000.

4 Boden, J., Palkes, M. and Thompson, D, 'A study on CO_2 Capture from a Gas-Fired Boiler by Oxyfuel Combustion without Flue Gas Recycle', The 2001 Joint AFRC/JFRC International Combustion Symposium, 10–13 September, 2001, Kauai, Hawaii.

5 Schoenfield, M., presentation at the IEA-GHG First OxyFuel Combustion Conference, Cottbus, Germany, 10 September, 2009.

6 Ochs, T., Brian, P., *et al.*, 'Oxy-Natural gas Firing of Jupiter Oxygen Oxy-Fuel Test Facility', 34th International Technical Conference on Coal Utilization and Fuel Systems, 1–4 June, 2009, Clearwater, Florida.

7 Paz, N.A., Plumly, A.L., Chow, O.K. and Chen, W., 'Waterwall Corrosion Mechanisms in Coal Combustion Environments', *Materials at High Temperatures*, Vol. 19, No. 3, 2002.

8 Kobayashi, H., Silver, L.S., Kwan, Y. and Chen, S. L., 'NO_x Emission Characteristics of Industrial Burners and Control Methods Under Oxygen Enriched Combustion Conditions,' 9th Members Conference, International Flame Research Foundation, Noordwijkerhout, The Netherlands 24–26 May, 1989.

9 Kobayashi, H., 'Advances in Oxy-Fuel Fired Glass Melting Technology', Proceedings of XX International Congress on Glass, 26 September–1 October, 2004, Kyoto, Japan.

10 Hottel, H.C. and Sarofim, A.F., *Radiative Transfer*, McGraw-Hill, Inc., 1967.

11 Kobayashi, H. and Tsiava, R., 'Oxy-Fuel Burners', Chapter 21 of *Industrial Burners Handbook*, pp 693–723, CRC Press, 2003.

13

High pressure oxy-fuel (HiPrOx) combustion systems

B. CLEMENTS, R. POMALIS, L. ZHENG and T. HERAGE,
CanmetENERGY, Natural Resources Canada, Canada

Abstract: High pressure oxy-fuel (HiPrOx) is a derivative of basic oxy-fuel technology that has the potential to overcome some issues associated with ambient pressure oxy-fuel systems. It has application both in the power generation and the industrial sectors. Its development and interest have been sparked by requirements for CO_2 capture and storage. This chapter describes the motivation and efficiency benefits of using pressure and technical approaches that are currently being considered. The central thrust of this chapter is to present the key concepts and focus on coal-fired Rankine and Brayton cycle power systems.

Key words: advanced power cycles, carbon capture and storage, direct contact steam generation, gasification, high pressure combustion, HiPrOx, *in situ* heavy oil production, steam assisted gravity drainage.

13.1 Introduction

Oxy-fuel combustion has been identified as a key technology for greenhouse gas (GHG) mitigation because it is capable of producing a concentrated CO_2 stream suitable for sequestration. This technology, as applied to pulverized coal Rankine cycle power systems, has recently been brought to a near-commercial status with several worldwide demonstrations. A barrier to its adoption has been the large additional auxiliary power required for oxygen production and CO_2 compression which results in low overall system efficiency.

High pressure oxy-fuel (HiPrOx) is a derivative of basic oxy-fuel technology that has the potential to overcome some of the issues associated with ambient pressure oxy-fuel systems. It has application both in the power generation and the industrial sectors. Its development and interest have been sparked by requirements for CO_2 capture and storage (CCS).

This chapter describes the motivation and efficiency benefits of using pressure as well as various technical approaches that are currently being considered in the development of these systems. The central thrust of this chapter is to present the key concepts and focus on coal-fired Rankine cycle power systems. These basic concepts are then extended to Brayton cycle power systems as well as industrial thermal processes.

13.2 Rankine cycle power systems

13.2.1 Coal-fired power systems

Existing coal-fired Rankine cycle power generation systems that use ambient pressure air-fired boiler technology are in common use throughout the world. The typical efficiencies (based on high heating value) of these systems are approximately 34% for subcritical cycles and 39% for supercritical cycles. With increased temperature and pressure on the steam side and other process enhancements, these cycle efficiencies are gradually increasing. Limitations on peak cycle efficiency are mostly due to material constraints. By today's standards, these systems typically have no control of CO_2 emissions and simply exhaust CO_2 into the environment. Although a number of technologies have been identified to address the GHG control problem, there are large energy penalties associated with each method for capturing CO_2 from these power generation stations.

Only a few basic configurations of thermal power generating systems are commonly used throughout the world. Within the power generation sector, this commonality of equipment and the sector's overall contribution to the GHG problem have made it an obvious choice for early technology development of CO_2 capture solutions such as oxy-fuel combustion.

CCS constraints for these processes mean that additional system requirements now exist that result in significant efficiency penalties. The additional requirements include production of concentrated CO_2 streams, removal of impurities (particulate matter, SO_2 and NO_x emissions) from that stream, and pressurization of the stream to a typical pipeline pressure of 110 bar or to a pressure suitable for sequestration. Some flexibility to these considerations exists based on the application for the captured CO_2, the distance from the source to the sequestration site, and pipeline constraints.

As a result of recent climate change concerns associated with GHG emissions from the power generation sector, three technology directions that show promise in capturing and sequestering CO_2 from coal-fired stations have become the focus of research and development. These technologies are oxy-fuel combustion, gasification, and post-combustion capture using chemical solvents.[1] Among these, oxy-fuel combustion is viewed to be a relatively simple solution that can be applied to either new boiler systems or the retrofit of existing installations.[2] Many of these basic technology concepts are also applicable for adaptation into the industrial sector for CO_2 control.

13.2.2 Ambient pressure oxy-fuel power systems

The basic problems associated with ambient pressure oxy-firing are the lower efficiencies, and higher capital and operating costs due primarily to the need for an air separation unit (ASU) and a CO_2 product recovery train (PRT). Within the power generation sector this leads to higher costs of electricity and the associated reluctance of many utilities to adopt this method for the economic generation of power.[3]

Cryogenic air separation is a mature technology and can meet the oxygen requirements in terms of capacity and purity for an oxy-fuel system. Most studies conclude that the optimal oxygen purity for oxy-fuel combustion is approximately 95%. Oxygen production imposes the largest energy penalty on the oxy-fuel system. Generally, the energy required to produce oxygen for ambient oxy-fuel system is about 200 to 220 kWh/t oxygen.[4] This means that up to 17% of the gross output of a power generation station is needed for oxygen production.[3,5]

The typical CO_2 purity for enhanced oil recovery (EOR) application is approximately 95%, although this can vary considerably. A multi-stage compression and refrigeration system is required[6] to purify, compress, and liquefy the final flue gas consisting mostly of CO_2 from an oxy-fuel system. The energy consumption of the PRT is very intensive due to the nature of the compression and refrigeration operations. The amount of energy required by the PRT can be as much as 10% of the gross output of a power generation system.[3,5]

The basic configuration of an oxy-fuel power system consists of an ASU supplying oxygen to the combustion system. Use of oxy-firing results in very high flame temperatures and, therefore, moderation of flame temperature is usually required. Moderation of the flame temperature within the furnace using flue gas recirculation (FGR) is a widely accepted method to achieve this control. In this manner, flame temperatures are maintained similar to those typically encountered in air-fired systems. This, in turn, results in similar radiation heat fluxes to the furnace walls. As a result, using this technique, oxy-firing can be accommodated within typical furnaces using traditional cooling rates and materials within the furnace walls.

This basic philosophy as applied to a typical power generation cycle is shown in Fig. 13.1. To attain control of main and reheat steam temperatures it is necessary to control the total heat transfer within the superheaters and reheaters. This can be accomplished by adjusting the FGR rate to obtain conditions matching both the furnace outlet temperature and mass flow through the heat transfer sections[7] with those of an air-fired system. The furnace temperature and mass flow must be adjusted using a single control variable (that is, the flue gas recycle rate); since the mass flow and temperature tend to be interrelated, adequate control becomes possible. Thus, with the various boiler systems studied, appropriate operating conditions can be attained within the control limits of the boiler systems.[7,8]

Ultimately, the flue gas is cooled to a state where the moisture is condensed. In preliminary configurations of an ambient pressure oxy-fuel system, the heat from the condensed moisture was considered for recuperation and use elsewhere within the system to improve the overall efficiency. Although a substantial amount of heat is available from this source, the temperature of this condensate is approximately 50°C to 55°C making it too low-grade for most useful applications within the system. Consequently, in these configurations this heat is simply

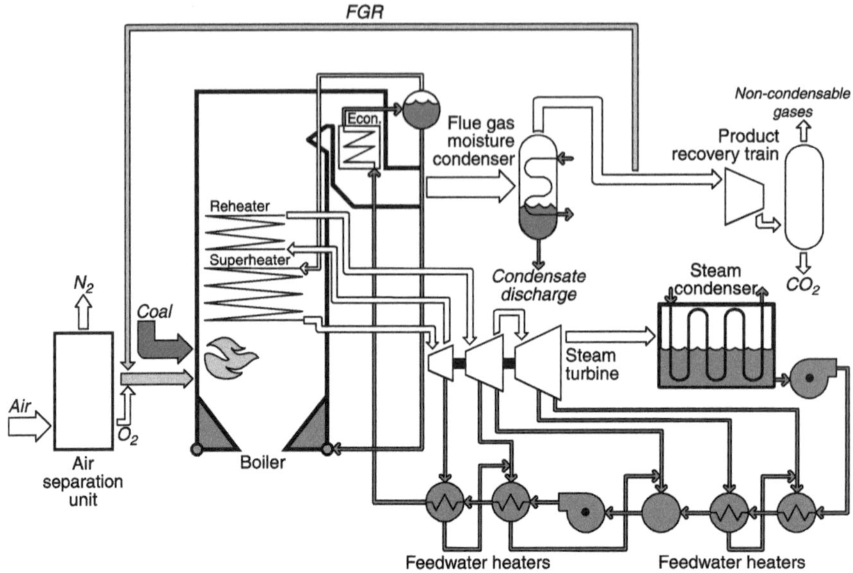

13.1 Ambient pressure oxy-fuel system.

rejected and the attainable thermal efficiency is very similar to air-fired systems where the moisture in the flue gas is not condensed.[7]

Infiltration air has a significant impact on the cost of the oxy-fuel system. Most air-fired furnaces are operated slightly below ambient pressure. For typical boiler systems it is estimated that roughly 3% of the total combustion air leaks into the boiler through various openings in the casing. Consequently, it is essential to minimize air infiltration in an oxy-fuel system to increase the flue gas CO_2 concentration and to reduce the energy penalty of the PRT associated with compression and separation of oxygen, nitrogen, and argon.

It was initially thought that ambient pressure oxy-fuel systems might hold some promise of furnace size reduction and reduced capital cost. This was due to the higher heat transfer rates attainable with increasing furnace temperatures and smaller volumetric flows of flue gas due to the absence of nitrogen. Although these avenues have not been exhaustively studied, a number of concerns about reducing furnace size became quickly evident.[9]

One major problem associated with furnace size reduction is that it presents difficulties in operating the furnace easily with air. The ability to operate these systems using air-firing was considered a big advantage and it would take a substantial reduction in capital cost to compensate for an inability to air-fire the systems.

Another issue exists primarily within the power generation sector: many existing power plants have traditionally been subcritical in design and the need for increased cycle efficiency has led power utilities to purchase supercritical units for new installations. Supercritical boilers, due to their higher steam pressures, have higher water temperatures within the waterwalls. The once-through nature of supercritical designs results in much less circulating water and a concomitant lower capability to cool the furnace walls. These reduced coolant flow and increased temperature conditions require the use of high grade materials within the furnace walls. This means that supercritical boilers are already limited by the maximum operating temperature of existing materials. Ultrasupercritical furnace designs offer cycle efficiency benefits but require even higher material grades that, as yet, have not had wide-spread commercial application. Many consider the thought of increasing flue side temperatures within the furnace impractical because it would increase the material requirements and costs with only minor cycle efficiency benefits. It is generally felt that as higher grade materials become commercially available they could be used as vehicles to increase steam temperatures and pressures, resulting in direct Rankine cycle efficiency gains.

Smaller furnaces would require handling ash related issues associated with various coals and this could mean developing and adopting completely new designs based on wet slag furnace systems or dry ash systems (fluidized beds). Such an approach is commercially risky compared with the experience and breadth of knowledge that has been gained over the years with traditional pulverized fuel furnaces.

Certainly, within the power generation sector, ambient pressure oxy-fuel systems have reached the demonstration phase; however, experience with these systems is still somewhat limited. In Canada, the USA and Europe, major developments have been underway for the demonstration of this technology at the industrial level. In 2008, Vattenfall Europe AG successfully converted a 30 MW_{th} unit in Germany to oxy-fuel operation with CO_2 capture. Jupiter Oxygen Corporation of the United States has started to retrofit a 22 MW_e at Orrville, Ohio for oxy-fuel operation as well.

13.2.3 High pressure oxy-fuel power systems

Due to the low efficiency and high auxiliary power consumptions associated with ambient pressure oxy-fuel systems various organizations have been exploring methods to increase the system efficiency and reduce capital costs. A very promising concept is the use of pressurized combustion. The use of pressure increases the power requirement of the air separation system; however, pressurization also increases the process efficiency while decreasing the power consumption of the CO_2 compression system. This, in turn, decreases the overall auxiliary power consumption with a net improvement in system efficiency. The use of pressure will also affect the performance and size of each piece of equipment within the system.

13.3 Uses of pressure in power systems

13.3.1 A brief history

Throughout history, many examples of the use of high pressure combustion systems exist in the fields of rocketry, internal combustion engines, and gas turbines. The common element within these systems is that they use a working fluid within a thermodynamic cycle which is first pressurized in its cold state, heated by means of chemical reaction of the fuel and oxidant (combustion), and finally expanded to create thrust or other mechanical work.

The most relevant of these systems for discussion within this chapter are within the realm of turbine engines using Brayton cycles and especially those adaptations suitable for use with solid fuels. The thermal efficiency of these cycles improves with increasing turbine inlet temperature. The maximum to minimum pressure ratio within the cycle affects both the efficiency and the maximum work output of the cycle and is therefore optimized for each specific application.[10] Typical maximum practical operating pressures for gas turbines are as high as 42.5 bar.[11]

Most turbine applications are open systems oriented towards gaseous and liquid fuels; however, some experience exists in expanding the applicability to solid fuels such as coal. The major technologies which allow this solid fuel direction would be integrated gasification combined cycles (IGCCs), closed Brayton cycles, or pressurized fluidized bed combustors (PFBCs).

IGCCs are typically oxy-fired to convert solid fossil fuel, such as coal, into synthesis gas (syngas) in a high-pressure and high-temperature gasifier. The resulting syngas is then cooled and cleaned, and fired in a gas turbine. The heat from the hot gas turbine exhaust together with heat from the gasifier, and gas cooling and cleaning process, is used to generate steam that drives a steam turbine. IGCCs are known to have very high process thermal efficiencies, due to the incorporation of both Brayton and Rankine cycles, and very low air pollutant emission levels.[12]

Closed Brayton cycles can be used to operate with a range of fuels including solid fuels such as coal and can also use a variety of working fluids. The disadvantages of these cycles are that the operating temperatures for the heat exchanger are high, presenting difficulties with respect to material limits, and the equipment size may be large. These cycles also require cooling of gases prior to introduction to the compressor, potentially lowering efficiency.[13] While the working fluid in these cycles is pressurized, the combustion system may not be pressurized.

PFBC systems were developed to use coal within a Brayton cycle, by first combusting the coal in an air-fired pressurized boiler, cleaning the ash from the combustion products, and then passing those products through a turbine. PFBC technology was the subject of intense research in the 1990s, although technical problems arose due to turbine damage by impurities in the combustion products. The role of pressure in PFBC is to improve the Brayton cycle efficiency, as well as to improve the reaction kinetics of coal combustion.[14] PFBC and IGCC power systems have been investigated across a lower operating range of pressurization, specifically 1 to 30 bar.[2]

13.3.2 Advantages and disadvantage of pressurization

Pressurization of a combustion system requires that both the fuel and oxidant streams (either air or pure oxygen) be first pressurized in their cold state and then be introduced into the furnace. Any other process streams must also be brought to the combustor pressure before introduction into the chamber. The major motivations for using high pressure combustion systems other than the traditional reasons are to increase the temperature of waste heat within the flue gas condensate, produce a flue stream that is pressurized, reduce equipment size, increase heat and mass transfer rates, and increase chemical reaction rates.

Higher pressure operation changes the temperature condition at which the gas to liquid phase change occurs. As a consequence, water vapor in the flue gas can be condensed at much higher temperatures. Moisture in the flue gas condenses in the range of 150°C to 200°C at 80 bar as opposed to 50°C to 55°C in ambient pressure oxy-fuel systems (Fig. 13.2). This makes the condensate a suitable heat source for use within the process for applications such as partially heating the boiler feedwater or condensate return within Rankine cycles, drying and preheating fuel, preheating oxidant streams (air or oxygen), drying and preheating product streams entering an industrial furnace, supplying high value waste heat sources to bottoming power cycles such as organic Rankine cycles or binary cycles (for example, Kalina cycles), and using the energy directly within the flue gas stream (industrial use).

The use of the waste heat in this manner can increase the thermal efficiency of boilers or furnaces by 10% to 40% depending upon the process. Another interesting artefact of this basic effect is that the relative thermal efficiency gain increases with the amount of moisture initially in the flue gas stream and increases generally with the amount of moisture in the fuel being fired. Thus, for higher moisture applications (that is, coal-water slurries, biomass, lignite coal, and so on) these concepts have increased value.

In a study of a coal-fired Rankine cycle the potential sinks for placement of waste heat back into the cycle were evaluated (Table 13.1). In this table the relative amounts of heat available and the temperature range are shown. The sink

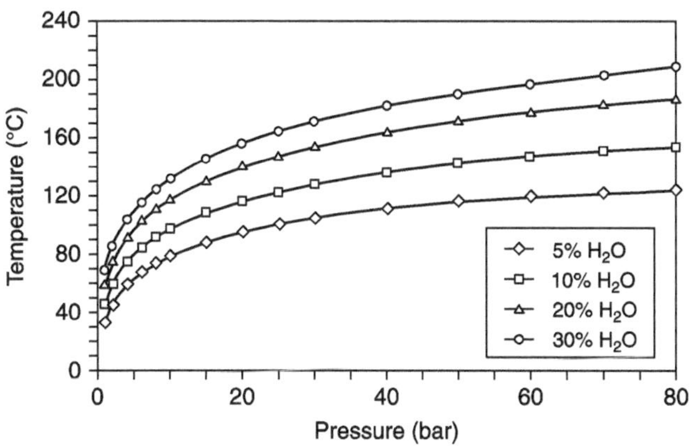

13.2 Flue gas condensation temperature versus pressure for various water contents.

Table 13.1 Potential heat sinks in a steam Rankine cycle

Potential sink	Inlet temperature (°C)	Outlet temperature (°C)	Amount (kJ/kg coal)
Oxygen	20	260	82
Flue gas recycle	20	300/226	357
Feedwater makeup	20	230	27
Condensate return	32	230	1079
Economizer water	230	296	467

with the most potential in terms of both heat quantity and temperature range is the condensate return (feedwater heating). In a common regenerative strategy used in many systems, 5 or 6 points of steam extraction are taken from the steam turbines to preheat feed and de-aerate water in a cascading manner using a combination of both direct and indirect heat exchangers. Use of waste heat in this manner replaces approximately half of these extractions and is readily applied to the indirect heat exchangers of the feedwater heaters. Less extraction from the various stages of the turbine allows the flow of steam to increase through various sections of the turbine, producing about 8% more gross power.

The usual reasons for requiring a pressurized flue stream are either for expansion to produce work in a turbine or for sequestration in a reservoir. Flue gas streams from ambient pressure oxy-fuel Rankine cycle power systems must be pressurized using a PRT in order to bring it to suitable pipeline or sequestration ready conditions. This requirement results in significant power consumption and is usually accomplished by a series of compression and refrigeration steps.[6] Using HiPrOx technology, since the entire system is pressurized, CO_2 is delivered at pressure resulting in about 50 MW savings out of 500 MW produced. Consequently, liquid CO_2 can be produced at ambient temperatures (Fig. 13.3). As a result of system

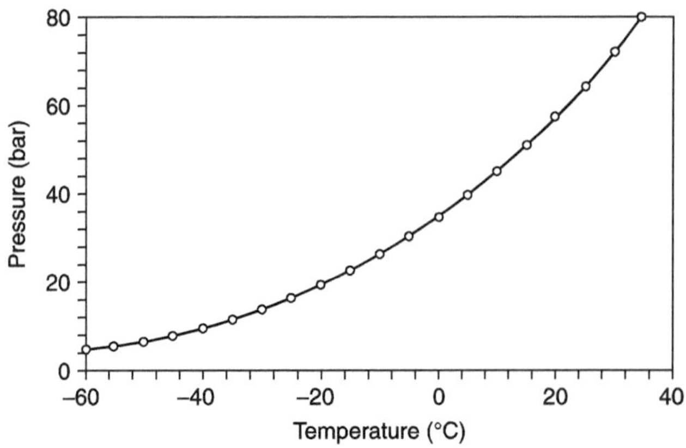

13.3 Pressure versus temperature required to produce liquid CO_2.

pressurization upstream of the coal combustion, fewer concerns exist with exposure of compressors to particulate matter and acidic species.

The down-side is that system pressurization requires that oxygen be supplied at working pressure, thus increasing the auxiliary power consumption associated with the ASU. Higher pressure delivery within an ASU is accomplished by first producing liquid oxygen, pumping the liquid to pressure and then reheating it to generate high pressure gaseous oxygen. The power consumption from a two-column ASU producing 1.6 bar 97.5% oxygen is about 230 kWh/t contained oxygen. This rises to about 330 kWh/t at 30 bar and 350 kWh/t at 70 bar.[4]

13.4 Equipment and operational considerations

13.4.1 Equipment size reduction

The sizing of furnace and air pollution control equipment in power systems is based to a large extent on residence time. Residence time is inversely proportional to the volumetric flow through the device. The volumetric flow is inversely proportional to the density of the flue gas, which varies directly with the pressure but is also influenced by the temperature throughout the system. Comparing a system at 80 bar with one operating at ambient pressure (1.014 bar) there can be a density ratio difference between 80 and 130. Thus, equipment sized on a residence time basis could be dramatically reduced in size. Naturally, there are practical limitations to the size reduction. As a result, it is likely that residence times will be considerably greater for high pressure systems when compared with those of ambient designs. It should also be noted that combustion reaction rates increase with higher O_2 partial pressures. Consequently, HiPrOx is expected to lead to improved fuel burnout as a result of longer residence times and faster combustion kinetics.

The furnace chamber is sized based on several different parameters including the residence time. Residence time is required for char burnout of coal particles and is typically around two seconds in many ambient pressure systems. A sufficient residence time is lower for easy to burn chars such as those encountered in lignite and sub-bituminous coals and increases for difficult chars such as some lower volatile bituminous and anthracite coals, and petroleum cokes. The dramatic increase in residence time for HiPrOx will correspond to a potential reduction in size; however, this opportunity should be tempered with the thought that it may mean that other constraints (ash flow and deposition rates, maximum heat flux achievable, etc.) dictate the size of the furnace. Certainly, the possibilities exist for furnace size reduction or the use of coarser, granular as opposed to pulverized, feed coal.

13.4.2 Improved heat transfer

Increased heat transfer results in reduced heat exchanger sizes for efficient transfer of heat between the combustion side of the system to the working fluid (steam, organic fluid, or other) or to any indirect heater such as is found within reformers or process liquid heaters. This may potentially result in decreased equipment costs although savings might be offset by more expensive materials depending upon temperature and pressure limits.

Although heat transfer in ambient pressure combustion systems contains both radiative and convective components, radiation predominates in the higher temperature sections and the convective aspect becomes increasingly predominant in the lower temperature sections. Both radiation and convection increase as a function of pressure. Radiation increases due to increased gas-side emissivity resulting from increased partial pressures of emitting species (CO_2, H_2O) and, to a lesser extent, increased density of solid matter (char, soot and particulate matter in general).

The rate of convection to the outside of a boiler tube, proportional to the Nusselt number, Nu, increases due to the increased density of flue gases at higher pressures.

$$Nu = C \times Re^{m} \times Pr^{n} \quad [13.1]$$

where the C, m and n are constants and Re and Pr are the Reynolds and Prandtl numbers, respectively. The value for m is typically around 0.6. This means that for the range of density ratio variation of 80 to 130, Nu varies from 13 to 19.

The outside heat transfer coefficient (h_o) is proportional to Nu and defined as:

$$h_0 = \frac{Nu \times k}{d} \quad [13.2]$$

where k is the fluid thermal conductivity and d is the characteristic diameter. Overall heat transfer through a boiler tube (Table 13.2) is largely dictated by the rate determining coefficient, which is the heat transfer from the flue gas stream to the outside surface of the tube (that is, the outside film coefficient). As furnace pressure increases, the outside film coefficient increases (proportionally to Nu) thus increasing the overall heat transfer.

Although both radiation and convection do increase with higher pressure, convection increases at a much faster rate and, therefore, convection will tend to predominate as operating pressure is increased (Table 13.3).

Due to the elevated pressures, all components (furnace, heat exchangers, air pollution control) are drastically reduced in size compared with ambient pressure

Table 13.2 Approximate convective heat transfer coefficients (W/m^2.°C)

System	Inside	Wall	Outside	Fouling	Overall
Air-fired at 1 bar	2576	9428	61	5678	59
Oxy-fuel at 1 bar	2576	9459	68	5678	65
HiPrOx at 26 bar	2436	9459	488	5678	354
HiPrOx at 80 bar	2641	9437	1113	5678	618

Table 13.3 Radiation and convective heat transfer contributions for various boiler technologies

System	Non-luminous radiation (%)	Convection (%)
Air-fired at 1 bar	23.4	76.6
Oxy-fuel at 1 bar	15.4	84.6
HiPrOx at 26 bar	1.4	98.6
HiPrOx at 80 bar	0.8	99.2

units. This results in a capital savings even when the need for a pressure vessel shell to contain the furnace is considered.

13.4.3 Temperature moderation with flue gas recirculation (FGR)

Furnace temperature moderation will be required for the HiPrOx case, as in the ambient pressure situation. The adiabatic flame temperature in the HiPrOx case is similar to the ambient pressure oxygen operations. FGR in the pressurized case is more difficult to achieve because of gas density and the increased pressure drop through the system. Less moderation is required for the fuel slurry-fired pressurized situation because the water in the slurry reduces the flame temperatures for typical bituminous coals from 3400°C to approximately 2900°C. In other words, a typical dry fed ambient pressure oxy-fuel system requiring approximately 60% FGR would need only 50% FGR need for the same moderation effect with a wet slurry feed. The efficiency penalties associated with moisture (such as slurry) for HiPrOx systems are less than those of ambient pressure oxy-fuel systems due to improved capability to recover the latent heat.

13.4.4 Pressure vessel construction

Although there appear to be a number of thermodynamic benefits, a number of mechanical hurdles exist in bringing this technology to reality. First, the fuel feed system must overcome the pressure of the system and this will probably require feeding of the fuel as a slurry. Second, the ash within the system must be managed when firing solid fuels and must be removed from the system periodically. The furnace design must be thoroughly examined. Two approaches can be considered for furnace design and some inspiration for these furnaces may be gained from various present-day gasifier and pressurized fluid bed designs.

The simplest approach is a slurry-fed reactor in which the ash is ‘wet’ (slagging) and runs continuously from the unit. The temperature inside the system must be controlled and this could be accomplished by FGR, cooled condensate recirculation or some combination of these two concepts. It is likely that the 'wet' furnace design would be the least expensive option.

Another approach that could be employed is the fluidized bed reactor. This is somewhat reminiscent of the PFBCs; however, HiPrOx systems may operate at higher pressure. In a PFBC, a lower reactor temperature may be more easily controlled using combinations of FGR, bed material recirculation and in-bed heat exchangers. The lower temperature operation has several advantages. The use of bed material results in a larger thermal capacitance within the system, making it more tolerant to upsets that can result in temperature excursions. This consideration is very important when operating pressurized furnaces with pure oxygen. Although this is probably a more expensive avenue, it may deliver a more reliable and robust system. Firing in this way will be less fuel specific, allowing for wider quality ranges, and moisture and ash contents. However, this approach involves additional feeding and removal of solids to and from the system, adding complexity and further cost.

CanmetENERGY has explored variants of HiPrOx power systems. One specific technology was studied in conjunction with ThermoEnergy Corporation and is the TIPS (ThermoEnergy Integrated Power System) process.[15] CanmetENERGY prepared a report[3,5] that outlined in considerable detail the technical potential of TIPS technology and its associated economics. Details of TIPS analysis are summarized below.

13.4.5 ThermoEnergy Integrated Power System

In 2001 CanmetENERGY began evaluating the use of HiPrOx power systems for use with coal-fired Rankine cycles. As noted above, one specific cycle, TIPS, was assessed in conjunction with ThermoEnergy Corporation.

The TIPS process is an advanced concept of high pressure oxy-fuel combustion, and as such, is quite different from atmospheric pressure oxy-fuel combustion. By pressurizing the entire process, TIPS is able to not only utilize the latent heat of the fuel, but more importantly can also condense the CO_2 in the flue gas at ambient heat sink temperatures thereby minimizing the need for multi-stage compression and refrigeration (Fig. 13.4). This type of cycle presents a number of advantages and disadvantages.[16]

One advantage is that the use of waste heat in flue gas condensate results in increased thermal efficiency of the boiler. Further, waste heat is used to replace regenerative extraction from the turbine, resulting in improved power output and increased steam-side efficiency. Another process benefit is that energy intensive flue gas compression and refrigeration for CO_2 capture is not required, precluding the large auxiliary power consumption requirement associated with ambient pressure systems. The increased pressure results in decreased physical sizes of volumetric devices such as furnaces and scrubbers. Pressurization also eliminates infiltration air, thus reducing nitrogen contamination of the flue gas. Further,

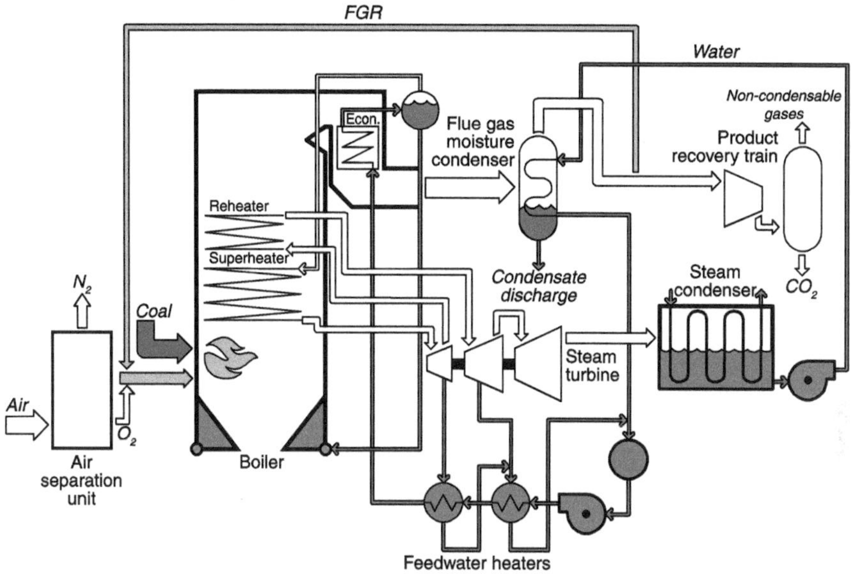

13.4 HiPrOx configuration of a Rankine cycle power station in the TIPS configuration.

increased gas density leads to higher convective heat transfer with concomitant heat exchanger component size reductions. Last, higher mass transfer and chemical reaction rates result in improved air pollution control device efficiencies.

Of course, there is a price to be paid for pressurization. The requirement for high pressure oxygen feed increases auxiliary power consumption as compared with ambient pressure oxygen production. Temperature moderation using the FGR approach is more difficult to achieve physically. Unlike ambient pressure oxy-fuel systems, high pressure variants do not have the flexibility to operate under both air and oxy-fuel conditions. Finally, the overall mechanical construction of the system requires it to be a pressure vessel or, to be contained within a pressure chamber.

In a study performed by CanmetENERGY, a common single reheat condensing steam turbine rated at 3600 + 3600/3600 rpm was selected. The main steam conditions were specified to be 540°C and 103.5 bar while those of the reheat steam were 540°C and 25.3 bar. The turbine consisted of high pressure (HP), intermediate pressure (IP), and low pressure (LP) sections. The design exhaust steam pressure of the LP section was 0.048 bar with a vapor fraction of about 90%. The net turbine output in the air case was rated at 100 MW_e.

The feedwater heater (FWH) system consisted of five steam-water heaters. For each heater, steam drawn (termed bleed steam) from the turbine is used to heat the condensate return. Three of the five feedwater heaters were replaced with heat from the condensed moisture in the flue gas.

The auxiliary power consumption for the ASU is more than the ambient pressure oxy-fuel scenario due to the need to supply oxygen to the system at an increased pressure. With system pressurization, there is no need for a PRT and its associated power consumption. Additionally, air pollution control equipment requirements are lowered primarily due to reduced flue gas volume. Tables 13.4 and 13.5 show the results for 500 MW_e (gross) power plants using various technologies. The boiler efficiency is increased due to the use of condensed moisture from the flue gas, while the steam side efficiency increases because of a lower extraction requirement for feedwater heating leaving more steam flow through the turbine. This, in turn, results in a higher overall efficiency than that obtained in an ambient pressure oxy-fuel system; however, there is always a penalty to pay for CO_2 capture.

In February 2009, ThermoEnergy Corporation teamed up with Babcock Power Corporation to commercialize the TIPS system. According to ThermoEnergy's web site, 'Babcock and ThermoEnergy engineers will begin work immediately to finalize the data needed to design, construct and operate a large-scale pilot plant at a host site.'

Table 13.4 Approximate power consumption (MW_e) for various power cycles

System	Boiler	FGR fan	ASU	PRT	Air pollution control
Air-fired (without capture)	28.8	0.0	0.0	0.0	18.0
Oxygen-fired (1 bar)	28.8	10.7	81.0	55.1	10.0
TIPS	26.8	0.1	121.5	0.0	5.0

Table 13.5 Overall system performances

System	Net output (MW_e)	Gross output (MW_e)	Boiler efficiency (%)	Steam side efficiency (%)	Net efficiency (%)
Air-fired (without capture)	453	500	89	38	34
Oxygen-fired (1 bar)	312	500	89	38	24
TIPS	347	500	97	41	29

13.5 Other high pressure power generation systems

13.5.1 Enel System

The Enel Group (Ente Nazionale per l'Energia eLettrica) is a major Italian utility company with operations in Canada, USA, Europe, Latin America, and South America. Enel, a company with more than 97,000 MW generating capacity, has committed to clean coal technology and plans to build 5000 MW_e clean coal generation capacity. In 2006, Enel began work in the area of pressurized oxy-fuel combustion with ITEA S.p.A. (recently acquired by Sofinter group). By 2007, the company had built a small pilot-scale plant (5 MW_{th}) based on a HiPrOx technology called Isotherm Pwr®. Isotherm Pwr® is a novel pressurized oxy-combustion technology, patented and jointly developed by the two Italian companies, Enel and ITEA S.p.A.[17]

This technology is currently being tested at a facility based in Gioia del Colle in southern Italy. It is noteworthy that ITEA S.p.A. developed the first experimental plant for liquid waste incineration in 1998 and scaled the technology up to 5 MW_t in 2003. In addition, Enel has recently completed the executive design of a 48 MW_t high pressure demonstration plant.

The Isotherm Pwr® process works at pressures above 4 bar. The combustion and the heat transfer processes occur in two separate units (Fig. 13.5), working at the same pressure. Cryogenically produced oxygen is the oxidant and the combustion temperature is controlled by recycling a portion of the flue gas from the boiler outlet. Consequently, only oxygen, carbon dioxide, and water are the main flue gas components. A portion of the recycled flue gas is mixed with oxygen entering the combustor, and a second flue gas recycle is used to control the temperature at the boiler entrance, where it is kept close to 800°C to avoid slagging problems.

The reactor section works under wet slagging conditions. Since the process allows for a relatively constant temperature profile over the entire length of the combustor, at temperatures above those required for the molten slag to flow, the ash removal efficiency is very high (greater than 95%). Molten slag is quenched in a water bath and coal ash is removed as vitrified inert material.

The main features of the Enel combustion process can be summarized as follows. The combustion system has the ability to burn low grade, cheap fuels, such as coals with high ash contents, tar sands, or other fuels with low heating

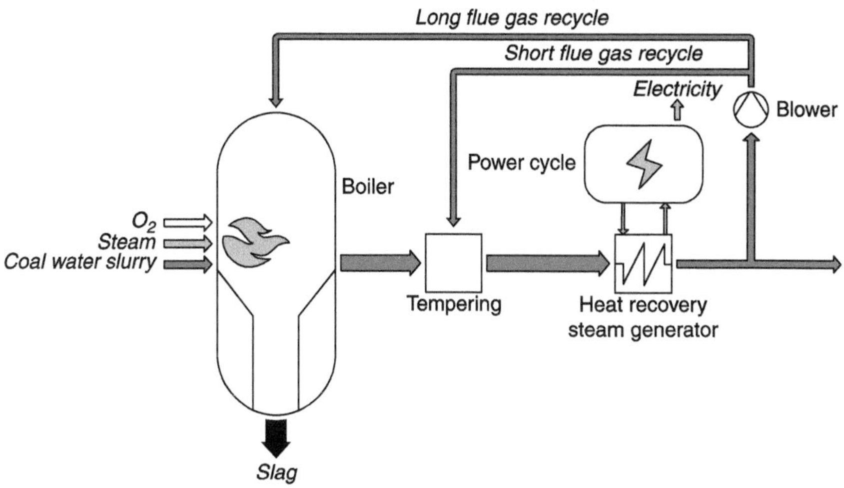

13.5 The Enel Isotherm Pwr® system.

values. The resulting increased heat transfer rates on the heat recovery steam generator (HRSG) lead to smaller convective heat transfer surfaces, and improved thermal efficiencies on large scales. The ash can be removed with high efficiency. Current research efforts are focusing on scaling up the technology.

13.5.2 Brayton cycles

For the most part to date, combustion systems pressurization has been employed to effect thermodynamic conditions suitable for Brayton cycles. For air-fired gas turbines the pressure ratio gradually rose from 5 to 10 during the 1960s. Since then, the pressure ratio has reached a value of about 42.[11,18] Recently, many of these systems have been adapted to use oxy-firing to allow for CO_2 concentration and capture.

Brayton cycle applications are most typically open systems that generally use a flue gas stream generated from natural gas-fired pressurized combustors. To moderate temperatures and improve the Brayton cycle thermodynamics, water or steam can be introduced into the combustor. Due to the nature of Brayton cycles and gas turbine technology, these systems are most commonly used with relatively clean fuels such as natural gas. However, some similar systems that fire liquid fuels do exist.

As mentioned previously, the main systems using open Brayton cycles and solid fuels are the IGCC and PFBC technologies. Closed Brayton cycles have been identified as a means of using solid fuels within a Brayton cycle system.[18]

If pure oxygen is used, the flue gas stream produced from these systems primarily consists of H_2O and CO_2. The water can be condensed to create a relatively pure CO_2 stream that may require further compression. The condensed water may be recirculated back to the combustor for re-use. A number of variations exist on this basic cycle.

Clean Energy Systems Corporation (CES) has been a pioneer in the area of pressurized oxygen use in gas turbine engines. As early as 2003, CES acquired an idle 5.5 MW_e biomass power plant located in California. Initially, this plant was used as a demonstration facility for their 20 MW_{th} gas generator firing natural gas and pure oxygen (Fig. 13.6) with an inlet pressure between 50 and 100 bar.[19] This system has also been used to demonstrate the use of this technology for a variety of fuels including syngas, glycerine, and a stabilized emulsion of heavy petroleum refining residuum in water. The combustor temperature is moderated by injection of water or steam.[20]

CES has been offering a commercial first generation form of this technology using a gas generator at 170 MW_{th} and modest gas inlet temperature of 760°C. Future technology generations are planned in which this temperature will be increased first to 1260°C and then to 1760°C.

Various gas turbine cycles with CO_2 capture concepts have been benchmarked in modeling studies. In studies conducted by Kvamsdal *et al.* and Bolland *et al.*, a

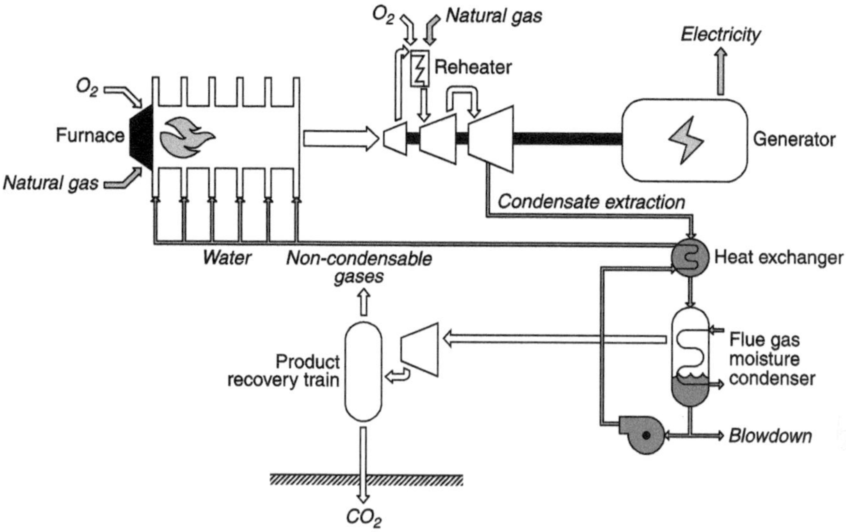

13.6 The CES power system.

number of thermodynamic cycles were modeled and compared.[21,22] Cycle variations on the HiPrOx theme exist and include: oxy-fuel combined cycle, water cycle, Graz cycle, advanced zero emissions power plant (AZEP) cycle, and the Matiant cycle.

13.6 The industrial sector

Within Canada, the overall CO_2 emissions from the industrial sector account for about 25% of the total GHG emissions[22,23] and combustion systems are responsible for approximately 64% of this figure.[24] Worldwide, mineral processing, pulp and paper making, smelting, iron and steel production, cement manufacture, lime production, and fertilizer manufacture sectors are examining possible technology pathways to reduce GHG emissions.

The diversity of equipment and processes within industry has been an impediment to early adoption and development of CO_2 emissions solutions. Focus has been primarily on post combustion technologies such as amine scrubbing. As a result, a philosophy of limited change in the basic processes exists, limiting the development of GHG reduction technologies and the associated risks. Although this is certainly a more conservative approach for the time being, it ignores possible synergies between certain types of industrial processing and techniques such as ambient pressure oxy-firing and HiPrOx.

The use of pressure is not solely limited to power systems. Industrial combustion systems have often used furnaces with low to high levels of pressurization. High pressure oxy-firing may be a good pathway to obtain a relatively pure CO_2 stream needed for sequestration. Possible process synergies with oxy-firing depend on the nature of the industrial process being considered. Some processes already use pure oxygen because of the need for high temperatures or when high reductant (CO, H_2) partial pressures within the flue gas stream are needed to promote chemical reactions such as conversion of metal oxides to their base metals.

Aside from many of the same benefits attributed to the power generation sector, the industrial sector can often benefit from preheating or drying feed material, increasing mass and heat transfer rates to loads or heat exchangers, and enhancing low grade waste heat to produce power. Examples of preheating or drying of feed materials include preheating of ores or preheating of limestone in the manufacture of lime or cement. Increased mass and heat transfer rates resulting from high pressure oxy-firing in thermal mineral processes may lead to equipment scale down and increased production efficiencies.

Power generation from low grade waste heat sources can be a significant benefit to an operation, especially within industrial sectors. In these situations, waste heat is used within an organic Rankine cycle (ORC) or a binary cycle (for example, a Kalina cycle) to produce power. The magnitude and quality of the waste heat in these situations dictates the economics of this type of installation and the use of HiPrOx technology may overcome some of the economic barriers by increasing the quality of the waste heat sources.

The use of HiPrOx systems for direct contact steam generation is another interesting adaptation of this basic technology. In these systems, steam can be generated at very high thermal efficiencies for applications within many industries. This is a particularly attractive option for the heavy oil extraction industry.

The oil sands and carbonate heavy oil extraction industries require vast quantities of high pressure steam, traditionally supplied by boilers. Boilers indirectly heat treated boiler water. The thermal efficiencies of these devices tend to be in the neighborhood of 80%. The 20% loss consists of sensible heat associated with the dry flue gases and latent heat associated with the uncondensed moisture exiting through the stack.

Industrial boilers generally produce process steam in an open cycle necessitating large amounts of chemicals and energetically expensive feedwater treatment. Most industrial boilers employ a double drum arrangement that allows for the purging of concentrated solids by means of a blowdown extraction from the lower drum. This blowdown results in a heat loss from the system, as well as wastewater that may require subsequent treatment before release into the environment.

Boilers used in the tar sands industry currently are an adaptation of the industrial boiler that allows for increasingly poorer feedwater. These once-through steam generators (OTSGs) generally produce 80% quality steam, meaning that there is 20% saturated water included in the product. Since all the product is not steam, the water concentrates the solids formed allowing them to be flushed through the system.

Extraction of bitumen from oil sands, and in particular *in situ* production, uses large quantities of pressurized steam injected into wells. The two major processes that are employed to effect bitumen extraction are steam assisted gravity drainage (SAGD) and cyclic steam stimulation (CSS).

These oil sands operations have resulted in a significant environmental problem associated with pools of wastewater that are typically contaminated with hydrocarbons and are laden with total dissolved solids (TDS). This industry tends to use OTSGs because of their ability to employ feedwater of minimal quality. The use of HiPrOx direct contact steam generators (DCSGs) has been identified as a niche application. In these systems a hydrocarbon fuel can be fired with oxygen and water to produce a flue stream consisting mostly of steam with a CO_2 component.

Although OTSG boilers require very large amounts of water, waste pools and produced water are not suitable candidates because they are too contaminated for use as boiler water. The result is that clean water sources are being used and more contaminated water sources are being created furthering the environmental problem. The use of DCSG can alleviate some of these constraints and allow for the use of poorer quality water.

Direct contact air-fired steam generators have been used for a number of years and several demonstrations within this industry have been carried out with relatively positive results.[25,26] Advantages of direct steam generation compared with conventional steam generation include smaller size and greater portability, lower capital costs, higher energy efficiency, and the ability to use lower quality water.

Disadvantages of direct contact steam generation compared with conventional steam generation include production of lower quality steam due to dilution with nitrogen, non-condensability of the nitrogen fraction, and production of carbonic acid leading to potential corrosion problems. The use of oxy-firing maintains the same advantages, and at the same time eliminates the disadvantages associated with low quality steam and non-condensable nitrogen.

Advantages of oxy-fuel direct contact steam generation compared with direct contact air-fired steam generation are further size reduction and greater portability, much lower capital costs, ability to sequester CO_2, high quality steam production, and the ability to use waste water. The only major disadvantage of oxy-fuel direct steam generation is that an air separation unit is required. A number of technologies that use this concept exist, although to date most of these systems use gaseous or liquid fuel sources.[27–29]

13.7 Future trends

HiPrOx technology is a relatively new area of development with only a few demonstrations. Opportunities exist in this area not only for power technology development but also for the industrial and oil and gas sectors. Most demonstrations have been focused on gaseous or liquid fuels with the exception of the Enel facility. With looming GHG regulations, the opportunities within this area are definitely increasing and it is expected that increased interest in HiPrOx will result in the future.

13.8 Acknowledgements

The authors would like to convey thanks to the Federal Panel on Energy R&D (PERD) and CanmetENERGY, Natural Resources Canada for support in this subject over the years. We would also like to thank Mr Alex Fassbender (formerly of ThermoEnergy Corporation) for his support in developing the TIPS process. Thanks are extended to Dr Kevin Fogash of Air Products and Chemicals Corporation for information regarding cryogenic air separation units. The International Technical Conference on Coal Utilization and Fuel Systems (Clearwater Coal Conference) has been an interested and supportive partner in allowing a forum for fostering developments in this field. Last, we would also like to thank Mr Marco Gazzino of Enel for contributions and active discussions in this area.

13.9 References

1 Electric Power Research Institute (EPRI) *Evaluation of Advanced Coal Technologies with CO_2 Capture*, Report 000000000001004880, April 2004.
2 Zheng, L., Tan, Y. and Wall, T. *Some Thoughts and Observations on Oxy-fuel Technology Developments*, The 22nd International Pittsburgh Coal Conference, Pittsburgh, PA, 12–15 September 2005.
3 Pomalis, R., Zheng, L. and Clements, B. *ThermoEnergy Integrated Power System Economics*, The 32nd International Technical Conference on Coal Utilization and Fuel Systems, Clearwater, FL, 10–15 June 2007.
4 Fogash, K., personal communication, September 2007.
5 Zheng, L., Pomalis, R. and Clements, B. *Technical Feasibility Study of TIPS Process and Comparison with other CO_2 Capture Power Generation Process*, The 32nd International Technical Conference on Coal Utilization and Fuel Systems, Clearwater, FL, 10–15 June 2007.
6 Zheng, L. *Product Recovery Train Development for CO_2 Capture in Oxy-fuel Environment*, The 30th International Technical Conference on Coal Utilization and Fuel Systems, Clearwater, FL, 18–22 April 2005.
7 Zheng, L., Clements, B. and Douglas, M. *Simulation of an Oxy-Fuel Retrofit to a Typical 400 MW_e Utility Boiler for CO_2 Capture*, The 26th International Technical Conference on Coal Utilization and Fuel Systems, Clearwater, FL, 5–8 March 2001.
8 Zheng, L., Clements, B. and Runstedtler, A. *A Generic Simulation Method for the Lower and Upper Furnace of Coal-fired Utility Boilers Using Both Air Firing and Oxy-Fuel Combustion with CO_2 Recirculation*, The 27th International Technical Conference on Coal Utilization and Fuel Systems, Clearwater, FL, 4–7 March 2002.
9 Zheng, C., Clements, B. and Zheng, L. *The Feasibility of Decreased Furnace Size with Reduced Flue Gas Recirculation in Coal-Fired Boiler Designs*, The 30th International Technical Conference on Coal Utilization and Fuel Systems, Clearwater, FL, 18–22 April 2005.
10 Faires, V.M. *Thermodynamics*, 5th edition, McMillan, 1970.
11 Brooks, F.J. *GE Gas Turbine Performance Characteristics*, GE Power Systems, 2000.
12 Williams, A., Pourkashanian, M., Jones, J. and Skorupska, N. *Combustion and Gasification of Coal*, Taylor and Francis, New York, 2000.

13 Petchers, N. *Combined Heat, Cooling and Power Handbook*, Fairmont Press, 2003.
14 Cuenca, A. and Anthony, E. *Pressurized Fluidized Bed Combustion*, Blackie Academic and Professional, 1995.
15 Fassbender, A. United States Patent No: US 6,196,000 B1, 6 March 2001.
16 Clements, B., Zheng, L. and Pomalis, R. *Oxy-Fuel Firing: The Transition to Pressurized Systems for Cycle Efficiency Optimization*, Proceedings of the 8th European Conference on Industrial Furnaces and Boilers, Algarve, Portugal, 25–28 March 2008.
17 Gazzino, M. and Benelli, G. *Pressurised Oxy-Coal Combustion Rankine-Cycle for Future Zero Emission Power Plants: Process Design and Energy Analysis*, ASME 2008 2nd International Conference on Energy Sustainability, Volume 2, Jacksonville, FL, 10–14 August 2008.
18 Sawyer, J. *Sawyer's Gas Turbine Engineering Handbook*, Turbomachinery International Publication, 1985.
19 Clean Energy Systems web site, www.cleanenergysystems.com, 2009.
20 Anderson, R., MacAdam, S. and Viteri, F. *Adapting Gas Turbines to Zero Emission Oxy-fuel Power Plants*, Proceedings of ASME Turbo Expo 2008, 2008.
21 Bolland, O. and Kvamsdal, H. *A Thermodynamic Comparison of the Oxy-fuel Power Cycles, Water-Cycle, Graz Cycle and Matiant Cycle*, Proceedings of the International Conference POWER Generation and Sustainable Development, Association of Engineers from the Montefiore Electrical Institute (AIM), Liege, Belgium, 2000.
22 Natural Resources Canada. *Canada's Emission Outlook: An Update*, 1996.
23 Natural Resources Canada. *Canada's Energy Outlook*, 2006.
24 Kvamsdal, H., Jordal, K., and Bolland, O. *A Quantitative Comparison of Gas Turbine Cycles with CO_2 Capture*, Energy, **32**, 2007.
25 Godin, M. *Direct Contact Steam Generation – Presentation*, PTAC, 2008.
26 Meyer, R. and Steele, C. *The Future of Heavy Crude Oils and Tar Sands – Chapter 61 – Innovative Approaches to Facilitate Production of Heavy Crudes, Havlena*, The Future of Heavy Crude Oils and Tar Sands International Conference Proceedings, 1979.
27 Eisenhawer, S., Mulac, A.J., Donalson, A.B. and Fox, R.L. *Steam Generation Having a High Pressure Combustor with Controlled Thermal and Mechanical Stresses and Utilizing Pyrophoric Ignition – United States Patent 4,648,835*, 1985.
28 Rao, D. *Liquid Vortex Gas Contactor – United States Patent 4,604,988*, 1984.
29 Rodwell, L. *Steam Generation from Low Quality Feedwater – United States Patent 4,398,603*, 1981.

14

Chemical-looping combustion for power generation and carbon dioxide (CO_2) capture

H. JIN and X. ZHANG, Chinese Academy of Sciences, P. R. China

Abstract: Chemical-looping combustion (CLC) is regarded as a promising technology for CO_2 separation with near-zero energy penalty. This chapter first discusses the characteristics of CLC using energy and exergy analysis. Then, CLC development is reviewed, including the looping materials and energy systems with CLC, and future trends are indicated.

Key words: chemical-looping combustion, CO_2 capture, near-zero emission.

14.1 Introduction

It appears that we face the potentially serious problem of rapid climate change due to anthropogenic emissions of greenhouse gases, mainly CO_2. As concerns about climate change grow, different technological solutions for mitigating the impacts of CO_2 emission due to energy generation and utilization are being explored. Although renewable energy offers some potential for reducing greenhouse gas emissions, fossil fuels are expected to remain a major part of the world's energy mix for the foreseeable future (Azar *et al.*, 1999). One of the options for controlling greenhouse gas emission is CO_2 capture and storage.

A number of technologies for CO_2 capture have reached industrial demonstration level: (a) pre-combustion, in which the fuel is de-carbonized prior to combustion; (b) oxy-fuel combustion, where pure oxygen is used for combustion instead of air, thus producing a CO_2-enriched flue gas ready for sequestration once the flue gas is cleaned; and (c) post-combustion capture, where a chemical solvent is employed to scrub the CO_2 from flue gas. Significant amounts of energy are required for all of these technologies, resulting in large energy efficiency penalties and major increases in the cost of electricity production.

The overlap between energy science and environmental science calls for novel ideas and approaches that go beyond the traditional 'treatment after pollution' and 'environment after energy conversion' considerations and the thermal cycle for CO_2 separation is clearly one of the major challenges here. Breakthroughs in these fields will generate an embranchment of burgeoning disciplines among the energy, environmental, and chemical industries. Therefore, from the viewpoint of scientific crossover between energy and environment, it is imperative to use innovative ideas to find a practical approach to resolving energy utilization and environment pollution issues simultaneously (Jin *et al.*, 2005).

Research has shown that there are close interactions between energy utilization and CO_2 separation, i.e. that they are not independent. At present, the major irreversibility in the power system takes place in the fossil combustion process where the chemical energy of fuel is converted into physical energy. It was found that almost 30% of the work availability of the chemical energy in the fuel was lost in this process. Another important point is that the combustion process is also the source of CO_2 generation alongside the chemical energy conversion process. Therefore, this suggests that there is great potential in the combustion process itself to simultaneously solve energy utilization and environmental pollution problems, and this could result in a breakthrough in CO_2 capture in power systems (Jin *et al.*, 2008).

Chemical-looping combustion (CLC) could fundamentally change the way fuel is utilized. This technology could significantly reduce exergy destruction during the combustion process and separate CO_2 with very little energy penalty.

Since both oxy-fuel combustion and chemical-looping combustion processes take place in an almost nitrogen-free environment, CLC is therefore, in a broad sense, a form of oxy-fuel combustion. The major differences relate to how the oxygen is generated and how the combustion is conducted.

This chapter reviews the mechanism of energy release in CLC, the development of oxygen carrier materials and reactors, and the energy systems. Despite the fact that CLC has attracted extensive attention in recent years and much work has been done in this area, the technology is still in its infancy. Development of oxygen carriers with the appropriate reactivity and stability is still the main challenge to the progression of advances in CLC. There is also a need to fully understand the interactions between cascade utilization of chemical energy and the decrease in energy penalty for CO_2 separation in CLC, the reactivity of CLC with liquid or solid fuel, and the design principles and methodologies of fluidized bed reactors, etc. In addition, the role of CLC in improving thermodynamic performance and the principles of energy systems based on CLC should be explored in the renewable energy and chemical industries. Finally, construction and operation of large-scale CLC demonstration plants are needed before this technology can be used commercially.

14.1.1 Chemical-looping combustion overview

From the viewpoints of both energy utilization and the environment, it has been found that the combustion process not only causes the largest exergy destruction but also generates large amounts of CO_2. Hence, there is a tremendous potential to resolve both energy utilization and environmental problems in the combustion process simultaneously. Furthermore, it is worth noting that energy consumption in CO_2 capture is significant. The question here is what will bring a breakthrough in CO_2 recovery and how we may effectively use high-quality energy in combustion for power generation and CO_2 capture. It is therefore desirable to ensure that the development of next-generation power plants is directed at integrating combustion and the CO_2 recovery process to decrease exergy destruction in the combustion process while taking the environmental impact into account.

CLC was first proposed by Richter and Knoche (1983) as a means of decreasing exergy destruction in combustion; at the time, CO_2 capture was not considered. In 1994, Ishida and Jin originally proposed a novel gas turbine cycle with CLC for inherent CO_2 separation (Ishida and Jin, 1994) and explored a new approach that integrated energy conversion and CO_2 separation; this was patented in the United States (US patent No. 5,447,024 (Ishida *et al.*, 1995)). Figure 14.1 illustrates the gas turbine cycle with natural-gas-fired CLC. The solid oxygen carrier is circulated between the air and fuel reactors. The gaseous fuel is fed into the fuel reactor where it is oxidized by the oxygen from the metal oxide.

A generalized description of the CLC process can be summarized as shown below.

Fuel reactor:

$$C_xH_y + (2x + y/2)MeO = xCO_2 + (y/2)H_2O + (2x + y/2)Me \qquad [14.1]$$

Air reactor:

$$(2x + y/2)Me + (x + y/4)O_2 = (2x + y/2)MeO \qquad [14.2]$$

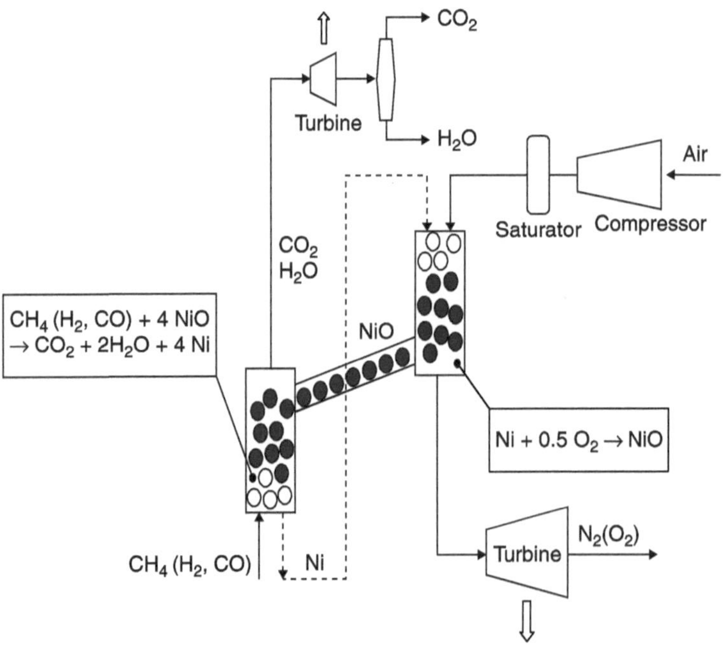

14.1 Schematic illustrating the concept of CLC.

Since no air is used for combustion in the fuel reactor, CO_2 and water vapor are the main products in the flue gas. A high-purity CO_2 stream can be recovered readily by condensing the water vapor, therefore eliminating the need for an additional energy-intensive CO_2 separation process. As in oxy-fuel systems, the combustion reaction takes place in the absence of nitrogen, giving rise to extremely low NO_x emissions.

The CLC system for power generation and CO_2 capture is expected to have a much higher thermal efficiency than most other combustion–gasification fuel utilization systems with CO_2 capture. This is mainly due to the fact that CLC minimizes the loss of chemical energy through lowering the energy level of the fuel and therefore greatly reduces the energy penalty for CO_2 capture (Jin *et al.*, 2005).

14.1.2 Development of chemical-looping combustion

This novel and advanced technology was mentioned in the highly influential report by the Intergovernmental Panel on Climate Change (IPCC) (Metz *et al.*, 2005). In the Fifth Framework Programme (FP5) projects of the European Union, CLC was ranked as one of the best in cost evaluations of the CO_2 Capture Project (CCP). This assessment was confirmed in FP6 and FP7 projects, where the cost of CO_2 capture by CLC was estimated to be 40–50% cheaper than that of post-combustion capture using amine scrubbing. (FP6: Chemical Looping Combustion CO_2-Ready Gas Power.)

Since the pioneering work of Ishida and Jin (1994), and in response to growing concerns over CO_2 emissions, major CLC research projects have been launched worldwide. For example, Chalmers University of Technology of Sweden (Lyngfelt *et al.*, 2001; Lyngfelt and Thunman, 2005), Southeast University of China (Xiang *et al.*, 2004), the New Energy and Industrial Technology Development Organization of Japan, Alstom of USA, and CanmetENERGY Technology Centre (Wang and Anthony, 2008) in Canada have all been working in this field. Since 1992, there has been a ten-fold increase in the number of published papers on the topic of CLC.

The US Department of Energy (DOE) has also initiated a number of research projects on CLC. These projects mainly focus on chemical looping for combustion and hydrogen production, fluidized bed chemical looping applications, and oxygen carrier development for CLC. Japan has also started a series of projects on CLC, with participants including NEDO (New Energy and Industrial Technology Development Organization) and RITE (Research Institute of Innovative Technology for the Earth).

The International Flame Research Foundation (IFRF) has stated that 'These [CLC] high efficiency levels can be achieved while simultaneously providing sequestration-ready CO_2'. Professor János Beér of the Massachusetts Institute of Technology (MIT) mentioned in his report (Ansolabehere *et al.*, 2007) that 'Novel separation schemes such as chemical looping should continue to be pursued at the process development unit (PDU) scale'.

The United States DOE pointed out that CLC would be regarded as one of the advanced technologies of CO_2 capture. In August 2008, the DOE also announced significant funding for the 15th CO_2 Capture Research Program, with CLC as the first and foremost technology to be investigated (http://www.fossil.energy.gov/programs/sequestration/cslf/index.html).

The integrated CLC combustion–gasification power system, proposed by Alstom and supported by the DOE, is worth mentioning in particular (http://www.netl.doe.gov/publications/factsheets/project/Proj293.pdf). The objective of this work is to design, construct, and operate a pilot facility to demonstrate this novel CLC system. The process diagram for the Alstom system is shown in Fig. 14.2. Calcium compounds are used to carry oxygen and heat between the various reaction loops. The chemical loop uses CaS and $CaSO_4$ reactions to gasify the coal. The dominant gas, CO, enters the shift reaction chamber where steam is used to convert the CO into CO_2 and H_2. The CO_2 is then removed from the gas using another chemical loop based on CaO and $CaCO_3$. These compounds are then directed to another reactor where a 'thermal' loop, using a bauxite heat transfer medium, drives off the CO_2 for use or sequestration. This system has the potential to achieve near-zero CO_2 emissions, to meet or exceed integrated gasification combined cycle (IGCC) efficiency, and could cost less than US$800 per kilowatt without CO_2 capture and less than US$1000 per kilowatt with CO_2 capture.

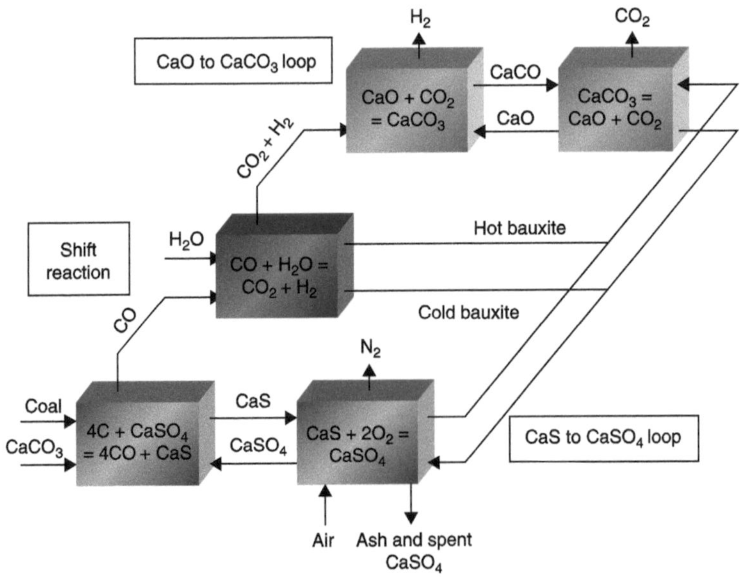

14.2 Schematic of the integrated CLC combustion–gasification developed by Alstom.

In order to avoid the carbon entering into the air reactor and the influence of ash on the system, measures should be taken to separate the oxygen carrier from the unburned carbon particles and ash. Chalmers University of Technology (Lyngfelt *et al.*, 2001; Lyngfelt and Thunman, 2005), Southeast University of China (Xiang *et al.*, 2004), and CanmetENERGY Technology Centre (Wang and Anthony, 2008) in Canada have carried out research in this area.

The largest prototype chemical-looping combustor has been in operation at Chalmers University of Technology (Lyngfelt and Thunman, 2005) and utilizes a 10 kW_{th} fluidized bed; a 50 kW_{th} CLC has also been constructed in Korea. The Chalmers work is supported by the BIGCO_2 of Europe and is shown in Fig. 14.3 (Tangen, 2008). Most recently, the SINTEF ER/NTNU EPT (SINTEF Energy Research/Norwegian University of Science and Technology, Department of Energy and Process Engineering) laboratory has designed and constructed a small, transparent cold demonstrator CLC rig.

14.2 Principle of systems integration for chemical-looping combustion

Reducing exergy destruction in a combustion process is mainly dependent on increasing the initial temperature (above 1400°C) of the thermal cycle. This approach requires raising the energy level of the energy acceptor, i.e. gas from the outlet of a turbine; however, thermal engine and material properties have put

14.3 First version of 100 kW rotating CLC reactor.

severe limits on the possibilities of this method. Further efforts to improve thermal efficiency should be directed towards innovative combustion processes and new approaches to energy conversion such as CLC.

14.2.1 Principle and theory of systems integration

In the CLC system the utilization of fuel is divided into two steps. In the reduction reactor, the chemical energy of the fuel is converted into the chemical energy of the metal, and at the same time the fuel is converted into CO_2 and H_2O. In the oxidation reactor, the chemical energy of the metal is converted into thermal energy, as the energy level of the oxidation reaction is obviously decreased. That is to say, the chemical energy of the fuel is stored in the metal fuel through the reduction reaction and then released in the oxidation reaction. The CLC system is therefore completely different from the traditional combustion process and thermal cycle.

Cascade utilization of the chemical energy of fuel

CLC has two distinguishing characteristics. One is that the integration of the energy conversion of fuel and the specific endothermic reaction would be expected to change completely the conventional simple and direct combustion of fuel. Namely, fuel does not directly contact the combustion air, but first carries out an endothermic reaction; after that, the resulting products proceed to an exothermic reaction with the combustion air. The second characteristic is that CLC may increase the thermal energy from combustion in comparison with direct combustion, and the increment in thermal energy is equal to that required in the endothermic reactions.

The high thermal efficiency of CLC is due to the energy-level matching in the oxidation and reduction reactions. In particular, the energy level of the reduction reaction and the level of the supplied heat are matched. In addition, the exergy destruction in the oxidation reaction is smaller than that in direct combustion because, in the oxidation reaction, the energy level of the metal oxidation with air is much lower than that of the direct combustion of fuel. The low-temperature heat sources with suitable temperature can decrease the exergy destruction in the reduction reaction. This favorable temperature of the heat sources leads to the improvement of the performance of the CLC system. On the other hand, the amount of heat absorbed by the reduction reaction in CLC significantly reduces the amount of heat that is released to the environment, which causes major exergy destruction in direct combustion.

Integration of combustion and carbon dioxide separation

The high energy penalty is one of the key problems when capturing CO_2 from conventional energy systems. When the CO_2 separation process is integrated with the cascade utilization of the chemical energy of fuel in CLC, this high energy penalty can be greatly reduced. This is mainly due to the fact that the reduction reactor and the oxidation reactor are separated and only solid granules are exchanged between them, the products of fuel oxidation are not diluted by nitrogen.

14.2.2 Mechanism of energy release in chemical-looping combustion

Decrease in the exergy destruction of combustion

The chemical energy of a fuel may be utilized, step by step, from energy level A_{ch1} to energy level A_{ch2}, as shown in Fig. 14.4. Here, the energy level A indicates the ratio of the exergy change to enthalpy change for a given energy conversion process. A_{ch} refers to the energy level of the chemical energy of the fossil fuel, while A_{Th} is the level of thermal energy, for example, the energy level of the high-temperature flue gas at the inlet of a gas turbine.

In CLC, the reduction of the metal oxide converts the chemical energy of the fossil fuel into the metal fuel energy. Since the energy level of the metal fuel is lower than that of the fossil fuel, the energy level is degraded from A_{ch1} to A_{ch2} in the chemical reaction. Subsequently, as the metal fuel carries out the next oxidation reaction, the energy level of chemical energy (A_{ch2}) is converted into that of thermal energy A_{Th}. In this way, the energy level difference between the chemical energy

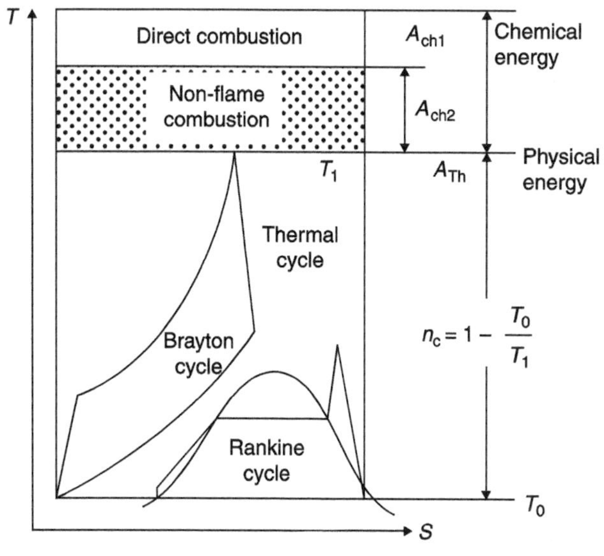

14.4 Schematic of mechanism of energy release of CLC.

and the thermal energy is decreased from ($A_{ch1} - A_{Th}$) to ($A_{ch2} - A_{Th}$), compared with direct combustion. In fact, the energy level from A_{ch1} to A_{ch2} as a 'driving force' is utilized to upgrade the energy level of the middle- and low-temperature thermal energy. That is, the part of the chemical energy corresponding to the energy level from A_{ch1} to A_{ch2} is firstly recovered by means of the chemical reaction, and then the remaining chemical energy corresponding to the energy level from A_{ch2} to A_{Th} is converted into thermal energy. Thus, the chemical energy of the fossil fuel may be indirectly released through the two reactions, rather than through the simple and direct combustion. In other words, CLC can lead to the effective use of chemical energy prior to combustion of fuel, bringing about a decrease in the irreversibility of the conversion of the chemical energy of fuel into thermal energy.

Increase in thermal exergy

Comparing with the stoichiometric equations of the conventional combustion reaction, the sum of the oxidation reaction and the reduction reaction of CLC results in the same combustion reaction. This means that both conventional combustion and CLC reaction systems may have the same input components and produce the same output components, but the metal/metallic oxide never leaves the CLC reaction system.

According to a simple energy balance:

$$\Delta H_{oxd} = \Delta H_{con} + \Delta H_{red} \tag{14.3}$$

where ΔH_{con} and ΔH_{oxd} represent the heat released from conventional combustion and oxidation reactions, respectively, and ΔH_{red} represents the heat absorbed by the reduction reaction. For CLC with an endothermic reduction reaction, the oxidation reaction usually generates more thermal energy at a high temperature through recovery of the thermal energy at a low temperature, compared with conventional combustion. From the viewpoint of the second law of thermodynamics, more work will be converted from the same amount of thermal energy. The potential for increasing work output by upgrading heat at the low- or middle-temperature regions is one of the key advantages of CLC systems. The increase in net work output, which refers to the maximum work output from the CLC power generation system compared with that of the conventional combustion power generation system, can be quantified as below.

The maximum work output W_{con} for a conventional combustion power generation system is:

$$W_{con} = \Delta H_{con} \eta_c \tag{14.4}$$

The maximum work output W_{oxd} for an oxidation reaction power generation system is (keeping the output temperature of combustion the same as for conventional combustion):

$$W_{oxd} = \Delta H_{oxd} \eta_c \tag{14.5}$$

The endothermic reduction reaction requires work input instead of yielding work, and the amount of work input should be subtracted from W_{oxd} to get the total maximum work output of the CLC power generation system. The work input W_{red} is:

$$W_{red} = \Delta H_{red}\eta_c' \qquad [14.6]$$

where η_c and η_c' represent the Carnot efficiency at the output temperature of combustion and at the temperature of the heat source for the reduction reaction, respectively. Consequently, the increase in net work output ΔW is calculated as:

$$\begin{aligned}\Delta W &= W_{CLC} - W_{con} = (W_{oxd} - W_{red}) - W_{con} = (\Delta H_{oxd}\eta_c - \Delta H_{red}\eta_c') - \Delta H_{con}\eta_c \\ &= (\Delta H_{oxd} - \Delta H_{con})\eta_c - \Delta H_{red}\eta_c' = \Delta H_{red}(\eta_c - \eta_c')\end{aligned} \qquad [14.7]$$

Equation [14.7] shows the difference in work output of the CLC power generation system and the conventional combustion power generation system. In reality, the difference in net work output will usually be lower than the result calculated by equation [14.7], which is based on Carnot efficiency. It can be seen that the temperature of the heat source for the reduction reaction is a key parameter for both ΔH_{red} and η_c': the increased net work output will be determined by the difference between the temperature of the heat source and the output temperature of combustion.

14.2.3 Integration of combustion and CO_2 separation

CLC has an inherent advantage for CO_2 separation. In the CLC system, an energy-intensive CO_2 separation process is not required because fuel and air never enter the same reactor. The CO_2 and the water formed at the reduction reactor are therefore never diluted with air as is the case in conventional combustion.

The net work output W_{net} of the system is calculated by subtracting the separation power consumption W_{sep} and the auxiliary power consumption $W_{auxiliary}$ of the system from the gross power output W_{gross}, expressed as:

$$W_{net} = W_{gross} - W_{sep} - W_{auxiliary} \qquad [14.8]$$

where $W_{auxiliary}$ is the total auxiliary power consumption mainly by compressors and pumps of the system.

The difference of net output, ΔW_{net}, between a CLC power generation system with CO_2 separation (the integrated system) and a conventional combustion power generation system can be expressed as:

$$\begin{aligned}\Delta W_{net} &= W_{net,CLC} - W_{net,non} \\ &= (W_{gross,CLC} - W_{gross,non}) + W_{sep,non} \\ &= \Delta W + W_{sep,non}\end{aligned} \qquad [14.9]$$

where the subscript 'non' represents a conventional system, and ΔW has the same meaning as that in equation [14.7]. For simplicity, the auxiliary power consumption is assumed to be the same for all systems.

Equation [14.9] indicates that the increasing net work output from CLC is due to the increasing power output without CO_2 separation and the decrease of exergy destruction in the combustion process. For example, a CH_4-fueled CLC system could have thermal efficiency at 55–62%, which could be four to eight percentage points higher than that of the combined cycle with CO_2 capture (Ishida and Jin, 1994).

14.3 Solid looping materials

The properties of the oxygen carriers are the most important factors determining the performance of CLC. Oxygen carrier materials are composed of an active metal oxide and an inert support/binder. The main function of the inert support is to provide high dispersion of the metal, by increasing the fluidization characteristics and mechanical strength of the oxygen carrier. It is believed that only the metal oxide phase is active in the combustion process, participating in both reduction and oxidation reactions.

A number of different transition state metals and their corresponding oxides have been investigated as possible candidates in experiments: Ca, Cu, Cd, Ni, Mn, Fe, and Co. In most cases, the thermogravimetric analyzer (TGA) method is employed in these experiments. In addition, scaled-down fluidized bed reactors have been used. Natural gas (CH_4), coal syngas (H_2, CO), hydrogen, etc. have been used as fuel.

High reactivity with fuel and air, and high resistance to attrition, fragmentation, and agglomeration, are the most desirable properties for oxygen carriers. It is also essential that the oxygen carrier is inexpensive and environmentally friendly. It should also be fluidizable and stable under repeated reduction/oxidation cycles at high temperature, i.e. have a high regeneration capability.

Because of the sintering or agglomeration problems associated with current carriers, operating temperatures of air reactors are limited to levels incompatible with those of high-efficiency gas turbines. This problem may be solved if carriers capable of withstanding temperatures of up to 1200°C or more can be developed. This problem can also be addressed by using the excess oxygen in the stream exiting the oxidation reactor to burn additional fuel and increase the gas temperature prior to the gas turbine.

Very little attention has been paid to the size of oxygen carriers. It is known that small particles can cause serious damages to the turbine blades, yet a gas clean-up system might lead to large pressure drops and low temperatures for the turbine.

At present, most research on oxygen carriers is mainly based on gas fuel. In future work it is important for researchers to prepare suitable oxygen carriers for solid fuel combustion and to develop cheap, environmentally friendly oxygen carriers.

14.3.1 Desirable characteristics of looping materials

The bulk of oxygen carrier research in the literature focuses on Ni-, Cu-, and Fe-based materials. In early research, Ni-based oxygen carriers were studied by several researchers (Erri and Varma, 2007; Ishida and Jin, 1996; Jin *et al.*, 1999; Villa *et al.*, 2003) because of their high reduction and oxidation rates and good repeatability. However, the Ni-based oxygen carriers are expensive; in addition, they also catalyze carbon formation. Cu-based oxygen carriers have better oxygen capacity; they are less expensive and do not catalyze the cracking of CH_4, which causes carbon formation. Cu carriers prepared by impregnation exhibit excellent regenerability and those prepared by co-precipitation can tolerate higher temperatures than other Cu-based oxygen carriers. However, Cu-based oxygen carriers are prone to agglomerate at high temperature, which limits their application in high-temperature reactions (Cho *et al.*, 2004; Chuang *et al.*, 2008; de Diego *et al.*, 2004). Fe-based oxygen carriers are relatively inexpensive and environmentally safe, but their reactivity is worse than that of the Ni-based carriers (Corbella and Palacios, 2007; Mattisson *et al.*, 2004). As a result of the fact that oxygen carriers based on Ni, Cu, Co, and Mn will inevitably leak into the environment to a certain extent and become secondary pollution sources, $CaSO_4$, which is much cheaper and more environmentally friendly, has been considered as a new type of oxygen carrier by some researchers (Shen *et al.*, 2007; Wang and Anthony, 2008; Zheng *et al.*, 2006).

The cost of oxygen carriers is one of the most important factors determining whether CLC can be commercialized. It is easy to see that the cost of producing synthetic oxygen carriers is too high. In addition, the lifetime of these oxygen carriers in a CLC system with solid fuels may be shortened by deactivation caused by fuel ash or by loss of material with the ash when it is separated from the oxygen carriers. Natural minerals such as ilmenite and perovskite have been considered as cyclic materials (Leion *et al.*, 2008a; Rydén *et al.*, 2008). Ilmenite has the advantages of excellent mechanical strength, low attrition rate, moderate reduction and oxidation rates, and high oxygen capacity. Compared with $Fe_2O_3/MgAl_2O_4$, using ilmenite could result in a 33% reduction in the amount of oxygen carrier needed. It has also been found that carbon formation does not occur in ilmenite during the long defluidization period. Ilmenite is an inexpensive oxygen carrier, costing about 100 times less than a Ni-based oxygen carrier. This could substantially compensate for the large quantity of ilmenite required because of deactivation by ash.

Among the candidate metal oxides, Co offers the highest oxygen carrying capacity. However, Ni, with the potential of transferring 0.5 moles of oxygen per mole of metal, is considered promising given its other favorable properties, as discussed below.

Pure metal oxide particles have been generally found to exhibit poor reactivity and mechanical durability, making them unsuitable for extended use in CLC systems. The bulk of the CLC research reported in the literature focuses on the porous support materials, which are thus used to improve particle porosity and surface characteristics, and regeneration ability, as well as to impart mechanical strength and reduce attrition. In some tests, carriers have been shown to undergo undesirable reactions with some support materials, making them incompatible for use in CLC. Inert materials tested include silica, alumina, yttria-stabilized zirconia (YSZ), kaolin, and various metal aluminates (Cho *et al.*, 2004; Ishida and Jin, 1997; Ishida *et al.*, 2002; Lyngfelt *et al.*, 2001).

Different particle production methods based on mechanical powder mixing, impregnation, and coprecipitation have been proposed (Jin and Ishida, 2001; Mattisson *et al.*, 2003; Ryu *et al.*, 2004; Villa *et al.*, 2003). To increase the reactivity and durability, Al_2O_3, YSZ, TiO_2, and MgO are added in the metal oxides (Jin *et al.*, 1998a, 1999). Table 14.1 shows results from some of the experiments. Results from the characterization of different types of carrier particles are discussed in the succeeding sections.

Ishida and Jin (1997) compared the reaction rates of NiO/YSZ, NiO/Al_2O_3, and Fe_2O_3/YSZ. The results demonstrate that NiO/YSZ has quite good reactivity, as shown in Fig. 14.5, and Jin's experiments show the effect of solid reactant and binder on reactivity (see Fig. 14.6) (Jin *et al.*, 1999). Mattisson and Jerndal (Jerndal *et al.*, 2006; Mattisson and Lyngfelt, 2001) reported on the thermodynamic characteristics of various oxygen carrier materials. Figure 14.7 shows plots of equilibrium constant K (logarithmic scale) versus $1/T$ for the reduction of some common materials using methane as the reducing agent (fuel), with a high log K value indicating potential for the metal oxide to react with methane. It is apparent that, under these conditions, MnO_2/Mn_2O_3, Mn_2O_3/Mn_3O_4, Co_3O_4/CoO, and CuO/Cu_2O have a greater tendency to react with methane than Fe_2O_3/Fe_3O_4 and NiO/Ni. However, MnO_2, Mn_2O_3, Co_3O_4, and CuO decompose into Mn_2O_3, Mn_3O_4, CoO, and Cu_2O, respectively, at low temperatures.

Apart from thermodynamic characteristics, some physical properties such as density, active surface area, pore volume, particle size, and crushing strength are important factors to be considered when selecting an oxygen carrier material (Adánez *et al.*, 2004; Cho *et al.*, 2004). The density and particle size not only determine the fluidizability of the oxygen carrier but may also affect the overall reaction rate due to their influence on mass and heat transfers. Oxygen carriers with particle sizes ranging from 0.08 to 2 mm are considered suitable for CLC.

Complete conversion of fuel is another important desirable characteristic for an oxygen carrier. In order to verify this characteristic, the degree of methane conversion to CO_2 was calculated using the method of minimization of Gibbs free energy by Jerndal and colleagues (Jerndal *et al.*, 2006.). In this calculation, CH_4, CO_2, CO, H_2O, H_2, and O_2 are considered as possible gaseous products. The conversion of methane to CO_2, the overriding chemical transformation property, is described in Fig. 14.8. This analysis shows that Mn_2O_3/Mn_3O_4, CuO/Cu_2O, Fe_2O_3/Fe_3O_4, and NiO/Ni carriers are able to convert methane to CO_2 almost completely.

Table 14.1 The research state of oxygen carrier particles

	Oxygen carrier	Fuel gas	Reduction temperature (°C)	Oxidation temperature (°C)	Dp (mm)
Nakano *et al.* (1986)	Fe_2O_3, Fe_2O_3/Ni Fe_2O_3/Al_2O_3	H_2 H_2/H_2O	700–900	800–1000	0.007
Ishida and Jin (1994)	NiO/YSZ, Fe_2O_3/YSZ NiO, Fe_2O_3/Al_2O_3	H_2	600, 800, 1000	700, 900, 1100	1–3
Ishida *et al.* (1996)	NiO/YSZ	H_2	600, 800, 1000	600, 800, 1000	1.8
Ishida and Jin (1996)	NiO, NiO/YSZ	H_2	600	1000, 1200	2
Jin *et al.* (1999)	NiO/YSZ, Co_3O_4/YSZ Fe_2O_3/YSZ, $CoO\text{-}NiO/YSZ$	H_2, CH_4	600	1000	1.8
Ishida *et al.* (1998)	NiO/YSZ, Fe_2O_3/YSZ, NiO/Al_2O_3, Fe_2O_3/Al_2O_3, NiO/TiO_2, Fe_2O_3/TiO_2	H_2/N_2 CO/CO_2	550, 600, 700, 800, 900		
Jin *et al.* (1999)	NiO/Al_2O_3, CoO/MgO, NiO/TiO_2, Fe_2O_3/Al_2O_3, NiO/MgO, Fe_2O_3/TiO_2, Fe_2O_3/MgO	H_2 H_2O/CH_4	600 700	1000 1000	2.1, 1.8
Hatanaka *et al.* (1997)	NiO	CH_4	400–700		0.07

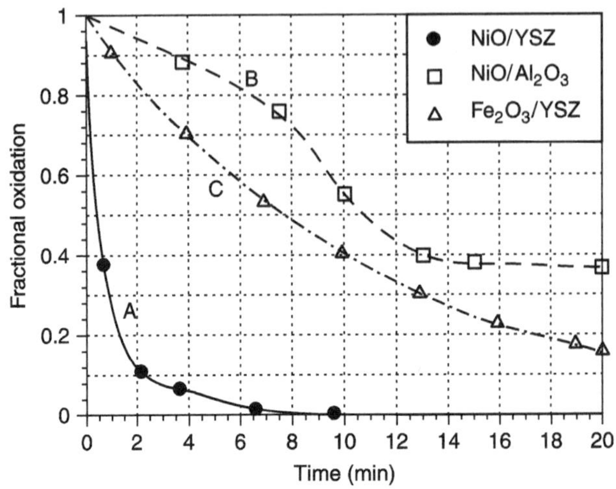

14.5 Comparison of reduction rates.

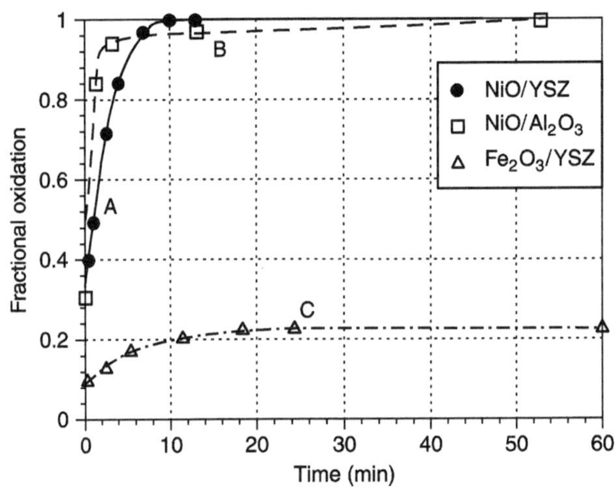

14.6 Comparison of oxidation rates.

Nickel (Ni)-based oxygen carriers

In general, reactivity of unsupported NiO would be reduced after repeated usage due to agglomeration, and therefore it is unsuitable for CLC. Apart from the commonly used materials, Ni/YSZ has shown excellent reactivity and regenerability given the abundant availability of NiO sites with no metal support complexes detected (Jin and Ishida, 2000a, 2001). It was found that NiO loading on YSZ provides high solid diffusivity for the NiO ion and helps to improve the

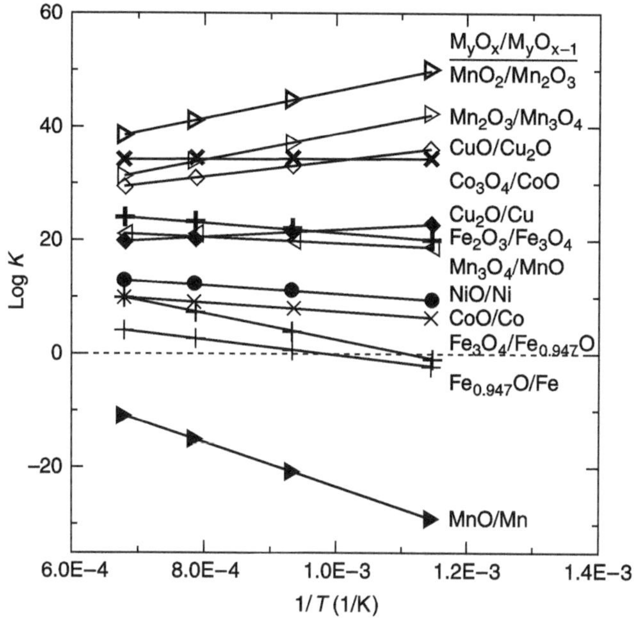

14.7 Log K as a function of $1/T$ in the temperature range 600–1200°C for different metal oxide systems combustion.

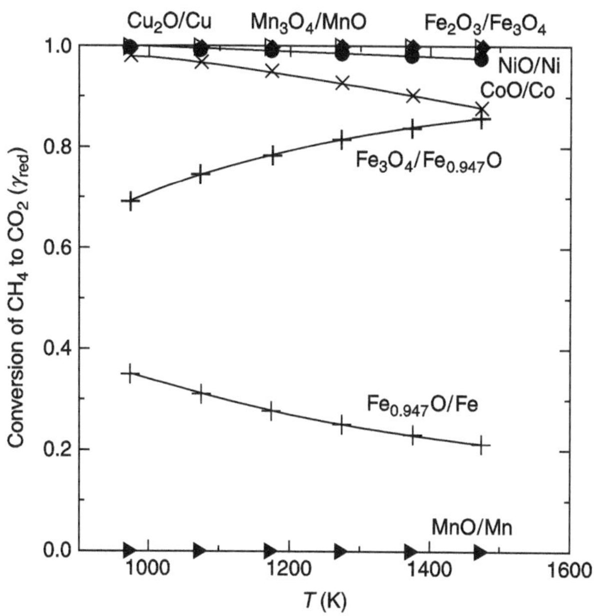

14.8 Conversion of CH_4 to CO_2 for different metal oxides.

composite material reactivity. With Al_2O_3 support, nickel could behave as a dispersed NiO phase. Dispersed phases are less prone to agglomerate after reduction. Consequently, supported nickel shows greater stability when exposed to repeated high-temperature CLC cycles than bulk NiO. In a Ni/Al_2O_3 system, there is always the concern of $NiAl_2O_4$ (nickel aluminate) formation as a result of the interaction between nickel and the alumina support. Nickel aluminate is known to be resistant to reduction below 1000°C (Cho *et al.*, 2005; Jin and Ishida, 2001). Hence, once nickel aluminate has formed on the oxygen carrier, it cannot then participate in the fuel combustion cycle. It is therefore essential to minimize metal/support contact in order to reduce nickel aluminate formation. X-ray diffraction (XRD) analysis of used samples shows that, during the initial cycles, the excess nickel reacts with Al_2O_3 forming $NiAl_2O_4$, and this helps to improve oxygen carrier stability.

Recently, Ni/$MgAl_2O_4$ has also been investigated as an alternative to Ni/$NiAl_2O_4$ (Johansson *et al.*, 2006; Villa *et al.*, 2003; Zafar *et al.*, 2006). The addition of Mg limits the sintering of NiO and stabilizes the Ni^{2+} in cubic (NiO) and spinel ($NiAl_2O_4$) phases. As a result, the oxygen carrier remains stable over repeated reduction and oxidation cycles even at temperatures above 1300°C.

Several researchers have also focused on Ni/TiO_2 materials (Adánez *et al.*, 2004; Jin *et al.*, 1999; Son and Kim, 2006). Cyclic reduction and oxidation of such materials shows lower reactivity than nickel supported on Al_2O_3. In fact, NiO was more prone to interact with TiO_2, forming $NiTiO_3$ (nickel titanate), which is less reducible than NiO. A disadvantage of this carrier is its tendency towards coke formation. Thus, the reduced carrier may contain coke, form CO_2 in the air reactor, and decrease the overall CO_2 capture efficiency.

Ni-based carriers tested in a small-scale fluidized bed exhibited good reactivity and mechanical durability, with virtually complete CH_4 conversions being achieved at 1050°C. Results also showed the carrier to be suitable for partial oxidation reforming of methane at 830–950°C. No difficulties with particle agglomeration or sintering were encountered (Copeland *et al.*, 2001). Despite some of the described limitations, nickel supported on alumina could hold significant promise as a potential oxygen carrier material for large-scale CLC application.

Iron (Fe)-based oxygen carriers

Iron-based oxygen carriers, such as oxides of magnetite (Fe_3O_4), hematite (Fe_2O_3), and wustite (FeO), are environmentally safe and more cost-effective than other oxides such as nickel (NiO) and copper (CuO) (Hossain and de Lasa, 2007). Due to their ready availability and low price, and the fact that they are environmentally friendly, iron oxides have attracted wide attention as oxygen carriers for use in CLC. It was confirmed that the reduction kinetics from hematite to magnetite ($Fe_2O_3 \rightarrow Fe_3O_4$) is the fastest while that of magnetite to ferrous oxide ($Fe_3O_4 \rightarrow FeO$) and ferrous oxide to iron ($FeO \rightarrow Fe$) is much slower.

For reduction with methane at 600°C, the Ni-based carrier exhibited reaction rates approximately 15 times faster than the Fe-based carrier. For oxidation at 1000°C, final conversion ratio of Fe-based particles was limited to about 0.25 (Ishida and Jin, 1997). Pure Fe_2O_3 was studied as an oxygen carrier at 720–800°C and was found to have excellent chemical stability and no loss of reactivity with cyclic redox. However, it started to agglomerate at 900°C, although the agglomeration rate was slow (Copeland *et al.*, 2002). Previous researchers have found that agglomeration and breakage of the particles could be avoided by adding Al_2O_3 (Cho *et al.*, 2000; Ishida *et al.*, 2005) prepared Fe_2O_3/Al_2O_3 composite particles and evaluated their applicability as solid looping materials. They found that two solid solutions, hematite(ss) and corundum(ss), were formed in Fe_2O_3/Al_2O_3 composite particles at temperatures above 1000°C. The mechanical strength of the Fe_2O_3/Al_2O_3 particles was improved by increasing the content of corundum(ss). Cho and colleagues (2004) compared Fe-, Ni-, Cu-, and Mn-based oxygen carriers for use in CLC. It was observed that oxygen carriers based on Ni, Cu, and Fe showed high reactivity, enough for their use in CLC systems to be feasible. However, samples of the Fe_2O_3/Al_2O_3 showed signs of agglomeration. The same research group developed Fe_2O_3-based oxygen carriers, together with various inert such as Al_2O_3, ZrO_2, TiO_2, and $MgAl_2O_4$, and found that Fe_2O_3/Al_2O_3 exhibited good reactivity (Mattisson *et al.*, 2004). An Fe-based oxygen carrier composed of 80 wt% Fe_2O_3 with 20 wt% Al_2O_3 was prepared using impregnation methods (He *et al.*, 2007a) and showed good reactivity in 20-cycle redox tests in a TGA reactor; 85% of the CH_4 was converted to CO_2 and H_2O during most of the reduction periods, with minor formation of CO and H_2. Abad *et al.* (2007) investigated the performance of Fe-based oxygen carriers in a continuously operating laboratory CLC unit, consisting of two interconnected fluidized beds using natural gas or syngas as fuel. The combustion of fuel gas was stable during the operation of the reactor. The combustion efficiencies of syngas and natural gas reached 99% and 94%, respectively. The reactivity and the crushing strength of the oxygen carrier particles were not affected significantly during operation. Agglomeration and carbon deposition were not observed and no mass loss of the solids was detected. It is clear that attrition and agglomeration of Fe_2O_3/Al_2O_3 oxygen carrier can be controlled under the right operating conditions, and therefore Fe_2O_3/Al_2O_3 is a promising oxygen carrier candidate for CLC.

Experiments with Fe-based carriers in a scaled-down fluidized-bed reactor indicate a particle life of the order of 10^6 cycles based on attrition rates. However, operating temperatures were limited to 800°C owing to agglomeration or sintering at higher temperatures. This restriction limits the efficiency of the CLC system unless additional fuel is used to reheat the flue gas or partially spent air from the oxidation reactor (Copeland *et al.*, 2001). Tests with natural hematite as the carrier have shown satisfactory results in terms of reactivity, with the particles showing some evidence of fragmentation after multiple cycles.

14.3.2 Role of binder in the looping materials

In addition to Ni- and Fe-based materials, research has been conducted on other materials such as perovskite, $CaSO_4$, and mixed-metal oxides. The development of $NiO/NiAl_2O_4$ and CoO–NiO/YSZ (Jin *et al.*, 1998b, 1999) has led to other research work (Tan *et al.*, 2006).

When using Al_2O_3 as an additive material and NiO as the reactant, a compound of the spinel type of $NiAl_2O_4$ is formed, which can reduce the content of the reactive component and hence the reduction conversion rate. Since $NiAl_2O_4$ is stable below 1173K, this has led to the synthesizing of a looping material of $NiO/NiAl_2O_4$. The reactivity of this new material has been compared with that of other materials. It is clear that both reduction and oxidation rates of this new material are much higher than those of NiO/YSZ. In particular, both reactions can be performed completely. Another advantage is that the cost of $NiAl_2O_4$ is less than 20% of the cost of YSZ.

By combining NiO with CoO, a new looping material was synthesized comprising a double metal oxide and added YSZ. This new looping material, CoO–NiO/YSZ, had excellent reactivity, no carbon deposition, and good repeatability. Figure 14.9 illustrates cross-sectional photographs of CoO–NiO/YSZ particles. The microstructure after the tenth cycle was similar to that of the fresh structure, with grains 0.5–2.0 mm in diameter, which indicated that CoO–NiO/YSZ was relatively stable over the course of multiple cyclic reactions compared with the single metal oxides.

Adánez *et al.* (2006) prepared mixed Ni–Cu oxides and examined their performance for methane CLC. The presence of CuO in the Ni–Cu oxygen carriers permits full conversion of CH_4 to CO_2 and H_2O with zero CO and H_2 emissions. Additionally, the presence of NiO in the Ni–Cu oxygen carrier allows the particles to operate at high temperatures (950°C). Hossain *et al.* (2007) developed a bimetallic $Co–Ni/Al_2O_3$ oxygen carrier for a fluidized bed CLC process. The $Co–Ni/Al_2O_3$ particles displayed excellent reactivity and stability. The study confirmed that the inclusion of Co in the double metallic $Co–Ni/Al_2O_3$ particles influences the state of the surface and minimizes the formation of nickel aluminate, which was believed to be contributing to the inferior reactivity of Ni-based oxygen carriers. The addition of Co also inhibits metal particle agglomeration during cyclic redox processes. The activation energy for $Co–Ni/Al_2O_3$ reduction was found to be less

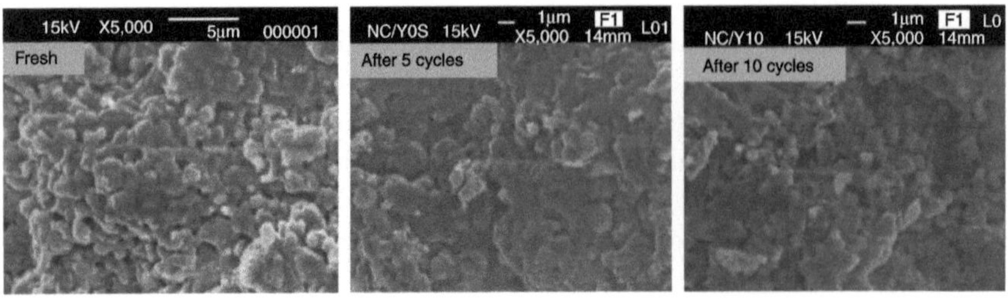

14.9 Cross-sectional photos of CoO–NiO/YSZ in the cyclic reaction.

than that of unpromoted Ni/Al_2O_3 samples. This suggested that doping with Co decreases the metal–support interaction and the binding energies between the metals and the fuel molecules (Hossain and de Lasa, 2007).

Readman *et al.* (2005) studied the feasibility of $La_{0.8}Sr_{0.2}Co_{0.2}Fe_{0.8}O_3$-$\delta$ as a potential oxygen carrier in a chemical-looping reactor. The results suggested that $La_{0.8}Sr_{0.2}Co_{0.2}Fe_{0.8}O_3$-$\delta$ has the redox properties required for chemical looping. Reduction and re-oxidation of the perovskite take place quickly enough for CLC. However, it displayed a low oxygen carrying capacity.

14.3.3 Carbon deposition considerations

Carbon formation may occur during the reduction period if carbon-containing fuels are used in CLC. Such formation could occur via two mechanisms: pyrolysis of methane and the Boudouard reaction:

$$CH_4 \rightarrow C+2H_2 \text{ (pyrolysis)} \qquad [14.10]$$

$$2CO \rightarrow C+CO_2 \text{ (Boudouard reaction)} \qquad [14.11]$$

The pyrolysis of methane is thermodynamically favored at high temperatures as it is an endothermic reaction. The Boudouard reaction is exothermic and thus is more likely to occur at lower temperatures. When Ni-based oxygen carriers react with carbonaceous fuel, fuel cracking causes carbon deposition, which affects the regenerability of oxygen carriers. In order to prevent carbon deposition, CH_4 was saturated at a ratio of $H_2O/CH_4 = 2.0$ (Ishida *et al.*, 1998; Jin *et al.*, 1999). The addition of water vapor may help the steam reforming and shift reactions (CH_4+H_2O=CO+$3H_2$; CO+H_2O=CO_2 + H_2), and it could therefore lead to far less carbon deposition.

In studying carbon deposition behavior for particles of NiO/YSZ and NiO/$NiAl_2O_4$ (Fig. 14.10), it is found that, before saturation, the particle weights were sharply increased due to carbon deposition after 200 s. However, it was found that NiO/$NiAl_2O_4$ had no weight increase after saturation of fuel. This indicated that carbon deposition can apparently be reduced by the addition of water into CH_4. Similar phenomena and conclusions have been reported by other researchers (Erri and Varma, 2007).

Chandel *et al.* (2009) suggested that both the pyrolysis and Boudouard reactions are slow in CLC without a catalyst. However, transition metals such as Ni and Fe can act as catalysts for methane decomposition. Metallic Ni in particular is well known to be a good catalyst for the thermal decomposition of hydrocarbons. These kinds of reactions are undesirable in CLC since they increase methane consumption, which competes with the main reduction reaction. The carbon formed is burned into CO_2 during the oxidation stage, resulting in lower efficiency of CO_2 capture. It was found that when using Ni-based oxygen carriers, carbon formation can be reduced by using mixed oxides as oxygen carriers or by adding steam to the fuel gas (Corbella *et al.*, 2006). Other parameters – such as oxygen

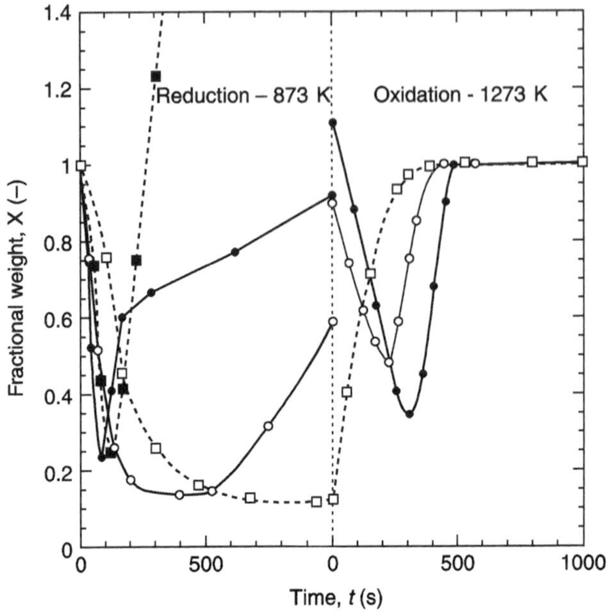

14.10 Effect of a double oxide particle on carbon deposition. Key: (–●–) NiO/YSZ(CH_4); (–○–)NiO/YSZ(H_2O/CH_4=2.0); (–□–)NiO/$NiAl_2O_4$(CH_4); (–■–) NiO/$NiAl_2O_4$(H_2O/CH_4=2.0).

availability of oxygen carriers, fuel conversion, temperature, and pressure – could affect carbon formation. Generally, carbon tends to be formed at low temperatures when the amounts of added oxygen are small.

Carbon formation should not be a problem under the conditions used in a CLC system where the conversion rate of the fuel is high. At a temperature of 950°C, no carbon formation is expected as long as more than 25% of the oxygen needed for complete oxidation of CH_4 is supplied (Villa *et al.*, 2003). If the CLC process is operated under optimum conditions, and a percentage of steam or CO_2 is added into the fuels, carbon formation is considerably inhibited.

14.3.4 Regeneration capability of looping materials

Regenerability is also important for looping materials. To date, researchers have examined NiO/YSZ, CoO/YSZ, and Fe_2O_3/YSZ particles as looping reaction materials and investigations have been conducted on reaction kinetics and carbon deposition, etc. by means of various particle preparation methods. While some of the materials have shown good reactivity, their regenerability and resistance to carbon deposition are not ideal. The material NiO/$NiAl_2O_4$, created by integrating NiO with the spinel-type metal oxide of $NiAl_2O_4$, gives complete conversion and fast rates for looping reactions (reduction and oxidation). In particular, this material has excellent regenerability in cyclic use. Carbon deposition can be completely avoided at a ratio of H_2O/CH_4=2.0. These promising results suggest that $NiO/NiAl_2O_4$ will provide outstanding performance as a solid looping material in CLC (Jin *et al.*, 1999).

NiO/YSZ has good reduction and oxidation rates, but its drawbacks are the higher carbon deposition rate, accompanied by an increase in both the grain and pore sizes, and the lower ability to regenerate. CoO/YSZ has good reactivity and a lower carbon deposition rate, but its regenerability is insufficient. The double metal oxides of NiO and CoO are promising for obtaining an excellent overall performance, with good reactivity, no carbon deposition, and significant regenerability over repeated cycles of reduction and oxidation (Jin *et al.*, 1998b).

14.4 Design of chemical-looping combustion systems

CLC requires perfect contact between the gas and solid oxygen carriers, as well as a significant flow of solid material between the reactors. It is essential to minimize gas leakage between the air and fuel reactors. A gas (fuel) leakage from the fuel reactor into the air reactor would cause CO_2 release into the atmosphere, reducing CO_2 capture efficiency. A gas (air) leakage from the air reactor into the fuel reactor would dilute the flue gas stream with N_2, adding extra cost to CO_2 separation. Most CLC reactor designs have focused on fixed bed and fluidized bed reactors (Adánez *et al.*, 2004; Bolt *et al.*, 1998; Corbella *et al.*, 2005; Jin and Ishida, 2000b; Johansson *et al.*, 2004; Mattisson *et al.*, 2006; Son and Kim, 2006).

Figure 14.11 is a fixed bed reactor for mass particles of solid–gas reaction operating at elevated high pressure (up to 9 atm) and high temperature (up to 1473 K) (Jin and Ishida, 2002). This reactor consists of a reaction tube (point 8 in the

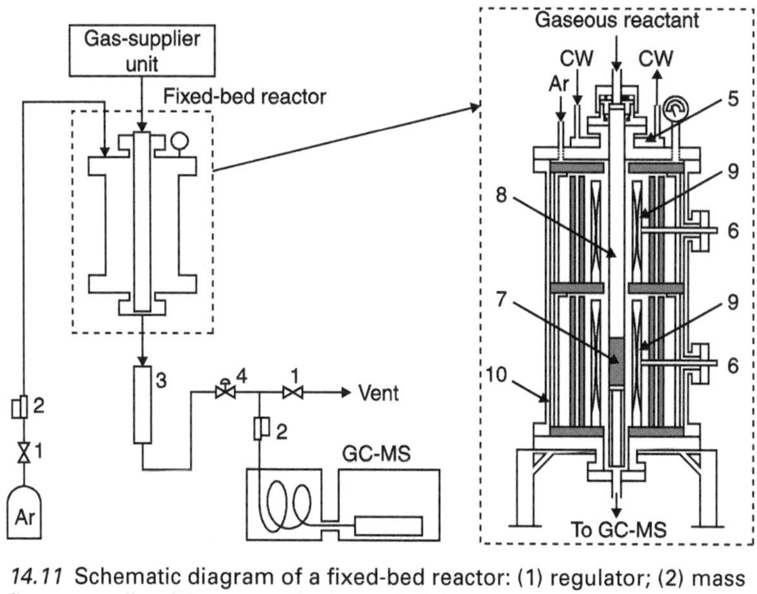

14.11 Schematic diagram of a fixed-bed reactor: (1) regulator; (2) mass flow controller; (3) vapor stripping unit; (4) auto pressure regulator; (5) water cooler; (6) type R thermocouple and controller.

figure), electric heater (9), and pressure shell (10). Pellet-shaped particles were supported on the Al_2O_3 disk (7) in a reaction tube (16 mm inner diameter) made of Al_2O_3. The length of the reaction tube was 900 mm, in order to preheat the gaseous reactants to a specified temperature. In this reactor, an auto-pressure regulator was used to control pressures. The main part of this apparatus is the gas chromatographic–mass spectrometric (GC-MS) system (JEOL GC-Mate), which can detect even very small amounts of gases exhausted from the reactor.

Figure 14.12 illustrates an interconnected fluidized bed CLC system. For this system, a high-velocity riser and a low-velocity bubbling fluidized bed are considered as the air and fuel reactors, respectively. The solid particles leaving the riser are recovered by a cyclone and sent back to the fuel reactor. In the low-velocity fluidized bed (fuel reactor), the oxygen is transferred from the carrier to the fuel. In this unit, particles circulate mainly by gravity and, as a result, the fuel reactor has to be placed at a sufficient height in the plant. In the high-velocity air riser reactor, the volumetric flow rate is approximately ten times greater than that

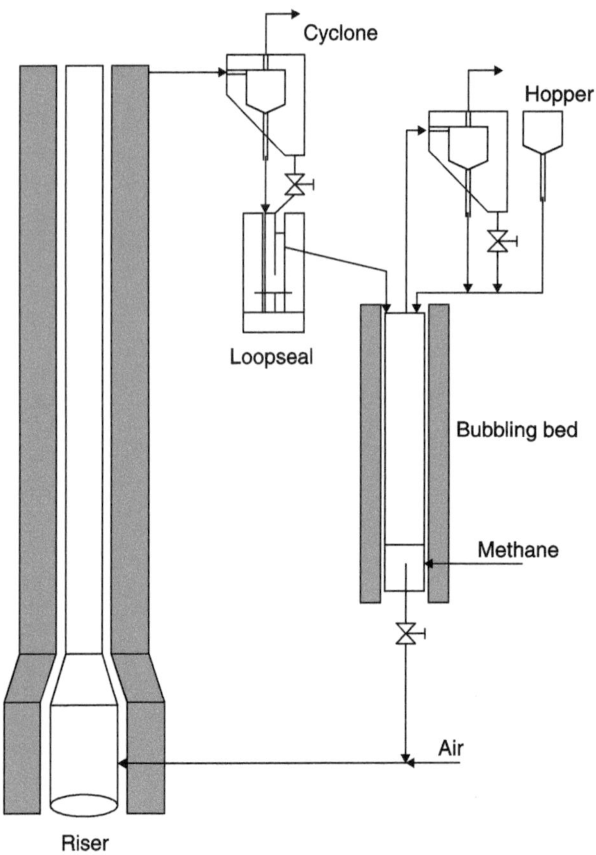

14.12 Layout of CLC process, with two interconnected fluidized beds.

in the fuel reactor. In order to maintain a comparable reactor size, a high velocity is employed for the air reactor. In addition, the high velocity in the riser (air reactor) provides the required driving force to circulate the particles between the two interconnected beds.

In order to minimize the capital and operating costs, Koronberger *et al.* (2005) proposed a two-compartment fluidized bed reactor configuration as shown in Fig. 14.13. This system has two adjacent fluidized beds, separated by a vertical

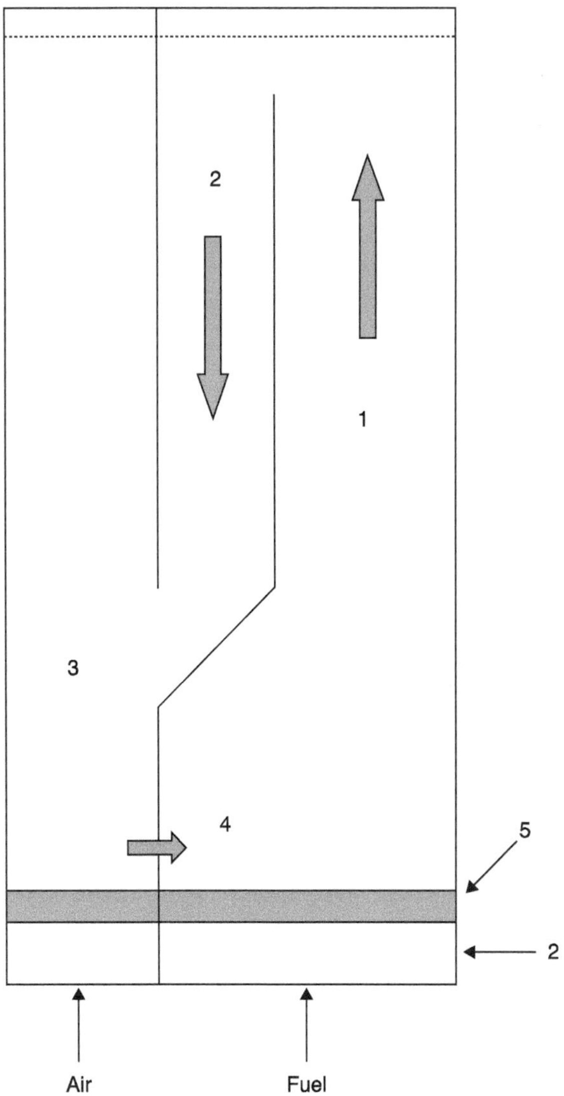

14.13 A two-compartment fluidized bed. (1) Air reactor; (2) downcomer; (3) fuel reactor; (4) slot; (5) gas distributor.

wall with two orifices. The high-velocity gas in the air reactor forces particles to travel upwards, with some of the particles falling into the fuel reactor.

Lyngfelt *et al.* (2001) proposed two loop-seal devices, one placed between the air reactor and the cyclone, the other located between the fuel reactor and the air reactor. A loop-seal device with steam injection into the downer (fuel reactor) can also help by creating a gas barrier, which minimizes gas leakage.

Several designs for small-scale fluidized bed and interconnected fluidized bed reactors have been proposed since 2001, and prototypes for CLC have been designed, built, and run. The interconnected fluidized bed reactor designed by Professor Anders Lyngfelt and colleagues of Chalmers University of Technology (Lyngfelt *et al.*, 2001; Lyngfelt and Thunman, 2005), is currently considered to be the best and largest reactor (Fig. 14.14) and uses a Ni-based oxygen carrier. The three main components of this system are: (1) a high-velocity riser; (2) a cyclone; and (3) a low-velocity bubbling fluidized bed. Tests conducted in a scaled cold model (Johansson *et al.*, 2003) suggested that typical leakage from the fuel reactor was 2%, i.e. a CO_2 capture efficiency of 98%. No leakage was detected between the cyclone and the fuel reactor. The typical leakage from the pot-seal into the fuel reactor was about 6%, which would dilute the CO_2 produced by approximately 6% air. However, this gas leakage can be avoided by using steam instead of air to fluidize the whole, or part, of the pot-seal.

A 10 kW CLC reactor was designed and built in 2002/2003 by Chalmers University under EU GRACE project funding. It is a circulating fluidized bed with an extra bubbling fluidized bed after the cyclone. It has two interconnected fluidized bed reactors, a fuel and an air reactor, a cyclone to separate gas and solid flow from the air reactor, and two loop seals. The gas velocity in the air reactor and riser provides the driving force for the circulation of particles between the two beds. Entrained particles are recovered in the cyclone and brought to the fuel reactor through a downcomer. The fuel reactor is a bubbling fluidized bed. Particles are circulated back to the air reactor due to gravitational force. The entire reactor system hangs on a scaffold and is guided vertically using rails below the air and fuel reactors. A supervision system has been installed to allow unmanned operation of the reactor. The supervision system permits operation at combustion conditions for long periods of time, e.g. 24 h, with continuous natural gas feed.

The CLC system at Chalmers University was further expanded with the addition of a 10 kW_{th} reactor that uses solid fuel. This significant part was designed by Berguerand (Berguerand and Lyngfelt, 2008a) and it is rather similar to the existing gas-powered 10 kW_{th} system that has been in operation, but important modifications to the fuel reactor chamber and the inclusion of an additional solids recirculation loop were made to adapt the system for solid fuels. The system consists of an air reactor where the metal oxides are oxidized, a riser leading to a cyclone where elutriated particles are separated, and a fuel reactor which was divided into three chambers. Two particle locks are located between solids inlets/ outlets of the air and the fuel reactors. They accommodate for the pressure

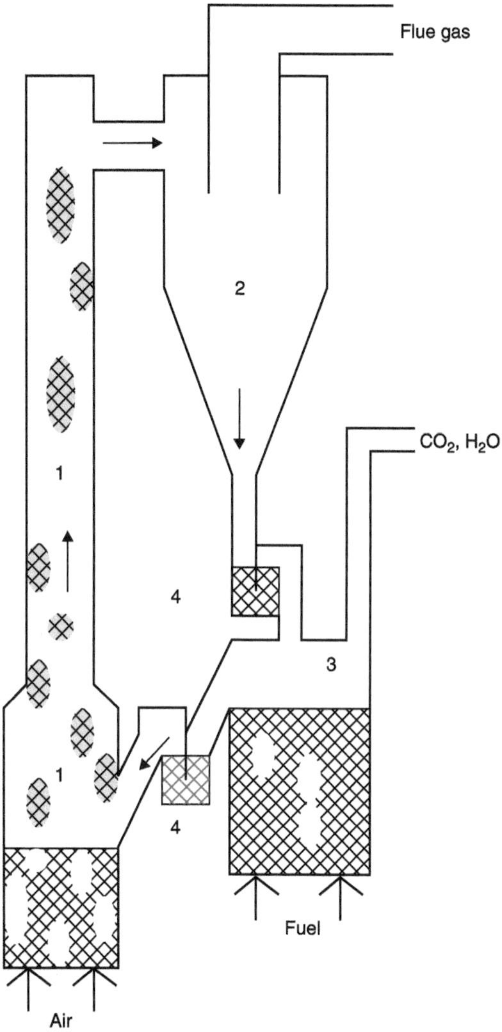

14.14 Layout of chemical-looping combustion process, with two interconnected fluidized beds: (1) air reactor; (2) cyclone; (3) fuel reactor; (4) high-velocity riser.

differences between the reactors and prevent gas mixing between the reactors. In the fuel reactor, steam is used as gasifying/fluidizing gas with low velocity and nitrogen is used in the carbon stripper with high velocity. It is also possible to fluidize the carbon stripper and the high-velocity part with steam, which would gasify the remaining coal to some extent and very likely improve the separation in the carbon stripper. The particle locks are fluidized with nitrogen although steam is another possibility.

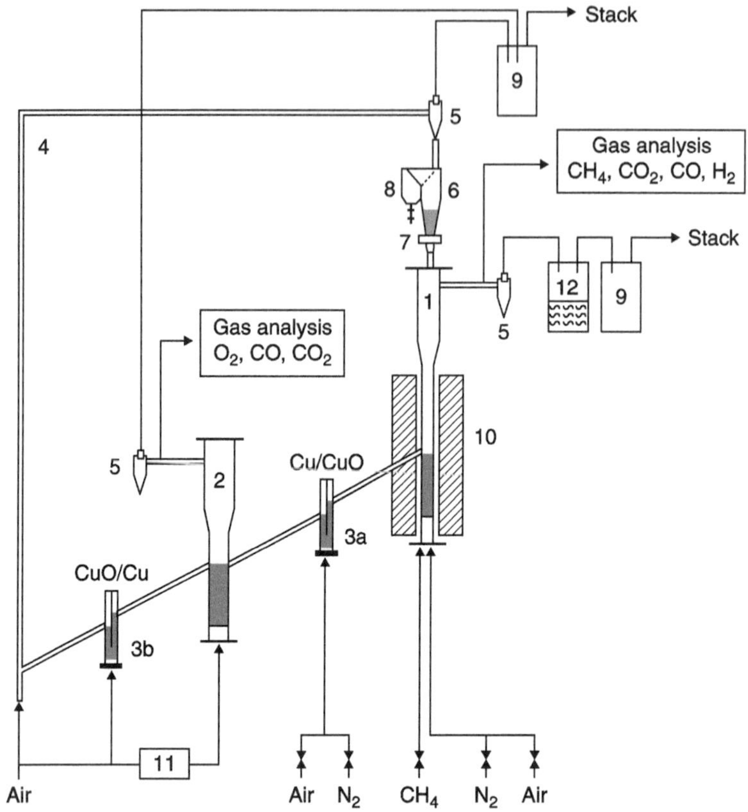

14.15 Schematic diagram of the CLC prototype: (1) fuel reactor; (2) air reactor; (3) loop seals; (4) riser; (5) cyclones; (6) solid reservoir; (7) solids valve; (8) diverting solid valve; (9) filters; (10) oven; (11) air preheater; (12) water condenser.

De Diego *et al.* (2007) of CSIC (Consejo Superior de Investigaciones Cientificas) in Spain have also successfully built a 10 kW_{th} CLC prototype. A schematic diagram of the 10 kW_{th} chemical-looping combustor used is shown in Fig. 14.15. The plant was designed to accommodate changes in the solid fuel flow rates while keeping the fuel-to-oxygen ratio constant. The fuel reactor is a bubbling fluidized bed (0.1 m inner diameter) with a bed height of 0.5 m and a freeboard of 1.5 m. A nearby furnace is used to provide heat to the CLC reactor during start-up and for accurate control of the operating temperature. Solids reduced in the fuel reactor are transported to the air reactor through a loop-seal fluidized bed reactor. Regeneration of the oxygen carrier takes place in the air reactor allowing residence times long enough to achieve complete oxidation of the reduced carrier. The regenerated oxygen carrier is returned to the fuel reactor by gravity from the solids reservoir located above a solids valve which controls the flow rates. A diverting solids valve located below the cyclone allows the

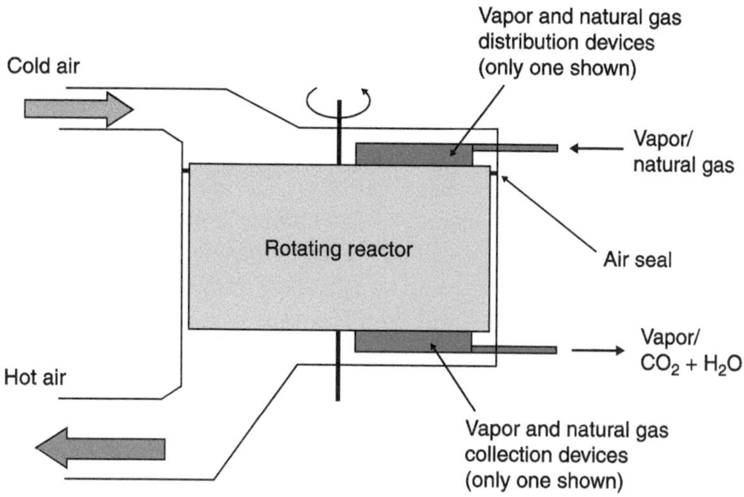

14.16 IFP rotating reactor.

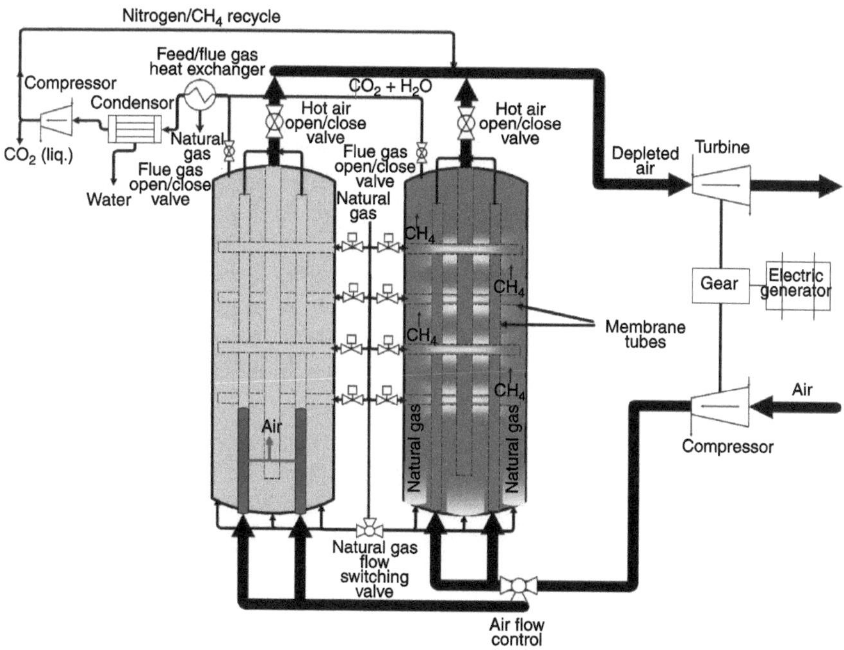

14.17 TNO membrane-assisted reactor.

measurement of solid flow rates at any time. The fine particles produced by fragmentation/attrition in the plant are recovered by the cyclones and the filters located in the fuel reactor and riser lines. There are two loop seals in the system to avoid solids back flow and gas mixing between reactors. A nitrogen loop seal prevents mixing of the fuel gas and oxygen while solids are flowing through it.

In ENCAP (Enhanced CAPture of CO_2) Sub-Project 4 (SP4), alternative concepts were also explored to evaluate the potential of novel fixed-bed reactor concepts in CLC. The study focused on natural gas as the primary fuel. Two studies were carried out in parallel: IFP energies nouvelles studied the case of rotating monolith-based CLC (Fig. 14.16) and TNO (http://www.tno.nl) developed the membrane-assisted CLC reactor (Fig. 14.17). In order to avoid particles from a fluidized bed CLC reactor entering the gas turbine, IFP proposed a rotating reactor that allows for continuous production of hot air on one side and CO_2 on the other side. The oxido-reduction reactions take place in a monolith coated with appropriate material. The reactor developed by TNO uses membranes to form a physical barrier between the active metal-containing particles and the oxidizing and reducing gas streams (Fig. 14.18). This barrier is a macro porous membrane allowing the gas streams to diffuse while the solid particles stay fixed. Some of the advantages are the smaller losses at the exhaust, the enhancement of the performance of the system due to high reactivity, and the immobilization of the particles, which results in no attrition and less stress on the particles. In addition, as the design does not contain any moving parts, high pressure could be applied between oxidation and reduction. A pilot reactor was set up at TNO and initial experiments (up to 1200°C) were run within the framework of the ENCAP project.

A rotating wheel CLC reactor rated at 100 kW_{th} is currently under development by SINTEF and NTNU. The SINTEF and NTNU CLC facility is the biggest in

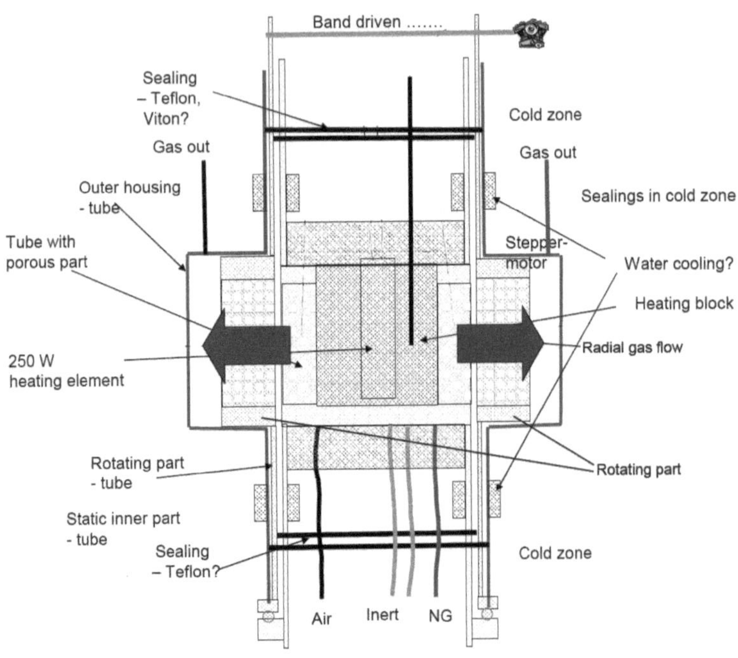

14.18 Layout of the rotating CLC reactor.

the world and is supported by the BIGCO2 Research and Development Platform. At present, the design and construction of a small transparent cold demonstrator CLC rig has been completed in the SINTEF ER/NTNU EPT laboratory and the rig is ready for testing (http://www.encapco2.org/publications/SP4_3_6and4_3_7_Summary_Report_final.pdf). The cold CLC rig will be used for gathering empirical input for reactor modeling and for identifying flow regimes; it will also serve as a basis for the final design of the hot rig.

Jung and Gamwo developed a reaction kinetics model for the fuel reactor and incorporated it into a multiphase hydrodynamic model (Jung and Gamwo, 2008). Simulation results revealed high weight fractions of unburned methane fuel in the flue gas along with CO_2 and H_2O. The low fuel conversion rate is partially due to fast and large bubbles rising through the reactor. In order to reduce the amount of unburned methane fuel in the flue gas, a nano-sized metal oxygen carrier might be preferable as it does not generate large bubbles and could increase the fuel conversion rate in the reactor.

Owing to the stringent requirements for heat transfer, reaction rate, conversion rate, and prevention of leaks, research on CLC reactors is still at an early and somewhat difficult stage. The problems center mainly on how to lower the reduction pressure, reduce the abrasion of the oxygen carrier, and prevent gas leakage. More insights and understanding will be gained if the theoretical models of reactors become more precise and optimized.

14.5 Chemical-looping combustion systems with different fuels

As mentioned above, most of the CLC cycles have used CH_4 as fuel. The results presented for CH_4 are highly relevant for common gaseous fuels such as natural gas and refinery gas. Other fuels, such as H_2 and syngas from coal gasification, have also been investigated for use in CLC.

14.5.1 H_2-fueled chemical-looping combustion system

The H_2/O_2 gas turbine cycle is a special thermal cycle with two serious areas of concern. First, from the exergy principle, since the heat exchange for steam generation is shifted into the combustor, the result is a significant energy level degradation from extremely high temperatures for hydrogen combustion (>2500°C) to very low temperatures for water evaporation (<350°C). The energy level degradation for steam generation in this cycle is much worse than that of the heat recovery steam generator (HRSG) from medium temperature (650°C) to low temperature (<350°C) in a combined cycle. The other issue is the cost of oxygen production. To address these problems, Jin and Ishida (2000a) proposed a novel gas turbine cycle with hydrogen-fueled CLC.

The hydrogen-fueled CLC consisted of two successive reactions: a hydrogen with metal oxide (reduction) reaction and the resulting metal with air (oxidation) reaction. The hydrogen fuel first reacts with the solid NiO in a reduction reactor, producing solid Ni and steam. In the next reactor, i.e. the oxidation reactor, saturated air at an elevated pressure reacts with the solid Ni produced by the first reactor, yielding NiO and high-temperature flue gas via strongly exothermic oxidation. Finally, the gases exiting from both reactors could be used as the working fluid to generate power in turbines. In this new gas turbine cycle employing hydrogen-fueled CLC, the exergy losses in combustion and power consumption were much lower than those in the H_2/O_2 combined cycle. As a result, the thermal efficiency of this new cycle with current-technology gas turbines (turbine inlet temperature of 1200°C) could be as high as 63.5%, a rise of at least 12 percentage points compared with that of the H_2/O_2 combined cycle (1350°C).

14.5.2 Coal-gasification chemical-looping combustion system

Jin and Ishida (2002, 2004) found that the reactivity of NiO with coal gas is much higher than that with natural gas (CH_4) in CLC when adopting CH_4 and coal gas as fuel and NiO/$NiAl_2O_4$ as the solid material. This is mainly due to the fact that the reduction of NiO with coal gas (CO and H_2) is a strongly exothermic reaction, whereas reduction with CH_4 (natural gas) is endothermic and has less force to drive the reaction. This is completely different from traditional combustion, in which natural gas is considered as an ideal fuel. It was also observed that there was no damage to the solid material samples after more than ten cycles of reduction and oxidation. This means that coal gas-fueled CLC has high reactivity, no carbon deposition, and high regenerability.

Based on this experimental result, a new integrated coal gasification combined cycle was proposed that incorporated the ideas of CLC and saturation of air (IGCLSA) (Jin and Ishida, 2000b) (see Fig. 14.19). The system consists of three main parts: a coal gasification and purification unit, a CLC unit, and a thermal cycle unit with saturated air. Purified coal gas mainly composed of H_2 and CO entered into the chemical-looping combustor and reacted with metal oxide ($H_2+CO+2NiO \rightarrow H_2O+CO_2+2Ni$). In the fuel reactor, the products H_2O and CO_2 were discharged from the upper part, and metal Ni exiting from the lower part entered into the air reactor and was oxidized into NiO by air ($Ni+0.5O_2 \rightarrow NiO$). As there was a strongly exothermic process taking place, the temperature of the gas increased sharply. The energy level of the Ni oxidation process was lower than that of direct combustion, and this provided the potential to reduce the energy level difference during the reaction process. Thermal performance analysis indicated that thermal efficiency of IGCLSA could reach 51.3%, net power output was 232.4 kJ/mol C, system efficiency was 5–10% higher than conventional IGCC, and the generation of CO_2 per kWh electricity could be significantly decreased from 0.52 kgCO_2/kWh to 0.36 kgCO_2/kWh.

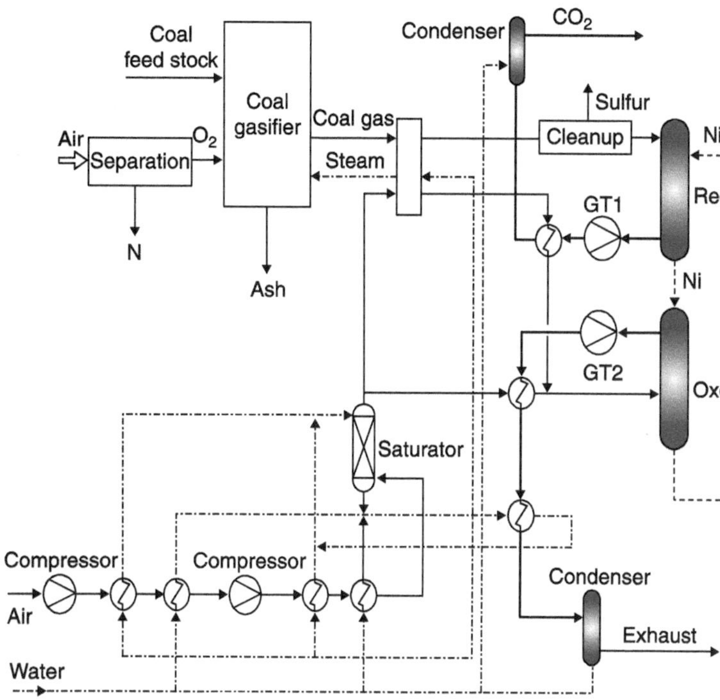

14.19 A process flow diagram for IGCLSA.

14.5.3 Solid coal with chemical-looping combustion

Most CLC research has focused on gaseous fuels. Since solid fuels are low cost and abundant, it is essential to explore the possibilities of using solid fuel in CLC. A 10 kW_{th} chemical-looping combustor for solid fuels was designed by Chalmers University of Technology (Berguerand and Lyngfelt, 2008a). The design for the 10 kW_{th} solid fuel pilot facility is rather similar to the existing gas-powered 10 kW_{th} facility at Chalmers, but important modifications in the fuel reactor chamber were made and an additional solids recirculation loop was included to adapt the system for solid fuels. Tests were conducted on this combustor with a South African coal and a petroleum coke as solid fuels, and with ilmenite as the oxygen carrier (Berguerand and Lyngfelt, 2008a, 2008b; Leion *et al.*, 2008b). The CO_2 capture ranged between 82.5% and 96% for the South African coal and from 60% to 75% for the petroleum coke. The latter was lower due to petroleum coke being a less reactive fuel than the South African coal. Although there are significant practical problems, mainly with 'external' functions such as fuel feeding and gas analysis, the tests showed that solid fuels could be used in CLC.

14.6 Future trends

Currently, the study of energy systems using CLC mainly focuses on the saturated air gas turbine cycle with syngas- or natural gas-fueled chemical looping. Several key scientific problems remain to be solved. For example, the mismatch in energy levels between the reaction heat and the thermal energy utilized in the thermal cycle has not been explored. Consequently, in most conceptual research, the combustion heat is used to drive the reduction reaction in CLC. From the viewpoint of cascade utilization of chemical energy, the performance of CLC has huge potential to increase and thus deserves more attention. It is interesting to note that efforts have been focused recently on integrating CLC with novel power cycles from the viewpoint of energy levels.

14.6.1 H_2 production with chemical-looping combustion

A chemical-looping process was recently proposed for the production of hydrogen. Chemical-looping hydrogen (CLH) generation, which originates from CLC, is a type of water-splitting process with redox of a metal oxide. As with CLC, the CLH system is composed of two reactors: a fuel reactor for burning fuels and a steam reactor for water decomposition, which is different from the air reactor in CLC. Fuel is introduced into the fuel reactor to reduce the metal oxide particles and the fuel is oxidized into CO_2 and H_2O simultaneously. The reduced metal oxide is transported to the steam reactor, where it decomposes water to generate H_2.

The outlet gas stream from the fuel reactor contains only CO_2 and H_2O following complete conversion of the fuel, while the exit gas from the steam reactor is composed of H_2 with excess H_2O. Therefore, pure H_2 and CO_2 can be obtained by condensing H_2O, without any further separation processes (Son *et al.*, 2009). Son *et al.* (2009) reported that 3.7 L of H_2 was generated per kilogram through the reaction between fully reduced Cu-based oxide and steam. Go *et al.* (2009) investigated H_2 production by chemical looping of methane in a fluidized bed reactor using a Fe-based oxygen carrier. It was found that pure H_2 with free CO_2 can be obtained from the reaction $FeO \rightarrow Fe_3O_4$ in a steam reactor at 900°C. Additionally, the authors proposed a continuous two-step steam methane reforming (SMR) process with double-loop solid circulation systems for fuel and steam. Solid fuels, such as coal char, could be used as reducing materials in CLC. Yang *et al.* (2008) investigated H_2 production in a steam-iron process with reduction of FeO by CLC of coal char. FeO was used to oxidize the coal char into CO_2 and H_2O leading to the reduction of Fe_2O_3 to FeO and metallic Fe in the fuel reactor. The reduced iron oxide, FeO and Fe, was oxidized by steam in the steam reactor, while H_2 was generated by water decomposition. Usually, FeO and Fe are oxidized to Fe_3O_4 in the steam reactor, therefore an air reactor is needed, where the Fe_3O_4 is oxidized to the original Fe_2O_3 by air. The authors proved that FeO and Fe produced from the Fe_2O_3 reduction of char could be used in H_2 production via the steam-iron process. The total H_2 produced was 1000 ml of H_2 per gram of tested coal char, and the energy efficiency was 50.2% with respect to the energy ratio of H_2/char.

Similarly, Chiesa *et al.* (2008) proposed a CLH production system with three reactors (Fig. 14.20). Compared with commercially available technologies, this CLH process shows similar efficiency but is much more environmentally friendly because of the inherent CO_2 separation. The authors concluded that CLH is a promising process for producing H_2. In using the steam-iron process to produce H_2, deactivation of the FeO is one of the problems during operation. Bleeker *et al.* (2009) considered that deactivation of the FeO is caused by a decrease in the surface area of the oxygen carrier particles, and that a higher conversion rate of the oxygen carriers in the redox cycles will give stronger deactivation.

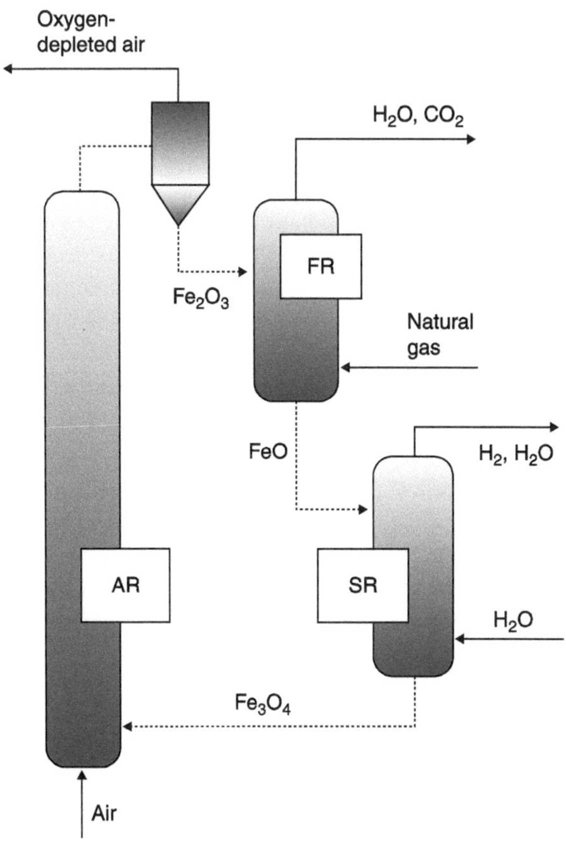

14.20 Conceptual scheme of the three-reactor CLH system. AR, air reactor; FR, fuel reactor; SR, steam reactor.

14.6.2 Solar hybridization power plant with chemical-looping combustion

Hybridizing fossil fuel energy and solar energy in solar thermal power plants is an attractive idea. This hybridization focuses on the endothermic processes, with fossil fuels being used exclusively as chemical reactants and solar thermal energy being used as a heat source to drive endothermic transformations. This would allow solar energy to be converted into chemical energy of syngas, which can be used in combined cycles to generate electricity. Most research has focused on utilization of concentrated solar thermal energy at temperatures above 1000°C to drive steam reforming of methane/coal gasification. The utilization of concentrators with high concentration ratios is very expensive currently. Furthermore, it is worth pointing out that the conversion and release of chemical energy from fossil fuel in this proposed process is still through direct combustion resulting in high irreversibility.

A methanol-fueled combined cycle that integrates CLC and solar thermal energy at around 200°C has been proposed (Hong *et al.*, 2006b), and preliminary experimental work has been conducted to validate the key components of the process (He *et al.*, 2007b). Figure 14.21 illustrates the cycle configuration. Methanol-fueled CLC consists of two successive reactions: (a) solar-driven Fe_2O_3 reduction with methanol fuel and (b) oxidation of the resulting FeO with air. The reduced solid product of the first reaction, FeO, is converted to Fe_2O_3 by oxidation in the second reaction. Here, Fe_2O_3 plays the role of an oxygen carrier between the two reactions. The first reaction of Fe_2O_3 particles and methanol fuel is endothermic ($CH_3OH+3Fe_2O_3 \rightarrow 6FeO+2H_2O+CO_2$) and is accomplished in the temperature range of 100–150°C at a pressure of 1 bar, so concentrated low-grade solar thermal energy can drive this reaction. The second reaction is an exothermic oxidation ($4FeO+3O_2 \rightarrow 2Fe_2O_3$) and releases heat in the high-temperature region of 1000–1300°C. The high-temperature exhaust gas from the oxidation reactor is fed to turbines to generate electricity. It was found that the net solar-to-electric efficiency of the proposed cycle could be 35%, superior to that of state-of-the-art solar thermal power plants (15%) and comparable to the advanced solar steam reforming plants (30%). Experiments on the reduction of Fe_2O_3 by methanol driven by solar thermal energy were analyzed using TGA. It was found that there were large grains and pores in the particles, and parts of the grains were broken down into smaller grains and agglomerated. These agglomerated grains might lead to a decrease in reaction rates.

A novel gas turbine cycle hybridizing solar thermal energy with natural-gas-fueled CLC has also been proposed (Hong *et al.*, 2006a). This system would allow concentrated solar thermal energy at around 1000°C to be lowered to 500°C, and provide a possible application in power systems hybridizing solar energy with natural gas. The CLC system could also be used for chemical storage. For example, some researchers (Otsuka *et al.*, 2003) found that, in principle, CLC could be used for the storage, transport, and supply of hydrogen.

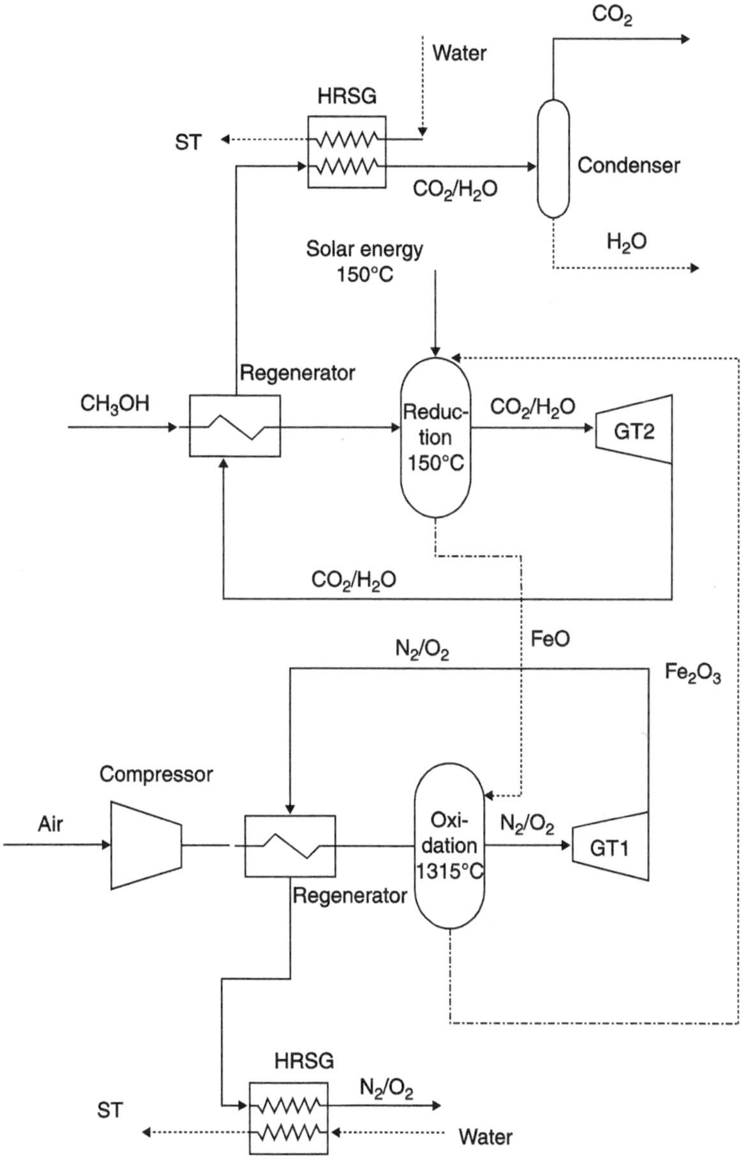

14.21 Simplified flow diagram for solar hybrid system with CLC.

14.7 Conclusions

CLC is an advanced technology for fossil fuel utilization and CO_2 capture. The basic principles of this technology are well understood and there are several small pilot scale test facilities where research programs are conducted. What is needed most in order to bring this very promising technology to commercialization is to have a well orchestrated R&D program that coordinates the international efforts which could lead to industrial demonstrations in the near future. Topics such as the interactions between cascade utilization of chemical energy and the decreasing of energy penalty for CO_2 separation in CLC, influence of microstructure of oxygen carrier on the chemical reaction, reactivity of CLC with liquid or solid fuel, and fluidized bed reactor design and operation in CLC are important issues that need to be further explored.

14.8 References

Abad A, Mattisson T, Lyngfelt A, Johansson M. 2007. The use of iron oxide as oxygen carrier in a chemical-looping reactor. *Fuel*, 86: 1021–1035.

Adánez J, de Diego L F, Garcia-Labiano F, *et al.* 2004. Selection of oxygen carriers for chemical-looping combustion. *Energy & Fuels*, 18: 371–377.

Adánez J, García-Labiano F, de Diego L F, *et al.* 2006. Nickel-copper oxygen carriers to reach zero CO and H_2 emissions in chemical-looping combustion. *Industrial & Engineering Chemistry Research.* 45: 2617–2625.

Ansolabehere S, Beer J, Katzer J, *et al.* 2007. *The future of coal: an interdisciplinary MIT study*. Massachusetts Institute of Technology, USA.

Azar C, Lindgren, K, Andersson B A, *et al.* 1999. *The role of carbon sequestration in a global energy future*. Mini-symposium on Carbon Dioxide Capture and Storage, Göteborg, Sweden.

Berguerand N, Lyngfelt A. 2008a. Design and operation of a 10 kW_{th} chemical-looping combustor for solid fuels-testing with South African coal [J]. *Fuel*, 87: 2713–2726.

Berguerand N, Lyngfelt A. 2008b. The use of petroleum coke as fuel in a 10 kW_{th} chemical-looping combustor. *International Journal Greenhouse Gas Control*, 2: 169–179.

Bleeker M F, Veringa H J, Kersten S R A. 2009. Deactivation of iron oxide used in the steam-iron process to produce hydrogen. *Applied Catalysis A: General*, 357: 5–17.

Bolt P H, Habraken F H P M, Geus J W. 1998. Formation of nickel, cobalt, copper, and iron aluminates form α- and γ-alumina-supported oxides: a comparative study. *Journal of Solid State Chemistry* 135, 59–69.

Chandel M K, Hoteit A, Delebarre A. 2009. Experimental investigation of some metal oxide for chemical looping combustion in a fluidized bed reactor. *Fuel* 88: 898–908.

Chiesa P, Lozza G, Malandrino A, *et al.* 2008. Three-reactors chemical looping process for hydrogen production, *International Journal of Hydrogen Energy*, 33: 2233–2245.

Cho P, Mattisson T, Lyngfelt A. 2000. *Reactivity of iron oxide with methane in a laboratory fluidized bed-application of chemical looping combustion.* Proceedings of 7th International Conference on Fluidized Bed Combustion, Niagara Falls, Canada; p. 599.

Cho P, Mattisson T, Lyngfelt A. 2004. Comparison of iron-, nickel-, copper- and manganese-based oxygen carriers for chemical-looping combustion, *Fuel* 83: 1215–1225.

Cho P, Mattisson T, Lyngfelt A. 2005. Carbon formation on nickel and iron oxide-containing oxygen carriers for chemical-looping combustion. *Industrial & Engineering Chemistry Research*, 44: 668–676.

Chuang S Y, Dennis J S, Hayhurst A N, *et al.* 2008. Development and performance of Cu-based oxygen carriers for chemical-looping combustion. *Combustion and Flame*, 154(1–2): 109–121.

Copeland R J, Alptekin G, Cesario M, *et al.* 2001. *A novel CO_2 separation system.* First National Conference on Carbon Sequestration. Washington, DC, USA.

Copeland R J, Alptekin G, Cesario M, *et al.* 2002. *Sorbent energy transfer system (SETS) for CO_2 separation with high efficiency*. Proceedings of 27th International Technical Conference on Coal Utilization and Fuel Systems, Clearwater, FL, USA.

Corbella B M, Palacios J M. 2007. Titania-supported iron oxide as oxygen carrier for chemical-looping combustion of methane. *Fuel*, 86: 113–122.

Corbella B M, de Diego L F, Garcia-Labiano F, *et al.* 2005. Performance in a fixed-bed reactor of titania-supported nickel oxide as oxygen carriers for chemical-looping combustion of methane in multicycle tests. *Energy & Fuels*, 19: 433–441.

Corbella B M, de Diego L, García-Labiano F, *et al.* 2006. Characterization and performance in a multicycle test in a fixed-bed reactor of silica-supported copper oxide as oxygen carrier for chemical-looping combustion of methane. *Energy & Fuels*, 20: 148–154.

de Diego L F, García-Labiano F, Adánez J, *et al.* 2004. Development of Cu-based oxygen carriers for chemical-looping combustion. *Fuel*, 83: 1749–1757.

de Diego L F, García-Labiano F, Gayán P, *et al.* 2007. Operation of a 10 kW_{th} chemical-looping combustor during 200h with a $CuO–Al_2O_3$ oxygen carrier [J]. *Fuel*, 86: 1036–1045.

Erri P, Varma A. 2007. Solution combustion synthesized oxygen carriers for chemical looping combustion. *Chemical Engineering Science*, 62: 5682–5687.

Go K S, Son S R, Kim S D, *et al.* 2009. Hydrogen production from two-step steam methane reforming in a fluidized bed reactor. *International Journal of Hydrogen Energy*, 34: 1301–1309.

He F, Wang H, Dai Y N. 2007a. Application of Fe_2O_3/Al_2O_3 composite particles as oxygen carrier of chemical looping combustion [J], *Journal of Natural Gas Chemistry*, 16(2): 12–18.

He P, Hong H, Jin H, *et al.* 2007b. The preliminary experimental study of the energy release principle in the energy system integrating methanol-chemical looping combustion and low-temperature solar thermal energy. *Journal of Engineering Thermophysics*, 28(2): 181–184.

Hong H, Jin H, Liu B. 2006a. A novel solar-hybrid gas turbine combined cycle with inherent CO_2 separation using chemical-looping combustion by solar heat source. *Journal of Solar Energy Engineering*, 128: 275–284.

Hong H, Jin H, Yang S. 2006b. A power generation system with inherent CO_2 recovery combining chemical-looping combustion with low-temperature solar thermal energy. *Journal of Engineering Thermophysics*, 27(5): 729–732.

Hossain M M, de Lasa H I. 2007. Reactivity and stability of $Co-Ni/Al_2O_3$ oxygen carrier in multicycle CLC. *AIChE. Journal*, 53(7): 1817–1829.

Hossain M M, Kelly E S, de Lasa H I. 2007. $Co-Ni/Al_2O_3$ oxygen carrier for fluidized bed chemical-looping combustion: desorption kinetics and metal-support interaction, *Chemical Engineering Science*, 62: 5464–5472.

Ishida M, Jin H. 1994. A new advanced power-generation system using chemical-looping combustion. *Energy*, 19(4): 415–422.

Ishida M, Jin H. 1996. A novel chemical-looping combustor without NO_x formation. *Industrial and Engineering Chemistry Research*, 35: 2469–2472.

Ishida M, Jin H. 1997. CO_2 recovery in a power plant with chemical-looping combustion. *Energy Conversion and Management*, 38(Suppl): S187–S192.

Ishida M, Yokohma M, Jin H. 1995. Chemical looping combustion power generation plant system. United States Patent, No. 5,447,024.

Ishida M, Jin H, Okamoto T. 1996. A fundamental study of a new kind of medium material for chemical-looping combustion. *Energy & Fuels*, 10: 958–963.

Ishida M, Jin H, Okamoto T. 1998. Kinetic behavior of solid particle in chemical-looping combustion: suppressing carbon deposition in reduction. *Energy and Fuels*, 12(2): 223–229.

Ishida M, Yamamoto M, Ohba T. 2002. Experimental results of chemical-looping combustion with $NiO/NiAl_2O_4$ particle circulation at 1200°C. *Energy Conversion and Management*, 43: 1469–1478.

Ishida M, Takeshita K, Suzuki K, Ohba T. 2005. Application of Fe_2O_3-Al_2O_3 composite particles as solid looping material of the chemical-loop combustor. *Energy & Fuels*, 19: 2514–2518.

Jerndal E, Mattisson T, Lyngfelt A. 2006. Thermal analysis of chemical-looping combustion. *Chemical Engineering Research and Design*, 84:795–806.

Jin H, Ishida M. 2000a. A novel gas turbine cycle with hydrogen-fueled chemical-looping combustion. *International Journal of Hydrogen Energy*, 25: 1209–1215.

Jin H, Ishida M. 2000b. Investigation of a Novel Gas Turbine Cycle With Coal Gas Fueled Chemical-looping Combustion. Proceedings of the ASME/Advanced Energy Systems Division-2000, Orlando, Florida, November 5–10, 2000

Jin H, Ishida M. 2001. Reactivity study on a novel hydrogen fueled chemical looping combustion. *International Journal of Hydrogen Energy*, 26:889–894.

Jin H, Ishida M. 2002. Reactivity study on natural-gas-fueled chemical-looping combustion by a fixed bed reactor. *Industrial and Engineering Chemistry Research, J Am Chem Soc*, 41: 4004–4007.

Jin H, Ishida M. 2004. A new type of coal gas fueled chemical-looping combustion. *Fuel*, 83: 2411–2417.

Jin H, Ishida M, Okamoto T. 1998a. Kinetic behavior of solid particle in chemical looping combustion: suppressing of carbon deposition in reduction. *Energy & Fuels*, 12, 223–229.

Jin H, Okamoto T, Ishida M. 1998b. Development of a novel chemical-looping combustion: synthesis of a looping material with a double metal oxide of CoO-NiO. *Energy Fuels*, 12(6): 1272–1277.

Jin H, Okamoto T, Ishida M. 1999. Development of a novel chemi-cal-looping combustion: synthesis of a solid looping material of $NiO/NiAl_2O_4$. *Industrial and Engineering Chemistry Research*, 38(1): 126–132.

Jin H, Hong H, Wang B Q, *et al.* 2005. A new principle of synthetic cascade utilization of chemical energy and physical energy. *Science in China Series E–Engineering and Materials Science*, 48(2): 163–179.

Jin H, Zhang X, Gao L, *et al.* 2008. Fudamental study of CO_2 control technologies and policies in China *Science in China Series E–Technological Sciences*, 51(7): 857–870.

Johansson M, Lyngfelt A, Mattisson T, *et al.* 2003. Gas leakage measurements in a cold model of an interconnected fluidized bed for chemical-looping combustion. *Powder Technology*, 134: 210–217.

Johansson M, Mattisson T, Lyngfelt A. 2004. Investigation of Fe_2O_3 with $MgAl_2O_4$ for chemical-looping combustion. *Industrial & Engineering Chemistry Research* 43: 6978–6987.

Johansson M, Mattisson T, Lyngfelt A, 2006. Creating a synergy effect by using mixed oxides of iron- and nickel oxides in the combustion of methane in a chemical-looping combustion reactor. *Energy & Fuels*, 20: 2399–2407.

Jung J K. Gamwo I. 2008. Multiphase CFD-based models for chemical looping combustion process: fuel reactor modeling. *Powder Technology*, 183: 401–409.

Koronberger B., Lyngfelt A., Loffler G., Hofbauer H., 2005. Design and fluid dynamic analysis of a bench-scale combustion system with CO_2 separation chemical-looping combustion. *Industrial & Engineering Chemistry Research*, 44: 546–556.

Leion H, Lyngfelt A, Johansson M, *et al.* 2008a. The use of ilmenite as an oxygen carrier in chemical-looping combustion. *Chemical Engineering Research and Design*, 86(9): 1017–1026.

Leion H, Mattisson T, Lyngfelt A. 2008b. Solid fuels in chemical-looping combustion. *International Journal of Greenhouse Gas Control*, 2: 180–193.

Lyngfelt A, Thunman H. 2005. Construction and 100 h of operational experience of a 10-kW chemical-looping combustor. *Carbon Dioxide Capture for Storage in Deep Geologic Formations – Results from the CO_2 Capture Project*, 1: 625–645.

Lyngfelt A, Leckner B, Mattisson T. 2001. A fluidized-bed combustion process with inherent CO_2 separation; application of chemical-looping combustion. *Chemical Engineering Science*, 56: 3101–3113.

Mattisson T, Lyngfelt A. 2001. *Capture of CO_2 using chemical-looping combustion.* In: Proceedings of First Biennial Meeting of the Scandinavian-Nordic Section of the Combustion Institute, 18–20 April, Göteborg, Sweden.

Mattisson T, Jardnas A, Lyngfelt A. 2003. Reactivity of some metal oxides supported on alumina with alternating methane and oxygen – application for chemical-looping combustion. *Energy and Fuels*, 17: 643–651.

Mattisson T, Johansson M, Lyngfelt A. 2004. Multicycle reduction and oxidation of different types of iron oxide particles – application to chemical-looping combustion. *Energy and Fuels*, 18: 628–637.

Mattisson T, Johansson M, Lyngfelt A. 2006. The use of NiO as an oxygen carrier in chemical looping combustion. *Fuel*, 85: 736–747.

Metz B, Davidson O, Coninck H, *et al.* 2005. *Carbon Dioxide Capture and Storage*. Special Report, Intergovernmental Panel on Climate Change, Cambridge University Press.

Nakano Y, Iwamoto S, Maeda T, *et al.* 1986. Characteristics of reduction and oxidation cyclic process by use of alpha Fe_2O_3 medium. *Iron & Steel Journal of Japan*, 72: 1521–1527.

Otsuka K, Kaburagi T, Yamada C, *et al.* 2003. Chemical storage of hydrogen by modified iron oxides. *Journal of Power Sources*, 122: 111–121.

Readman J E, Olafsen A, Larring Y, Blom R. 2005. $La_{0.8}Sr_{0.2}Co_{0.2}Fe_{0.8}O_3\text{-}\delta$ as a potential oxygen carrier in a chemical-looping type reactor, an in-situ powder X-ray diffraction study, *Journal of Materials Chemistry*, 15: 1931–1937.

Richter H J, Knoche K, 1983. Reversibility of combustion process, efficiency and costing: second law analysis of process. *ACS Symposium Series*, 235: 1129–1158.

Rydén M, Lyngfelt A, Mattisson T, *et al.* 2008. Novel oxygen-carrier materials for chemical-looping combustion and chemical-looping reforming; LaxSr1-xFeyCo1-yO3-δ perovskites and mixed-metal oxides of NiO, Fe_2O_3 and Mn_3O_4. *International Journal of Greenhouse Gas Control*, 2: 21–36.

Ryu H J, Bae D H, Jo S H, Jin G T, 2004. Reaction characteristics of Ni and NiO based oxygen carrier particles for chemical-looping combustor. *Korean Chemical Engineering Research*, 42: 107–114.

Shen L H, Xiao J, Xiao R, *et al.* 2007. Chemical looping combustion of coal in interconnected fluidized beds of $CaSO_4$ oxygen carrier. *Proceedings of the CSEE*, 27(2): 69–74.

Son S R, Kim S D. 2006. Chemical-looping combustion with NiO and Fe_2O_3 in a thermobalance and circulating fluidized bed reactor with double loops. *Industrial & Engineering Chemistry Research*, 45: 2689–2696.

Son S R, Go K S, Kim S D. 2009. Thermogravimetric analysis of copper oxide for chemical-looping hydrogen generation, *Industrial & Engineering Chemistry Research*, 48: 380–387.

Tan R, Santos S, Spliethoff H. 2006. *Chemical Looping Combustion for Fossil Fuel Utilization with Carbon Sequestration.* Study Report, International Flame Research Foundation, Velsen Noord, The Netherlands, January.

Tangen G. 2008. BIGCO2 R&D platform: closing the knowledge gaps of the CO_2 chain. *Greenhouse Issues*, 90: 20–21.

Villa R, Ctistiani C, Groppi G, *et al.* 2003. Ni based mixed oxide materials for CH4 oxidation under redox cycle conditions. *Journal of Molecular Catalasys A: Chemical*, 204–205: 637–646.

Wang J S and Anthony E J. 2008. A process for clean combustion of solid fuels. *Applied Energy*, 85: 73–79.

Xiang W G, Di T T, Xiao J, *et al.* 2004. Investigation of a novel gasification chemical looping combustion combined cycle. *Proceedings of the CSEE*, 24(8): 170–174.

Yang J, Cai N, Li Z. 2008. Hydrogen production from the steam-iron process with direct reduction of iron oxide by chemical-looping combustion of coal char. *Energy & Fuels*, 22: 2570–2579.

Zafar Q, Mattisson T, Gevert B. 2006. Redox investigation of some oxides of transition-state metals Ni, Cu, Fe and Mn supported on SiO_2 and $MgAl_2O_4$. *Energy & Fuels*, 20: 34–44.

Zheng Y, Wang B W, Song K, *et al.* 2006. The performance research on new oxygen carrier $CaSO_4$ used in chemical-looping combustion. *Journal of Engineering Thermophysics*, 27(3): 531–533.

15
Oxy-fuel combustion of gaseous fuel

N. ZHANG and W. HAN, Chinese Academy of Sciences, P. R. China

Abstract: This chapter examines the issues in utilizing gaseous fuel for power generation and CO_2 capture in an oxy-fuel combustion process. The major oxy-fuel cycles described in the literature, with both conventional and advanced air separation, are reviewed. Issues related to realizing the concepts at an industrial scale and major efforts of several demonstration projects are also addressed.

Key words: oxy-fuel combustion, gaseous fuel, CO_2 capture, thermal cycle.

15.1 Introduction

Natural gas has become an increasingly attractive choice for many applications, both as a fuel and as the raw material for chemical syntheses, e.g. methanol and ammonia. Over the past few decades, the natural gas reserves-to-production ratio has remained stable at about 60. The relatively abundant reserves keep the price stable. The percentage of natural gas as a proportion of world-wide energy consumption continues to grow. In 2008, natural gas consumption accounted for 24.1% of the total world energy consumption, taking third place behind oil and coal.[1] In Organization of Economic Cooperation and Development (OECD) countries, the share of natural gas in energy consumption surpasses that of coal, and it is the second largest primary energy source after oil.

Compared with coal, natural gas has several advantages across various aspects of combustion. Coal is known to be a less concentrated form of energy, and combustion of coal is difficult. Coal includes a large amount of non-combustible matter and water bundled with the hydrocarbon content, so it requires additional transportation capacity and elaborate combustion equipment and processes. In contrast, natural gas is very easy to burn and combustion efficiency is very high using simple combustion equipment. Hence, natural gas-fired generating units and chemical processes have lower capital and non-fuel operating costs (including transportation) than competing coal-fired units.[2] Furthermore, natural gas is much cleaner than coal because it contains less sulfur, ash, and nitrogen. The combustion products of natural gas contain few pollutants (such as SOx, NOx, and ash), and so do not require the type of clean-up units used in coal combustion. In addition, natural gas discharges half the amount of carbon dioxide (CO_2) per unit of electricity produced because of its high hydrogen/carbon ratio and the high thermal efficiency of gas-fired power plants.

Power generation accounts for about one-third of the CO_2 emissions from fossil fuel use, mainly from coal and natural gas, and thus measures to decrease CO_2 emissions in this sector could be highly effective.[3] Although natural gas has low carbon intensity, the quantity of CO_2 emissions derived from natural gas world-wide is extremely large, and new additions to generation capacity globally are dominated by natural gas-fired power plants. The key factors affecting the choice of power plants with CO_2 capture and storage (CCS) are the cost per output unit of electricity and the cost per unit of CO_2 emissions avoided. Efforts to improve CCS in natural gas-fired power plants are very necessary, as pointed out by Chiesa and Consonni,[4] 'Should those costs be lower for natural gas-fired systems, the optimal arrangement of the power generating system under a constraint on total CO_2 emission may well consist of a combination of natural gas-near-zero CO_2 emissions and plants fired with more carbon intensive fuels (like coal) with relatively large CO_2 emissions.'

In recent years, many innovative power cycles with low CO_2 emissions have been studied. These cycles are usually grouped into three alternatives:[4–19] post-combustion decarbonization, oxy-fuel combustion, and pre-combustion decarbonization; and they apply to coal-fired power plants. The adoption of one of these alternatives depends on the overall cost of the electricity (COE) produced, on the availability of safe long-term CO_2 storage options, and on the technological feasibility and reliability of the innovative and expensive equipment required, for example CO_2 gas turbines, high-temperature steam turbines, and O_2 production membranes.[10] Several studies[20–24] have compared the cost and energy penalties of these options for low emission power generation.

The CO_2 capture situation is different for natural gas-fired power plants. Most of the natural gas-fired power plants in the world utilize a gas/steam combined cycle. The combination of high efficiency, low investment costs, improved operating flexibility, short installation time, and low environmental impact has made the combined cycle very attractive for both medium and base load power generation, as well as for cogeneration of heat and power.

A combined cycle based on pre-combustion uses a reformer or a partial combustion reactor to convert natural gas to syngas, and then a shift reactor converts the CO in syngas to CO_2 before CO_2 capture. A power plant that incorporates pre-combustion is very complex and this method is only suitable for incorporation when building a new combined cycle power plant. Existing power plants can be adapted to utilize post-combustion and oxy-fuel methods. Post-combustion involves the addition of a CO_2 absorption unit at the outlet of the heat recovery steam generator (HRSG). The CO_2 concentration in the flue gas of a combined cycle gas-fired plant is only 3–5%, which is much lower than that in the flue gas of coal combustion (13–15%). This means CO_2 capture is more difficult and CO_2 separation will consume more energy. Moreover, to recover the same quantity of CO_2, the combined cycle plant must deal with 3–5 times more flue gas than in the coal-fired power plant, which makes the CO_2 capture equipment larger and significantly increases the investment cost. Oxy-fuel combustion is a feasible way of capturing CO_2 in a combined cycle, and is the only method that enables nearly 100% CO_2 capture. Combustion takes place in the absence of the large amounts of nitrogen present when air is used, and produces only CO_2 and H_2O. CO_2 separation is accomplished by condensing the water from the flue gas and therefore requires only a modest amount of energy. The disadvantage of this approach is the need for oxygen production (which is expensive both in terms of capital cost and energy consumption), and a complete redesign of the gas turbine.

Foy and Yantovski summarized the history and current state of the art of various zero emission power cycles based on oxy-fuel combustion.[25] Oxy-fuel combustion is already at the industrial demonstration stage for coal-fired plants, but the journey towards oxy-fuel combustion of gaseous fuel in any of the zero emission cycles proposed so far is taking considerably longer and remains uncertain, due to the need to develop new gas turbines. However, current cryogenic oxygen technology is showing continuing cost reductions based on improved component performance and scale enlargement. More efficient air separation technology, using ion transport membranes, is expected to reduce significantly the energy required and capital cost of separating oxygen from air. Several demonstration projects based on oxy-fuel combustion technology for gaseous fuel are being developed in Europe and the US. As these proceed and the technologies begin to achieve market penetration, it is expected that they will become competitive relative to alternative options based on pre- and post-combustion CO_2 capture. It is interesting to note that no particular CO_2 capture strategies seem to be winning so far for natural gas-fired electricity generation technologies.

This chapter examines the issues in utilizing gaseous fuel for power generation and CO_2 capture in an oxy-fuel combustion process. The hydrocarbon fuel does not have to be natural gas, but could be syngas made from coal, refinery residues, biomass, landfill gas, and so on. In the case of solid fuel applications, a fuel processing sub-system is required, which generally consists of a gasifier and clean-up systems to convert the raw fuel into clean syngas.

In sections 15.2 and 15.3 below, the cycles are grouped according to the oxygen production method they employ. Section 15.2 describes oxy-fuel cycles that use a separate air separation unit, COOPERATE, MATIANT, Graz, CES, that proposed by C. Gou, and COOLCEP. Section 15.3 describes oxy-fuel cycles with integrated air separation, mainly the advanced zero emission power plants (AZEP), in which an oxygen ion transport membrane is used to reduce the oxygen production efficiency penalty. The chemical looping combustion (CLC) system will not be addressed in detail as Chapter 14 by Jin and Zhang in this book describes CLC. Section 15.4 describes an integrated gasification combined cycle (IGCC) power plant with semi-closed cycle and oxygen-blown combustion, which uses solid fuel with gasification technology.

It should be pointed out that all the performance data in this chapter are taken from the referenced papers and thus were calculated by different authors using different assumptions, models, boundary conditions, and even performance criteria definitions. Some conclusions can, however, still be drawn from thermodynamic analyses comparing various gas-fired options. It is widely accepted that the conversion efficiencies of the MATIANT, Graz, and CES cycles with conventional air separation units (ASUs) are lower than those of advanced oxy-fuel concepts such as AZEP and CLC, yet these advanced processes require far more development and their prospects for commercialization are far from certain.

15.2 Thermodynamic cycles using conventional air separation technology

15.2.1 The carbon dioxide (CO_2) prevented emission recuperative advance turbine energy (COOPERATE) and MATIANT cycles

Yantovski *et al.* proposed the CO_2 prevented emission recuperative advance turbine energy (COOPERATE) cycle, with CO_2 re-circulation and gas combustion in a CO_2/O_2 mixture. This cycle is called quasi-combined, as it consists of a gas turbine recuperative cycle with multistage compression and a steam-like cycle with liquid CO_2 pumping and low temperature expansion, and the two cycles use the same working substance, CO_2.[26–28] Efficiency is reported to reach 52% when the pressure and temperature at the inlet of the steam-like turbine are 240 bar/800°C, 60 bar/1250°C at the second turbine inlet, and 15 bar/1250°C at the third. Yantovski[27] compared the COOPERATE cycle to a standard combined cycle, and concluded that the payback period is three years if using lubricants as fuel is considered as a profit.

A variant of this cycle is the MATIANT cycle proposed by Mathieu and Nihart[29] in 1999. The layout of the MATIANT gas cycle is shown in Fig. 15.1 and the corresponding *T-s* diagram in Fig. 15.2. It consists of a supercritical CO_2 Rankine-like cycle (2–3–4–5–6) combined with a regenerative CO_2 Brayton cycle with reheat (6–7–8–9–10–11–12–1–2).[29]

There are two combustion chambers. In the first combustion chamber (h in Fig. 15.1), the fuel is burned with O_2 and the working fluid CO_2 at pressure (P_2, Fig. 15.2) along the isobar (7–8). The combustion-generated mixture of CO_2/H_2O at 1300°C expands along (8–9), and is then reheated up to temperature (10) in the second combustion chamber, where additional fractions of fuel and O_2 are injected at (P_3). At the exit of the second combustion chamber, the fluid contains the combustion products, namely 8% CO_2 and 6% H_2O, in addition to the 100% vol. CO_2 circulating along the total cycle. This fluid is expanded in (10–11) to produce additional electricity. It is then cooled and the heat released is used to preheat the recycled CO_2 before and after the expansion (5–6) in a recuperator. At the recuperator outlet, water is condensed in a cooler and extracted in a CO_2/H_2O separator. The remaining CO_2 stream is compressed in a three-stage compressor with intercooling up to the CO_2 condensation pressure (70 bar) at about 30°C. CO_2 is condensed and pumped to the higher cycle pressure (P_1), the excess CO_2

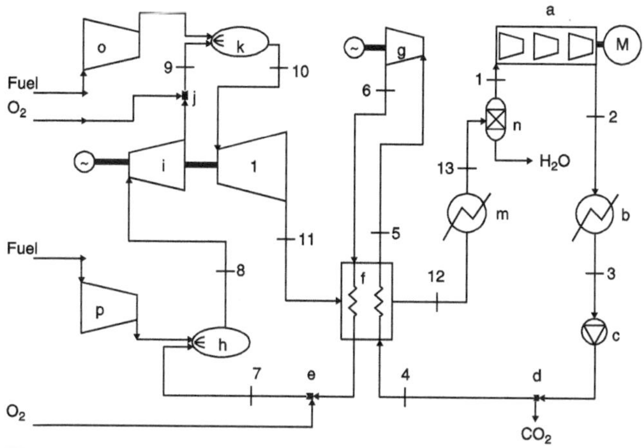

a: CO_2 compressor	b: Condenser	c: CO_2 pump
d: Splitter	e: Mixer	f: Recuperator
g: Uncooled high pressure expander	h: Combustion chamber 1	i: Medium pressure expander with internal cooling
j: Mixer	k: Combustion chamber 2	l: Low pressure expander with internal cooling
m: Cooler	n: Water separator unit	o and p: Fuel compressors

15.1 Layout of the MATIANT cycle.[29]

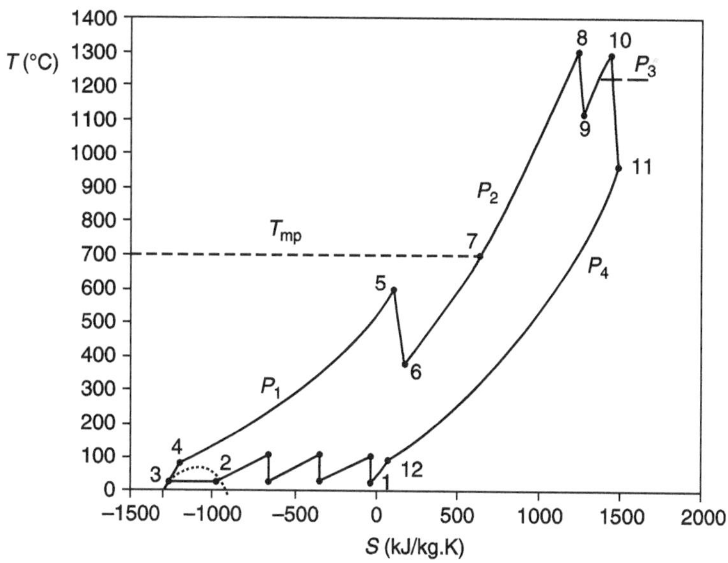

15.2 T–s diagram of the MATIANT cycle.[29] T_{mp} = temperature of middle pressure CO_2.

is removed in the liquid state. The rest is then heated by recuperated heat to state (5) and expanded along (5–6) in a steam-like turbine to generate electricity.

An O_2 purity of 99.5% is expected from an air separation unit (ASU) with a specific energy consumption of 0.28 kWh/kg O_2 at 5 bar. With the high pressure

(P_1) equal to 300 bar, (P_2) to 40 bar, (P_3) to 9.3 bar and (P_4) to 1 bar, the temperatures at the recuperator outlet (7) and the low pressure (LP) and medium pressure (MP) gas expanders inlet (8 and 10) are 700°C and 1300°C respectively, and the net cycle efficiency is reported as 44.3%.

In the MATIANT cycle, the non-condensable gases need to be vented out of the condenser to avoid accumulation, and some CO_2 inevitably escapes with them. To limit this CO_2 leakage, a later version[30] eliminates the condenser from the cycle, and the CO_2 stream after water removal is compressed in a four-stage compressor with intercooling above the CO_2 saturation line, directly up to a supercritical pressure. This cycle variant is more like an inter-cooled recuperative gas turbine cycle. The working fluid CO_2 is cooled along a supercritical isobar (80 bar) down to 30°C. The cycle efficiency was estimated to be almost the same as that with condensation. This eliminates the condenser and the associated cost; however, the pressure elevation must be achieved by compressors and thus requires more compression work.

Another thermodynamic analysis of trans-critical quasi-combined CO_2 cycles can be found in Fioravanti *et al.*[31] The common features of these cycles are the use of CO_2 as the working fluid and O_2, produced by an ASU, as the fuel oxidizer. With CO_2 condensation at a pressure of 60–70 bar (temperature 20–30°C), efficiencies of 35–49% were reported for plants based on such cycles, despite the additional power use for O_2 production and CO_2 condensation. Staicovici[32] proposed an improvement to these cycles by coupling them with thermal absorption technology to lower the CO_2 condensation below ambient temperature (30 bar, –5.5°C), and estimated a net power efficiency of 54%. To further lower the exothermal temperature of the cycle, Zhang and Lior proposed integrating it with liquefied natural gas (LNG) evaporation process as the cycle cold sink.[33,34] As a result, the cycle condensation process can be achieved at a temperature much lower than ambient. Recovery of LNG cold exergy and internal exergy produced net thermal and exergy efficiencies for a base-case cycle of over 65% and 50% respectively.

Some thermal parameters in the MATIANT cycle are very notable. For example, in the three-pass recuperator, the high pressure steam at 300 bar is heated to 600°C, the mid-pressure steam at 40 bar is heated to 700°C, and the inlet temperature of the hot side stream is as high as 930°C. This means that the recuperator has to work under very tough conditions, and brings challenges for both the recuperator design and material lifetime.

In the trans-critical, quasi-combined cycles found in the literature, the high pressure steam turbine inlet temperature is generally limited to 600°C to reduce the mechanical stresses on the materials, and the gas turbine uses a much higher inlet temperature of 1300°C to elevate the average heat absorption temperature. The combined cycle employs high pressure (with low compression power consumption) in the Rankine cycle, and high temperature (with high specific power output) in the Brayton cycle, thus taking advantage of both Rankine and

Brayton cycles to produce high efficiency. Some also include reheating in the Brayton cycle to further improve the thermodynamic performance. However, the configuration is complex and may bring problems for operation and control.

15.2.2 The Graz cycle

The basic principle of the Graz cycle was developed by Jericha in 1985.[35] Improvements and further developments by researchers at the Institute of Thermal Turbomachinery and Machine Dynamics at Graz University of Technology can be found in Jericha *et al.*[36, 37, 40] and Sanz *et al.*[38, 39] Several cycle configuration variants have been proposed with different steam content fractions in the working medium. Jericha *et al.* proposed a Graz cycle power plant with a cycle efficiency of 63%.[36]

The Graz cycle consists of a recuperated oxy-fuel gas turbine cycle with steam injection and a Rankine-type cycle. The oxy-fuel combustion produces a working fluid that is mostly CO_2 to power the high temperature turbine. After heat recuperation, the stream is further expanded in a lower pressure turbine to a condenser pressure of 0.25 bar. Water and CO_2 are separated by water condensation. The water is pumped, vaporized, and superheated to high pressure steam. After expansion in the high pressure turbine, the steam is used to cool the burners and the high temperature turbine hot stages. CO_2 from the condenser is compressed to atmospheric pressure; the combustion-generated CO_2 is then removed, and the remaining CO_2 is further compressed and fed to the combustor to cool the liners. The general layout of all components and thermodynamic details of this cycle for a 92 MW pilot plant can be found in Jericha *et al.*[36, 37]

The arrangement of the Graz cycle offers several thermodynamic advantages. It allows heat input at very high temperature, as well as expansion to vacuum conditions, and the high-pressure steam can be expanded to generate additional work in the high-pressure turbine (HPT), before being fed to the combustion chamber. It thus combines the advantages of a high-temperature Brayton cycle with a low back pressure Rankine cycle, and therefore offers high thermal efficiency.

Recent developments include the high steam-content Graz cycle (S-Graz cycle).[38, 39] The principle flow scheme as published[39] is shown in Fig. 15.3. The high-temperature Brayton cycle consists of compressors C1/C2, a combustion chamber, and a high temperature turbine (HTT); the low-temperature Rankine cycle includes a low-pressure turbine (LPT), condenser, HRSG, and HPT. Similarly to the original Graz cycle, the fuel is burned with a nearly stoichiometric mass flow of oxygen in the combustion chamber, which operates at a pressure of 40 bar. Steam and the recycled CO_2/H_2O mixture are added to cool the burners and the liner, and bring the mixture temperature at the exit of the combustion chamber to 1400°C. The mixture expands in the HTT to 1 bar for power generation. With the applied cooling steam, the exhaust gas from the HTT

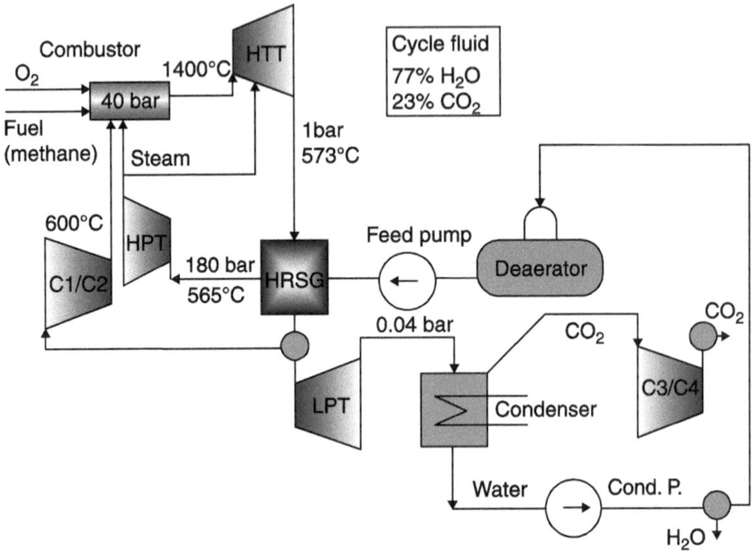

15.3 Principle flow scheme of the S-Graz cycle power plant.[39, 40] HTT, high temperature turbine; HRSG, heat recovery steam gen.; LPT, low pressure turbine; C3/C4, CO_2 compressors; C1/C2, cycle fluid compressors; HPT, high pressure turbine.

contains 77% steam and is used in the HRSG to vaporize and superheat steam for the HPT.

The major difference between this and the original Graz cycle is that only 45% of the mixture coming out of the HRSG further expands in the LPT to a condensation pressure of 0.04 bar absolute, significantly lower than for the original Graz cycle because of the higher steam content of the working fluid. Most of the cycle fluid after the HRSG is compressed by C1/C2 and fed to the combustion chamber with a maximum temperature of 600°C, avoiding release of the latent heat of steam vaporization by condensation. In the condenser, steam is condensed and separated from CO_2. After extracting the H_2O formed by combustion, the water is vaporized and superheated in the HRSG to steam at 180 bar and 565°C. The combustion-generated CO_2 is compressed to atmospheric pressure by C3 and C4 with intercooling and further water extraction for disposal.

Table 15.1 gives the power balance of the S-Graz cycle. An efficiency of nearly 70% is claimed when burning oxygen-blown coal gas, falling to 57.7% when oxygen production and CO_2 compression are taken into account. Based on these data, Statoil became interested in the project. The S-Graz cycle was re-evaluated in cooperation with Statoil, resulting in a net efficiency prediction of 52.7% for methane firing. The CO_2 mitigation costs were estimated to be \$20.7/ton CO_2 avoided.[39]

In the S-Graz cycle, only part of the working fluid is condensed; the major part is compressed in the gaseous phase, as in a Brayton cycle. The disadvantage of the increased compression work is compensated for by the high heat content of

Table 15.1 S-Graz cycle power balance[40]

HTT power (MW)	634.7
HPT power (MW)	48.0
LPT/LPST power (MW)	70.5
Total turbine power (MW)	753.2
C1 power (MW)	137.2
C2 power (MW)	90.2
C3 power (MW)	11.5
C4 power (MW)	4.8
Pump power (MW)	5.3
Total compression power (MW)	249.0
Net shaft power (MW) without mechanical losses	504.2
Total heal input Q_{zu} (MW)	758.6
Thermal cycle efficiency (%)	**66.47**
Electrical power output (MW) including mechanical, electrical, and auxiliary loss	490.3
Net electrical cycle efficiency (%)	**64.63**
O_2 generation and compression (MW)	74.7
Efficiency considering O_2 supply (%)	**54.78**
CO_2 compression to 100 bar (MW)	15.6
Net power output (MW)	**400.0**
Net efficiency η_{net} (%)	**52.72**

the steam taken back to the combustor. In addition, the high steam content in the working fluid enables much lower condenser pressure for the same cooling medium temperature, leading to a high specific power output from the LPT.

Further modification of the S-Graz cycle configuration includes condensation in the range of 1 bar by slight recompression, to alleviate the technical difficulty of condensing water out of a mixture of steam and non-condensable gases at very low pressures. This is integrated with a separated steam cycle with very low condensation pressure, and a high cycle efficiency of 53% had been claimed for this process.[40]

15.2.3 Clean energy system (CES) cycle

The Clean Energy System, Inc. (CES) has developed an oxy-fuel zero emission power plant (ZEPP), 'integrating proven aerospace technology into a conventional power system'.[41–45] It is similar to the Graz cycle in that steam is recirculated to the combustion chamber, but is less complex. The core of the CES process involves combustion of gaseous hydrocarbon fuel with oxygen in a gas generator adapted from rocket engine technology. This produces a mixed steam/CO_2 stream at high temperature and pressure. The plant configuration consists of four key sub-systems: fuel processing and compression, air separation and oxygen compression, power generation, and CO_2 capture and conditioning.

Clean gaseous fuel is compressed to the required combustor injection pressure. The air separation unit delivers a pressurized high purity gaseous oxygen stream to the combustor. In the configuration shown in Fig. 15.4,[43] the power generation system consists of oxy-combustors, turbines in series, and heat recovery equipment. The oxy-combustor operates at a pressure of 50–100 bar and its temperature is moderated by the injected water and steam. It produces a steam/CO_2 working fluid at a temperature of 500–600°C, to be delivered to the HPT. The composition of the working fluid varies according to fuel type, and water/steam injection amount and temperature, but is typically about 91% vol. steam and 9% vol. CO_2. Most of the HPT exhaust is reheated in a second combustor which burns additional fuel with oxygen.

The high-temperature intermediate turbine (IPT) necessitates advanced turbine material and cooling technology. Depending on the exhaust temperature, the IPT exhaust heat is recuperated in the HX to preheat the recycled feed water, or to generate steam to be injected into the combustor. Compared with the recuperator in the MATIANT cycle, the LPT turbine inlet temperature is much lower without combustion ahead of it. The HX in the CES cycle also works under more benign conditions, with much lower hot stream inlet temperatures and cold stream pressure. An HRSG is included downstream of the HX to condense the IPT exhaust, generating low-pressure steam for a separate low-pressure steam Rankine cycle, in which the pure LP steam is expanded to vacuum pressure in a conventional

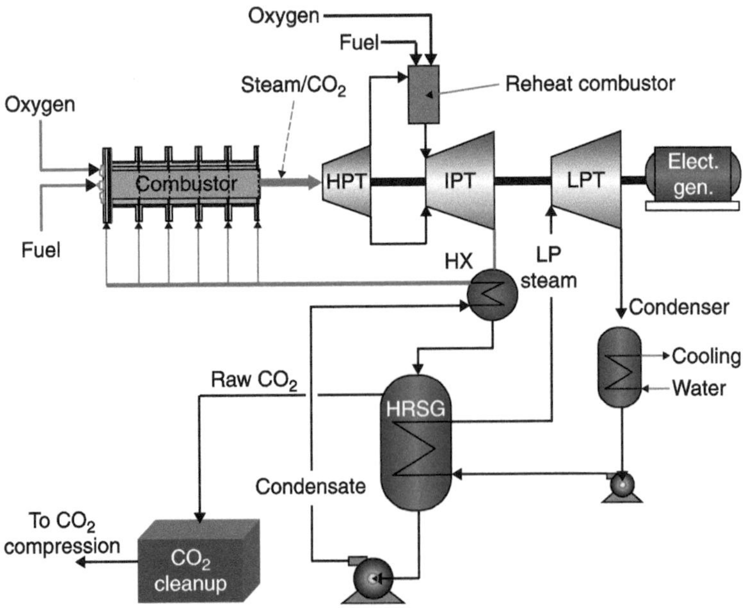

15.4 Conceptual layout of CES oxy-fuel cycle.[43]

LPT. The recovered CO_2 is conditioned and purified as appropriate for sequestration or utilization.

A 5 MW demonstration plant for this process began operation in KImberlina, California, in March 2005.[43–45] A 20 MWt oxy-fuel combustor has been tested successfully. Studies have shown that the system performance is enhanced greatly with increasing IPT inlet temperature. Although gas turbines that utilize high inlet temperatures are readily available, it requires significant efforts to adapt these machines to the H_2O/CO_2 working fluid. CES plans to develop and commercialize its oxy-fuel technology in stages, with each stage linked to a specific IPT with a progressively increasing turbine inlet temperature. For the first generation ZEPP, a General Electric J79 turbine has been adapted to integrate with a CES 170 MWt high-pressure oxy-fuel combustion system. A modest inlet gas temperature of 760°C was selected to eliminate the need for turbine cooling, the predicted net output power is 60 MWe at 30% efficiency (LHV), including the loads for O_2 production and CO_2 compression.[43]

15.2.4 The oxy-fuel power cycle proposed by Gou *et al.*[46]

Gou *et al.* proposed an innovative oxy-fuel power cycle with two configurations.[46] This cycle is also a quasi-combined one as it consists of a high-pressure steam Rankine-like cycle, and a high-temperature steam/CO_2 recuperative-reheat gas turbine cycle. The second configuration[46] is presented here. The layout and the corresponding *T-s* diagram are shown in Fig. 15.5 and Fig. 15.6, respectively.

Liquid water (1) is first pumped to 189 bar to be vaporized and superheated in the HRSG. After expanding in the HPT, the steam (3) enters the first combustor, which is operated at 40 bar. Fuel (methane in this study) is burned with a stoichiometric mass flow of O_2 to produce combustion gas at a mean temperature of 1300°C at the exit of the combustor. After expanding in the IPT, the flow (10) is reheated again to 1300°C by firing with additional fuel and O_2 in the second combustor at a pressure of 2 bar. The stream (13) is then expanded to 0.11 bar in the LPT. The fluid (14) enters the HRSG to preheat the feed water (2) and to be cooled down. The steam (15) after the HRSG is split into two branches. One branch (27), 36% by mass, is compressed by C4 and fed back to the second combustor. The rest of the steam (16) enters the condenser, where the liquid phase is separated from the gaseous phase. The gas fluid (17) from the condenser contains 85.5% CO_2 by mass, equal to the combustion generated CO_2, and is compressed with intercooling and water extraction to the correct pressure for liquefaction or storage. The combustion-generated water (30) is extracted out of the system from the condenser, and the rest (1) is delivered back to Pump 1 as recycled working fluid. A simulation with Aspen Plus estimated the system efficiency to be 50.6% with an optimized P_3 (the second combustor pressure) of 2 bar.

The authors compared this process with the MATIANT and CES cycles, the results showing that the proposed cycle has better performance in terms of system

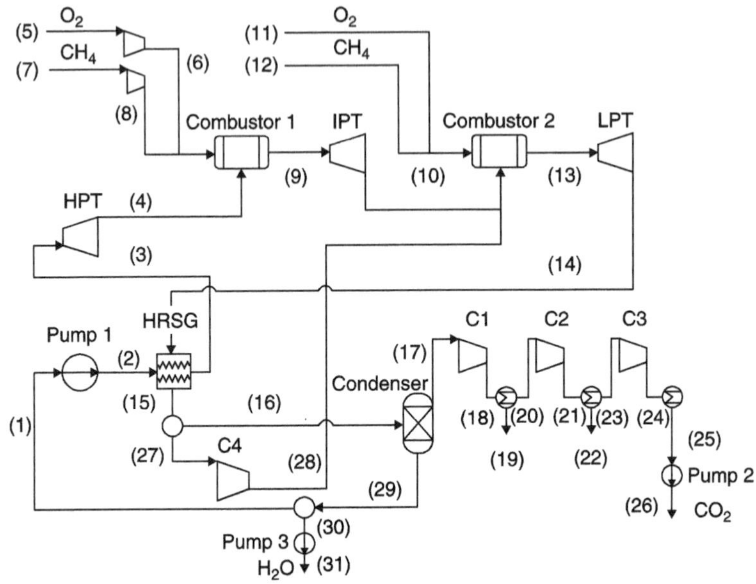

15.5 Flowsheet diagram of the power cycle proposed by Gou *et al.*[46]

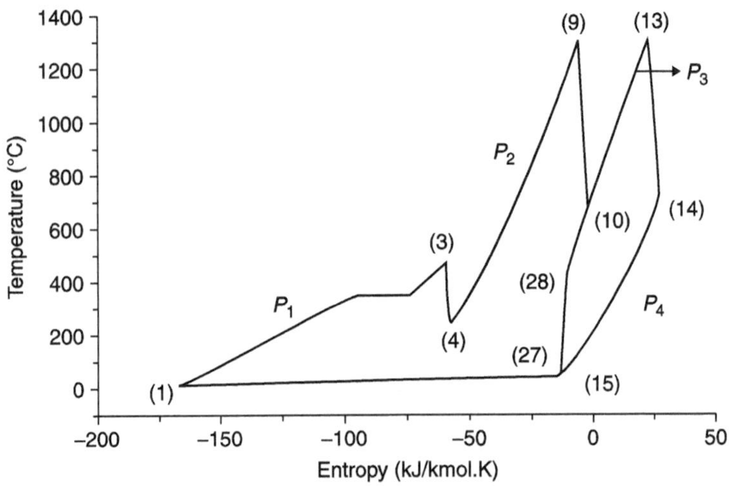

15.6 *T-s* diagram of the oxy-fuel power cycle proposed by Gou *et al.*[46]

efficiency and specific work. The efficiency of the proposed system was six percentage points higher than that of the MATIANT and CES cycles. A fraction of the working fluid is compressed and added to the IP turbine. This results in more working fluid with lower heat capacity on the hot side in the recuperator, and less working fluid with higher heat capacity on the cold side, and helps to reduce the exergy loss related to heat transfer.

The Gou *et al.* cycle, however, is more like a thermal dynamic exercise than an engineering solution, because of the ideal assumptions, such as stoichiometric and perfect combustion, in the analysis. Similarly to the MATIANT cycle, it employs two combustors to produce high temperature fluids at 1300°C for the IPT and LPT. Reheating ahead of the LPT elevates its exhaust (HRSG hot stream inlet) temperature to a high level of 730°C, and therefore is capable of directly producing high temperature steam for expanding in the HPT. Unlike the MATIANT cycle, this cycle uses steam as the major working fluid, which eliminates the compression process and offers a higher specific power output. The splitting of the stream before condensation, while it improves the thermal match in the HRSG, results in a more complicated configuration, and imposes engineering challenges.

Gou and co-workers also proposed a hybrid oxy-fuel power system integrated with low temperature thermal heat, which can either be the waste heat from industrial processes or solar heat. The external, low-temperature heat source is used to generate steam as (part of) the working fluid, which reduces fossil fuel consumption and consequent emissions.[47, 48]

15.2.5 Cool clean efficient power (COOLCEP)

LNG has a temperature of about 110K, much lower than that of the ambient air or water, and thus retains a large amount of cold exergy. At the receiving terminals, instead of simply evaporating the LNG using the heat from ambient seawater or air, as is often done in practice, it is possible to extract the cold exergy from LNG evaporation by investing it in a process that recovers it for some useful application. One way to achieve this is by incorporating it into a properly designed thermal power cycle that uses the LNG evaporator as its cold sink.

Zhang *et al.* proposed and analyzed oxy-fuel cycles integrated with LNG cold exergy recovery and utilization.[33, 34, 49, 50] These cycles integrate the LNG evaporation process as the heat sink, and thus take advantage of the low temperature that the stored LNG offers, convert its coldness into power generation, evaporate the LNG for further use, and at the same time capture the combustion-generated CO_2. One example is the COOLCEP (cool clean efficient power) cycle, based on a concept first proposed by Deng *et al.*[51] It is basically a cogeneration (power and refrigeration) recuperative Rankine cycle with CO_2 as the main working fluid. Two configurations have been proposed, namely COOLCEP-C and COOLCEP-S.[49, 50]

Figure 15.7 shows the layout of the COOLCEP-C cycle, which consists of a power subcycle and an LNG vaporization process. The interfaces between them are the CO_2 condenser CON, the heat exchanger HEX1, and the fuel feed stream 8.

The power subcycle can be identified as 1–2–3–4–5–6–7–8–9–10–11–12/13–13a–14–1. The low temperature (–50°C) liquid CO_2 as the main working fluid (1) is pumped to about 30 bar (2), then goes through a heat addition process (2–3) in the evaporator EVA1 and can thereby produce refrigeration if needed. The O_2 (4) produced in an ASU is compressed and mixed with the main CO_2 working fluid.

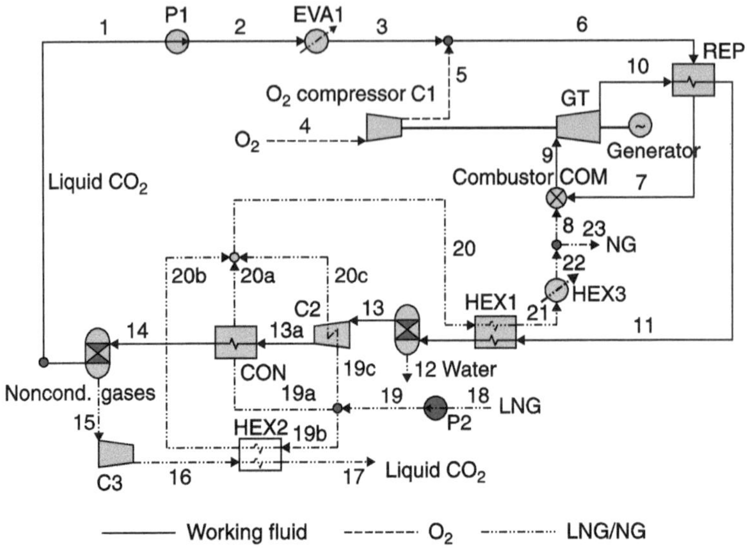

15.7 The process flowsheet of the COOLCEP-C system.[49]

The gas mixture (6) is heated (6–7) by turbine (GT) exhaust heat recuperation in REP. The working fluid temperature is further elevated in the combustor COM, fueled with natural gas (8) to its maximal value (the turbine inlet temperature, TIT) (9). The working fluid expands to near-ambient pressure (10) in the gas turbine (GT) to generate power and is then cooled (to 11) in the recuperator REP.

The gases in the mixture at the exit of REP (11) are separated, and the combustion-generated CO_2 component is condensed for ultimate sequestration, which is performed by further cooling in the LNG-cooled heat exchanger HEX1, where the H_2O vapor in the mixture is condensed and drained out (12). Afterwards, the remaining working gas (13) is compressed to the condensation pressure (13a), and one stage inter-cooling (19c–20c) is employed in the compressor to reduce the compression work. The CO_2 working fluid is condensed (14) in the condenser CON against the LNG evaporation, and recycled (1). The remaining working fluid (15), enriched with noncondensable species (mainly N_2, O_2 and Ar), is further compressed in C3 to a higher pressure level, under which the combustion-generated CO_2 is condensed and captured, ready for final disposal.

The LNG vaporization process is 18–19–19a/b/c–20a/b/c–20–21–22–23/8. LNG (18) is pumped by P2 to the highest pressure (73.5 bar) typical for receiving terminals which supply long-distance pipeline networks, and then evaporated using the heat from the power cycle. The evaporated natural gas (NG) may produce a small amount of cooling in HEX3 if its temperature is still low enough at the exit of HEX1, and thus contribute to the useful outputs from the system. Finally, the emerging natural gas stream is split into two parts, most of it (23) is sent to outside users and the rest (8) is used as the fuel in the combustor.

In this configuration, a compressor C2 is required to raise the expanded CO_2 gas (13) pressure to the condensation level, with the associated efficiency penalty due to the energy consumption of the compressor.

Noting the efficiency penalty of the compression work in COOLCEP-C, COOLCEP-S was configured so that the working fluid expands in the turbine, GT, only to the working fluid condensation pressure, thus eliminating the need for the gas compression process but at the expense of some power generation in the turbine. As a result, the turbine in COOLCEP-S has a higher back pressure compared with that in COOLCEP-C, and its exhaust is at a higher temperature (the regenerator REP hot stream inlet temperature). It is noted that the higher temperature in the heat exchanger REP requires careful attention to design. The working fluid pressure elevation is accomplished entirely by the process of pressurizing a liquid (process 1–2 in Fig. 15.7), which consumes much less energy.

Features of the COOLCEP system that contribute significantly to its performance include: the high turbine inlet temperature and turbine exhaust heat recuperation, which gives a high temperature level of added heat; and the heat exchange with LNG, which results in a heat sink at a temperature lower than the ambient and provides high power generation efficiency. Synergistically, these low temperatures allow condensation of the working fluid, thus the combustion-generated CO_2 is captured. Furthermore, the sub-critical re-evaporation of the CO_2 working fluid is accomplished below ambient temperature, so can provide refrigeration if needed.

The two configurations were simulated, and their performances are summarized in Table 15.2.[49] The COOLCEP-S cycle shows better performance, with a thermal efficiency of 59% at the turbine inlet temperature of 900°C. An economic analysis was also conducted, which found that the payback period is 8–10 years including the construction period, and the cost of electricity was estimated to be 0.34–0.37 CNY/kWh. The COOLCEP systems exhibit good thermal and economic performance. Their capacity needs to be matched with the temperature (coldness) of the LNG.

15.3 Thermodynamic cycles using advanced air separation technologies

15.3.1 Advanced zero emission power plant (AZEP) with oxygen (O_2) transport membranes (OTM)

Cryogenic air separation has a very negative effect on the efficiency of oxy-fuel systems. Novel oxygen generation technologies, such as oxygen transport membranes (OTM), could significantly reduce the energy penalty and lead to a new generation of power generation processes. Due to the high-temperature operating conditions and high investment costs, a stand-alone OTM might not

Table 15.2 COOLCEP cycle performance summary (Zhang *et al.*, 2009)

	COOLCEP-S	COOLCEP-C
Net power output, W_{net} (MW)	20	20
CO_2 condensation pressure (bar)	7/60[a]	7/60[a]
Turbine back pressure p_b (bar)	7.1	1.1
Combustor outlet temperature and pressure (°C/bar)	900/28	900/28
Turbine outlet temperature (°C)	701	474
LNG mass flow rate m_{LNG} (kg/s)	95.16	61.8
Fuel mass flow rate m_f (kg/s)	0.69	0.79
Main working fluid mass flow rate (kg/s)	101.6	60.45
Power consumption in ASU (MW)	2.33	2.67
Specific power output W_{sp} (kl/kg)	197	332
Refrigeration output, Q_c (MW)	56.7	26.6
Refrigeration exergy, E_C (MW)	8.90	4.09
Ratio of power/cooling energy, R	2.25	4.91
Thermal efficiency, η_e (%)	59.1	51.6
Exergy efficiency, ε (%)	39.8	37.3

[a] 7 bar is the condensation pressure for the main working fluid in the condenser; and 60 bar is the condensation pressure for a small fraction of the working fluid (stream 16)

offer much advantage over a cryogenic process. However, when integrated with the combustion process, it offers the possibility of oxygen production without a significantly adverse effect on efficiency, and probably at much lower cost than a cryogenic system.

Figure 15.8 shows a concept for integrating a pressurized OTM in a gas turbine cycle. According to this scheme, compressed air from a gas turbine compressor is heated to the required membrane temperature of 800–900°C by direct combustion of natural gas and is introduced into the OTM.[52] Oxygen is separated by passage through the OTM, and the oxygen-depleted air on the membrane feed side can then be fed to a second combustion chamber to obtain a high temperature, equivalent to that of modern conventional gas turbines (e.g. 1430°C).[53] The exhaust gases, after expansion through a turbine, can enter a heat recovery steam generator to generate steam for a steam turbine or for process steam. This conceptual system can cogenerate oxygen and electric power efficiently. It will, however, not be a zero emission power cycle, since the oxygen generated by the OTM is not used in the combustion of fossil fuel. This idea should therefore be seen as one of many possible ways of integrating membranes in an innovative way in power cycles.

A direct integration of the combustion chamber and an OTM membrane is employed in the advanced zero emission power plant (AZEP).[54] In this reactor, compressed air enters on the feed side, some of the oxygen permeates the membrane and is mixed with a natural gas flow, and a fuel oxidation reaction

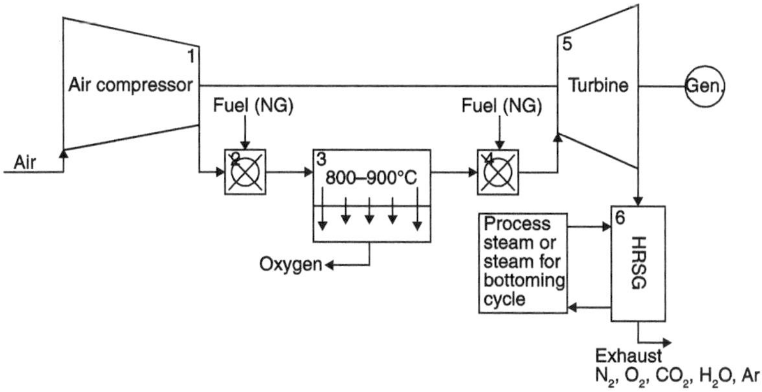

15.8 Concept for separating oxygen from air, based on OTM.[53]

takes place. The heat generated from this combustion provides the required heat for OTM operation, and heats the oxygen-depleted air on the air feed side.[55–57] There are two advantages to the integrated reactor. First, the oxygen on the permeate side is consumed quickly by the combustion reaction, which results in a low oxygen partial pressure on this side and in turn increases the oxygen flux and accelerates the air separation process, so the air separation unit can be smaller in size. Second, the combustion reaction provides the heat needed for air separation, so no additional heating equipment is required.

The AZEP plant substitutes the AZEP reactor for the combustion chamber of an ordinary gas turbine. The exhaust gas from the gas turbine is fed to a heat recovery steam generator to generate steam for electricity production through a steam turbine, as in an ordinary combined cycle. A simplified process scheme is shown in Fig. 15.9. A portion of the combustion products on the permeate side is re-circulated into the reactor, and the other portion is expanded in a so-called sweep turbine, for additional power generation. Of the exhaust gases, harmless oxygen-depleted air is the major fraction, while CO_2 is a small fraction and is easily separated by condensing the H_2O.[53]

The membrane operates at very high combustion temperatures. If the temperature is not high enough, the efficiency of the gas turbine will decrease. The top temperature limitation is determined by the material from which the OTM is constructed. Perovskites and fluorites are the two main types of ceramics with oxygen transport capabilities. The maximum operating temperature of perovskite materials is about 800°C–1000°C, and that of fluorite materials is far higher. Recently, research has focused on developing special OTM materials for combustion, and membranes made from modified fluorites may play a key role. Studies show that it might be possible to increase the maximum temperature up to 1400°C with a relatively high oxygen flux. Such high-temperature membrane materials are, however, still being researched.

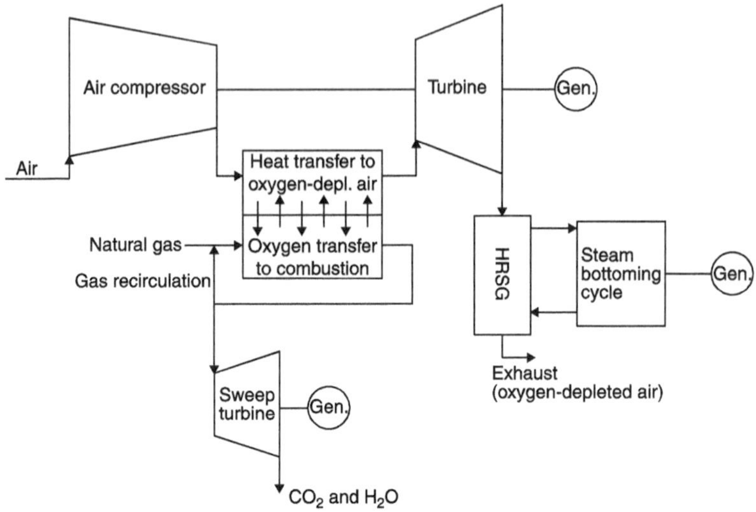

15.9 Simplified sketch of the AZEP concept.[53]

Recently, a new concept, ZEITMOP, was proposed,[58, 59] which could have a high turbine inlet temperature, low membrane operation temperature and near zero CO_2 emission. In this system, the combustor and membrane reactor are separated, and the oxidant for combustion comes from the pure oxygen generated in the low-temperature membrane. The ZEITMOP cycle presented in Fig. 15.10 consists of three main loops: circulating CO_2 (components 7, 9, 10, 11 and 12), a mixture of CO_2+H_2O (components 3, 6, 7, 10, 12, 13) and air/O_2-depleted air/O_2 branch (components 1, 2, 3, 4, 5, 7), together with the fuel supply system (components 7, 8). Ambient air enters compressor (1). After compression, it is heated up to around 800–900°C in heat exchanger (3) by the flue gases exiting

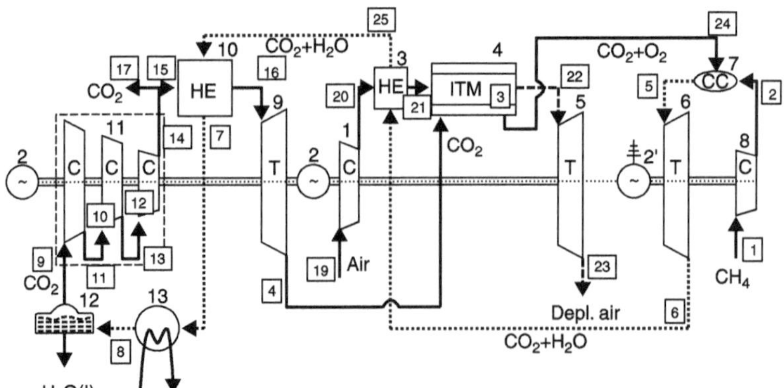

15.10 ZEITMOP power cycle with separate combustion and ITM reactor.[59]

turbine (6). The hot pressurized air then enters the OTM oxygen ceramic (4), which generates an oxygen stream from the air passing through the membrane, with the remainder being a relatively high pressure oxygen-depleted air stream. The CO_2 after (9) is mixed with oxygen in (4) before being directed to the combustion chamber (7), which is also fed with pressurized natural gas. The hot pressurized oxygen-depleted air stream leaving (4) is expanded in turbine (5) before being discharged to the atmosphere. The flue gas mixture of CO_2 and H_2O exiting the combustion chamber (at about 1300–1600°C) is expanded in the low pressure turbine (6), before being cooled in (3), (10), and (13). At ambient temperature, the water in the flue gas mix is liquid while the CO_2 remains gaseous. The bulk of the water is drained out of the cycle in (12). Almost-pure CO_2 enters the multi-stage inter-cooled compressor (11), from where a fraction (a few percent) of highly compressed, supercritical or liquid CO_2 is drained out of the cycle in (14) to be sequestered or used elsewhere. The major portion of the CO_2 is heated in (10) before being expanded in the high pressure turbine (9) down to about 15 bar. The CO_2 then enters combustor (7) via (4) in a mixture with oxygen.

At the highest cycle temperatures of 1300–1500°C, a cycle efficiency of 46–56% is estimated. The biggest technical obstacle is the high-temperature gas-to-gas heat exchanger (3), which is difficult to manufacture and the investment cost may be extremely high.

15.3.2 Chemical looping combustion (CLC)

CLC is a novel method for utilizing fuel and capturing CO_2.[60] The oxygen for combustion of the fuel is provided by a regenerateable metal oxide catalyst. Most of the work on CLC has focused on gaseous fuels. Chapter 14 outlines the technology and development of CLC.

15.4 Use of solid fuel with gasification technology

15.4.1 IGCC power plants with semi-closed cycle and oxygen-blown combustion

As the fossil fuel most widely used for power generation and the main producer of CO_2, coal has received a great amount of attention from researchers. Among current studies, the semi-closed cycle with oxygen-blown combustion based on coal gasification is an interesting idea because of its simple system configuration, such as no need for water-gas-shift reactors or a CO_2 separation unit.[61]

Figure 15.11 shows the conceptual configuration of the semi-closed cycle with oxygen combustion. There are two interesting features. One is that the exhaust from the turbine, which contains a high concentration of CO_2, is partly re-circulated back to the compressor, so the working fluid in this power cycle is highly enriched CO_2. Another feature is that the syngas is burned in pure or highly enriched

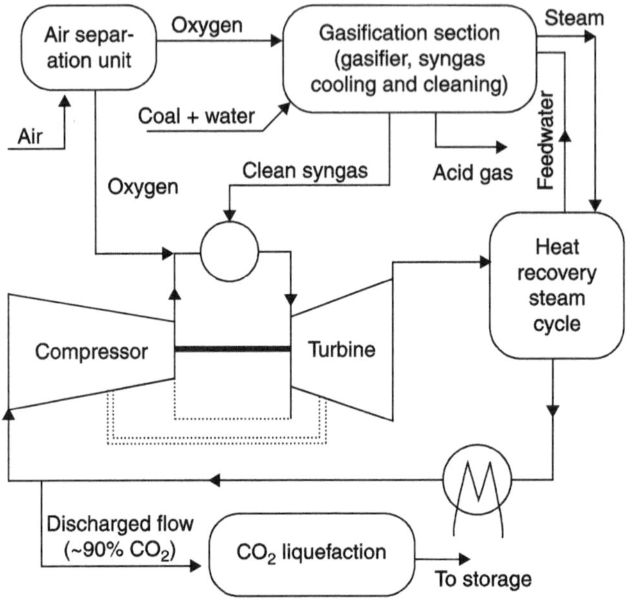

15.11 Concept overview of the semi-closed cycle with oxygen combustion.[61]

oxygen rather than air, so the flue gas is mainly made up of H_2O and CO_2, which are easy to separate from each other.

A complete flowchart of the semi-closed cycle with oxygen combustion is shown in Fig. 15.12. Air is separated in the ASU, which consists of an air intercooled compressor and a double separation column. The compressor is driven directly by the gas turbine. The pressure of the air when exiting the compressor is 0.48 MPa and the oxygen purity is approximately 95%. The O_2 separated here is split into two streams. One is sent to the gasifier, and the other goes to the combustor of the gas turbine.

The gasification system includes an entrained-flow slurry-feed gasifier, a syngas cooler (a radiative-convective heat exchanger), and a scrubber which cleans the syngas. Coal is gasified along with O_2 and H_2O in the gasifier, and the major components of the syngas are CO, CO_2, H_2, and H_2O. Then the syngas is cooled and cleaned sequentially. The gasification pressure and temperature are assumed to be 60 bar and 1600 K, respectively.

The power section consists of a topping cycle and a bottoming cycle. The key part of this unit is a gas turbine, which is assumed not to be affected by the change of working fluid. The heat recovery steam cycle (the bottoming cycle) is a conventional three-pressure reheat cycle. The HRSG supplies both steam to the steam turbine and feed water to the syngas cooler in the gasification unit to produce saturated steam with the heat retrieved from raw syngas.

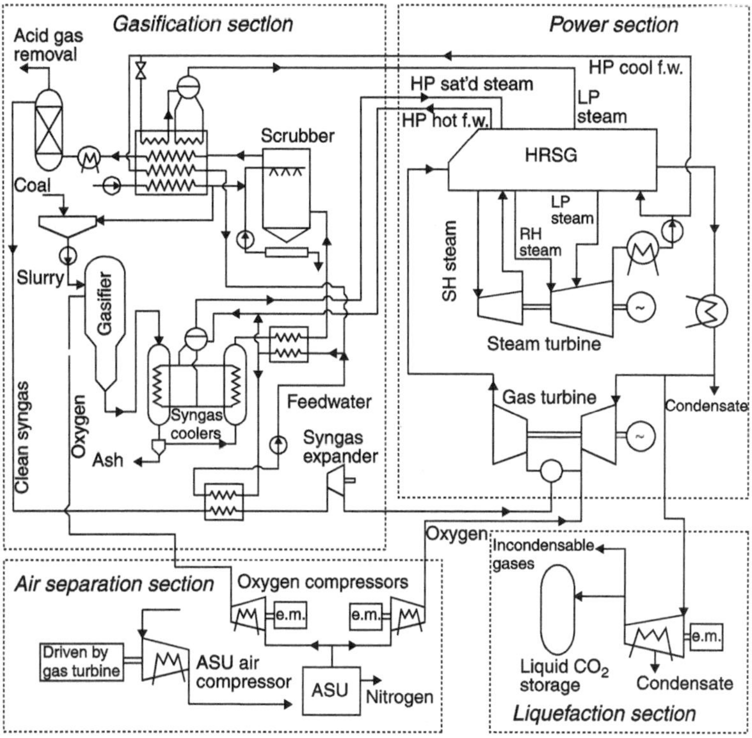

15.12 Flowchart of the semi-closed cycle with oxygen combustion.[61] LP, low pressure; HP, high pressure; RH, reheated; SH, superheated.

15.5 Future trends

There are several industrial scale oxy-fuel demonstration projects currently under development. Naturally, most of the activities are focused on coal as fuel, given that coal is widely used for power generation, and it releases approximately twice as much CO_2 as natural gas for each unit of electricity generated. Demonstration projects with natural gas as the fuel include the 30 MW Lacq Basin pilot project by Total of France, the Kimberlina Power Plant by Clean Energy Systems (CES) in California, USA, and the Norwegian ZENG (Zero Emission Norwegian Gas) project using the gas generator developed and demonstrated by CES.

French company Total launched a pilot project in the Lacq gas treatment complex in the south-west of France at the end of 2006.[62, 63] Aimed at testing the industrial-scale feasibility of an integrated chain of oxy-combustion for CCS, the project covers the whole chain from the boiler (CO_2 source) to the reservoir (CO_2 sink). An existing boiler built in 1957, rated at 30 MW_{th} and producing 40 t/h of steam (at 60 bar/450°C), is being retrofitted for oxy-fuel combustion. The oxygen requirement is about 240 t/h. During its first two years of operation, the pilot project will capture 150,000 t CO_2, at a rate of 200 t/day. The CO_2 stream will then be treated, compressed and transported via a pipeline to the Rousse gas field near Capelle de Rousse, where the CO_2 will be injected into a depleted natural gas reservoir.

The retrofit involves installation of a cryogenic air separation unit, which will be designed and supplied by Air Liquide, together with four oxy-burners. Air Liquide has demonstrated the performance of the oxy-burner concept with a downscaled (1 MW) design in a dedicated cold-wall test-rig with flue gas recycling.[63] High CO_2 concentration (94% vol. on a dry basis) was systematically achieved with excess O_2 at 2% vol. on a wet basis at the boiler exit. The downscaled 1 MW oxy-burner design needs to be upscaled to 8 MW_{th} to meet the requirement of the Lacq CCS project. Currently, the ASU is under construction, the boiler being modified and start-up is scheduled to begin soon.

Clean Energy Systems (CES), Siemens Energy, Inc., and Florida Turbine Technologies, Inc. (FTT) are jointly developing oxy-fuel combustion-based power systems, and have planned a stepped development strategy to advance the technology through an initial proof-of-concept phase and subsequent development of first and second generation power plant systems.[43–45]

The main principle behind the CES oxy-fuel combustion cycle involves replacing conventional steam boilers and exhaust gas cleaning systems with a gas generator that burns gaseous oxygen and natural gas under stoichiometric conditions in the presence of clean water. The resulting steam/CO_2 exhaust gas powers a combination of steam and advanced gas turbines before entering a heat recovery cycle and condenser/separator at low pressure. The recovered CO_2 is conditioned and purified (if required), while water is largely recycled back to the gas generator. The core of the CES zero-emission system is the oxy-combustion concept adapted from rocket engine technology. In the initial proof-of-concept phase, bench-scale research was performed to develop the primary combustion system. A 20 MW_{th} (10 MW_e) gas generator was designed and fabricated under the Vision 21 Program. The gas generator was used to explore operability and performance limits while operating on a variety of fuels over a wide range of conditions. It has completed more than 1500 hours firing with approximately 300 start and shutdown sequences. A 5 MW demonstration plant has been set up in Kimberlina, California. The 20 MW_{th} combustor was successfully tested by driving an existing turbine/generator set at the Kimberlina prototype power plant in early 2005. The Kimberlina plant is the first zero emission power plant in the world.

The work after the proof-of-concept phase is to develop and demonstrate first and second generation oxy-fuel power plant systems. The CES power generation system includes oxy-combustors and turbines in series, i.e. high, intermediate, and low pressure turbines. Both high pressure and low pressures turbines may be based on readily available steam turbine technology. The intermediate turbine (IPT) is critical, since the system performance is enhanced greatly by increasing the IPT inlet temperature. The high-temperature IPT thus makes advanced gas

Table 15.3 Turbine technologies for the CES demonstration power plant[44]

Generation	Temperature range	Candidate turbine
First	760–927°C	GE J79
Second	1080–1260°C	Siemens SGT900
Third	1650–1760°C	Advanced Design

turbine technologies necessary. CES planned a three-stage development process to advance and commercialize this technology, with each generation linked to a specific IPT with a progressively higher inlet temperature. The three generations of turbine technologies they defined are shown in Table 15.3.[44]

For the first generation demonstration plants, CES chose a modified GE J79 gas turbine because of the closely matched operating conditions. The engine modifications include the removal of the 17-stage axial flow compressor and connection of the CES gas generator to its three-stage turbine, adapting the 10 can-annular combustors for oxy-fuel combustion. This combustor functions as a reheater (RH) to heat the HPT exhaust steam/CO_2 working fluid from approximately 315 to 925°C.

A larger 170 MW_{th} oxy-combustor was designed and constructed to produce driving gases for the power generation system. All this equipment was assembled at the Kimberlina power plant. Commissioning tests of the combustion system were conducted during the latter half of 2008. Integration of the combustion system with the modified J79 gas turbine was completed in 2009.

The second generation focuses on power generation with increased efficiency and power output. The Siemens SGT-900 gas turbine was selected as the basis for the IPT. Conceptual design studies were performed to identify the modifications needed in order to adapt the hardware to function with the oxy-fuel gases, mainly involving the removal of the compressor and integration of the oxy-fuel reheater(s). Preliminary design of DGT-900 combustors adapted for oxy-combustion has already begun. The high turbine inlet temperature of 1260°C necessitates internal cooling for the turbine; this will be accomplished by using a fraction of the steam/CO_2 discharged from the HPT. The major challenges lie in the mechanical design configuration, and the thermal/structural behaviors in steam/CO_2 environments.

The Zero Emission Norwegian Gas (ZENG) Program is being co-developed by Lyse Energi AS, Nebb Engineering AS, Procom Venture AS, and CO_2-Norway. The Carbon Dioxide Capture and Storage Project is located in Risavika, Stavanger, Norway, and is timed into two phases.[64, 65] In August 2004, the ZENG project team completed Phase 1: concept and feasibility study for a 40 MW pilot and demonstration zero-emission power plant. As shown in Fig. 15.13, the 'Base Case' configuration was completely based on commercially available components

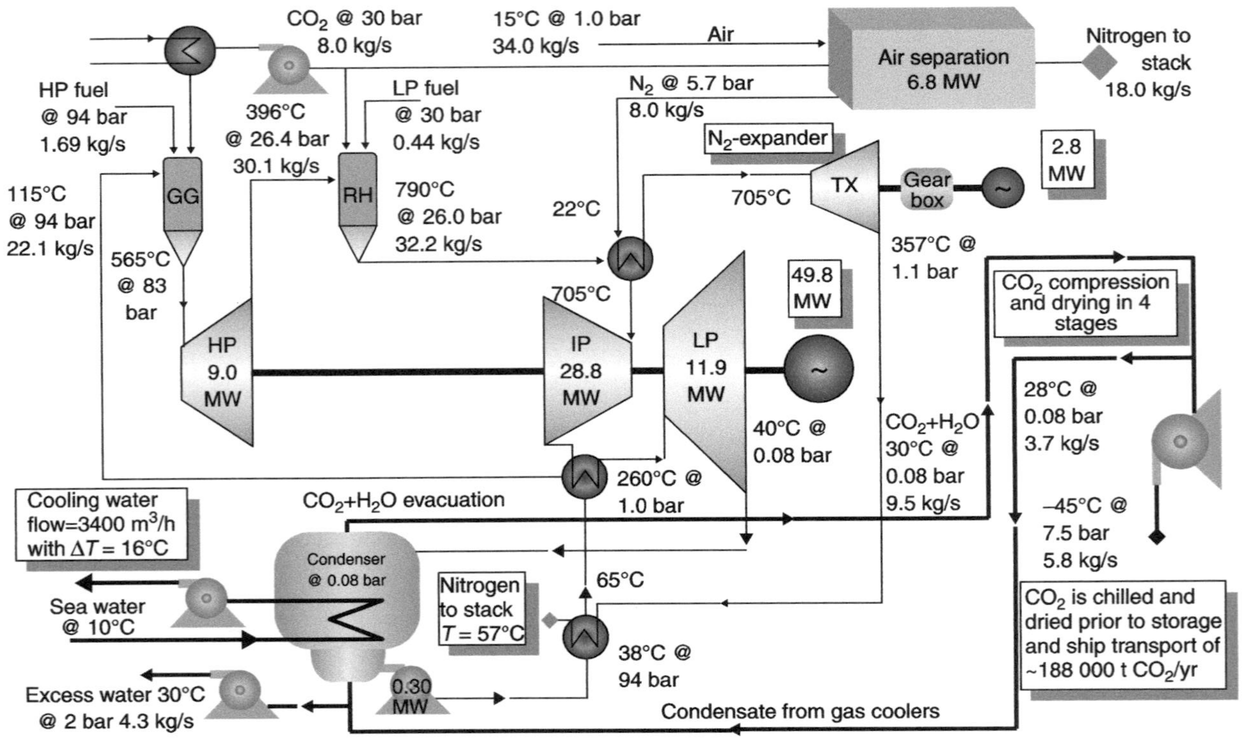

15.13 Process flow schematic for 'Base Case' configuration with 42 MWe net output and cycle efficiency of ~38%.[63]

to meet the schedule for plant commissioning in 2008. One feature of this new plant is the integration of nitrogen expansion from the ASU, with the nitrogen gas being expanded in a N_2-turbine expander to produce additional power. A thermal efficiency of 38% has been estimated based on LHV, which could be improved with further optimization. These potential improvements include: increased turbine inlet temperature for the IPT, integration with the ASU, and employing a low-pressure 'pure' Rankine steam cycle.

Phase 2 was scheduled as a concept definition study for a 70 MWe zero-emission demonstration plant (2007) and investment decision (2008). The plant has a targeted efficiency of ~48% based on LHV. The 70 MWe power module is based upon reconfiguring near-term commercially available components and early commercialization in niche industrial markets. Novel technologies to be explored include: a 170 MWth CES gas generator; a modified intermediary pressure turbine with blade cooling; the intermediary pressure oxy-combustion reheator; integration of ASU and nitrogen expansion; the low-pressure section of the process cycle; etc. For the medium-term (2010–2012), the project will target technology improvements and economies of scale towards attaining a 400 MWe commercial zero-emission power plant with ~55% efficiency. The Phase 2 study is also supposed to define the key technologies and develop risk assessments for realizing the next phase and longer term development goals.

Oxy-combustion with CO_2/steam recycling produces a working fluid different from that in air-based combustion. The thermodynamic properties of this gas are significantly different, leading to different performance characteristics in turbines. Therefore turbine technology needs to be developed substantially in order to utilize oxy-fuel gas as a working fluid. Basically, it means re-designing the turbine. Such an approach would be out of step with the trend towards higher efficiency combined cycles,[24] and gas turbine manufacturers appear to have little interest in developing oxy-fuel gas turbines as they have not been persuaded of the potential market in this area. Large-scale demonstrations of new technologies such as CES gas generators are thus very much needed, to prove both the feasibility and economics of the technology.

Oxygen flames are characterized by quicker kinetics than flames in air. This results in a rapid and dense heat release of chemical energy from the fuel, in the form of short and intense flames.[63] The heat transfer characteristics of oxy-fuel combustion products are also very different from those of air combustion products. In particular, the thermal conductivity of steam is known to be substantially higher than that of nitrogen. The primary impact of this property difference is to increase the convective heat transfer on both the hot and cold sides of the turbine components, thereby reducing the thermal resistance between the hot and cold mediums. The consequently increased heat flux and temperature gradients may lead to increased thermal stresses within the turbine components and thus need to be evaluated carefully.[44] In addition, the steam/CO_2 environments also pose questions about the behaviors of turbine materials. These impacts may increase the cost of retrofitting natural gas-fired boilers for oxy-fuel operation, and retrofitting turbine internal cooling systems.

Although the oxy-fuel process has been identified as having great technical potential, and many advanced thermal cycles have been proposed and studied, there is still a long way to go before gaseous fuels are employed on an industrial scale.

15.6 References

1 *Energy Policy Research* (in Chinese), 6, 2009.

2 A. Denny Ellerman, 1996, The Competition between Coal and Natural Gas, *Resources Policy*, 22: 33–42.

3 J. M. Beér, 2007, High Efficiency Electric Power Generation: The Environmental Role, *Progress in Energy and Combustion Science*, 33: 107–134.

4 P. Chiesa, S. Consonni, 2000, Natural Gas Fired Combined Cycles with Low CO_2 Emissions, *Journal of Engineering for Gas Turbines and Power*, 122: 429–436.

5 A. Corradetti, U. Desideri, 2005, Analysis of Gas-Steam Combined Cycles with Natural Gas Reforming and CO_2 capture, *Journal of Engineering for Gas Turbines and Power*, 127: 545–552.

6 G. Lozza, P. Chiesa, 2002, Natural Gas Decarbonization to Reduce CO_2 Emission from Combined Cycles – Part I: Partial Oxidation, *Journal of Engineering for Gas Turbines and Power*, 124: 82–88.

7 G. Lozza, P. Chiesa, 2002, Natural Gas Decarbonization to Reduce CO_2 Emission from Combined Cycles – Part II: Steam-Methane Reforming, *Journal of Engineering for Gas Turbines and Power*, 124: 89–95.

8 N. Zhang, N. Lior, 2008, Comparative Study of Two Low CO_2 Emission Power Generation System Options with Natural Gas Reforming, *Journal of Engineering for Gas Turbines and Power*, 130: 051701-1–11.

9 N. Zhang, N. Lior, 2008, Two Novel Oxy-Fuel Power Cycles Integrated with Natural Gas Reforming and CO_2 Capture, *Energy*, 33: 340–351.

10 R. Gabbrielli, R. Singh, 2005, Economic and Scenario Analyses of New Gas Turbine Combined Cycles with No Emissions of Carbon Dioxide, *Journal of Engineering for Gas Turbines and Power*, 127: 531–538.

11 J. M. Frias, S. M. Aceves, J. R. Smith, H. Brandt, 2004, Thermodynamic Analysis of Zero-Atmospheric Emissions Power Plant, *Journal of Engineering for Gas Turbines and Power*, 126: 2–8.

12 D. Fiaschi, L. Lombardi, L. Tapinassi, 2003, The Recuperative-Auto Thermal Reforming and Recuperative-Reforming Gas Turbine Power Cycles with CO_2 Removal – Part I: The Recuperative-Auto Thermal Reforming Cycle, *Journal of Engineering for Gas Turbines and Power*, 125: 933–939.

13 D. Fiaschi, L. Lombardi, L. Tapinassi, 2004, The Recuperative-Auto Thermal Reforming and Recuperative-Reforming Gas Turbine Power Cycles with CO_2 Removal – Part II: The Recuperative Reforming Cycle, *Journal of Engineering for Gas Turbines and Power*, 126: 62–68.

14 I. Garcia, J.-V.-M. Zorraquino, 2002, Energy and Environmental Optimization in Thermoelectrical Generating Processes – Application of a Carbon Dioxide Capture System, *Energy*, 27: 607–623.

15 O. Bolland, P. Mathieu, 1998, Comparison of Two CO_2 Removal Options in Combined Cycle Power Plants, *Energy Conversion Management*, 39: 1653–1663.

16 P. Chiesa, S. Consonni, 1999, Shift Reactors and Physical Absorption for Low-CO_2 Emission IGCCs, *Journal of Engineering for Gas Turbines and Power*, 121: 295–305.
17 A. Corrado, P. Fiorini, E. Sciubba, 2006, Environmental Assessment and Extended Exergy Analysis of a 'Zero CO_2 Emission', High-Efficiency Steam Power Plant, *Energy*, 31: 3186–3198.
18 E. I. Yantovski, K. N. Zvagolsky, V. A. Gavrilenko, 1992, Computer Exergonomics of Power Plants without Exhaust Gases, *Energy Conversion Management*, 33: 405–412.
19 G. Wall, E. I. Yantovski, L. Lindquist, J. Tryggstad, 1995, A Zero Emission Combustion Power Plant for Enhanced Oil Recovery, *Energy*, 20: 823–828.
20 K. Damen, M. van Troost, A. Faaij, W. Turkenburg, 2006, A Comparison of Electricity and Hydrogen Production Systems with CO_2 Capture and Storage. Part A: Review and Selection of Promising Conversion and Capture Technologies, *Progress in Energy and Combustion Science*, 32: 215–246.
21 K. Damen, M. van Troost, A. Faaij, W. Turkenburg, 2007, A Comparison of Electricity and Hydrogen Production Systems with CO_2 Capture and Storage. Part B: Chain Analysis of Promising CCS Options, *Progress in Energy and Combustion Science*, 33: 580–609.
22 H. M. Kvamsdal, K. Jordal, O. Bolland, 2007, A Quantitative Comparison of Gas Turbine Cycles with CO_2 Capture, *Energy*, 32: 10–24.
23 J. Davison, 2007, Performance and Costs of Power Plants with Capture and Storage of CO_2, *Energy*, 32: 1163–1176.
24 P. Freund, 2003, Making Deep Reductions in CO_2 Emissions from Coal-Fired Power Plant Using Capture and Storage of CO_2, *Proceedings of the Institution of Mechanical Engineers*, 217, Part A: *Journal of Power and Energy*, 1–7.
25 K. Foy, E. Yantovski, 2006, History and State-of-the-Art of Fuel Fired Zero Emission Power Cycles, *International Journal of Thermodynamics*, 9: 37–63.
26 E. I. Yantovski, K. N. Zvagolsky, and V. A. Gavrilenko, 1995, The COOPERATE-demo Power Cycle, *Energy Conversion and Management*, 36: 861–864.
27 E. I. Yantovski, 1996, Stack Downward Zero Emission Fuel-fired Power Plants Concept, *Energy Conversion and Management*, 37: 867–877.
28 E. I. Yantovski, J. Gorski, 2000, Further Elaboration of Quasi-combined Zero-Emission Power Cycle, *Proceedings of ECOS 2000*, Enschede, Netherlands, pp. 1083–1092.
29 P. Mathieu, R. Nihart, 1999, Zero-Emission MATIANT Cycle, *Journal of Engineering for Gas Turbines and Power*, 121: 116–120.
30 P. Mathieu, R. Nihart, 1999, Sensitivity Analysis of the MATIANT Cycle, *Energy Conversion and Management*, 40: 1687–1700.
31 A. Fioravanti, L. Lombardi, G. Manfrida, 2000, An Innovative Energy Cycle with Zero CO_2 Emissons, *Proceedings of ECOS 2000*, Enschende, Netherlands, pp. 1059–1070.
32 M. D. Staicovici, 2002, Further Research Zero CO_2 Emission Power Production: The 'COOLENERG' Process, *Energy*, 27: 831–844.
33 N. Zhang, N. Lior, 2006, Proposal and Analysis of A Novel Zero CO_2 Emission Cycle with LNG Cryogenic Exergy Utilization, *Journal of Engineering for Gas Turbines and Power*, 128: 81–91.
34 N. Zhang, N. Lior, 2006, A Novel Near-Zero CO_2 Emission Thermal Cycle with LNG Cryogenic Exergy Utilization, *Energy*, 31: 1666–1679.
35 H. Jericha, 1985, Efficient Steam Cycles with Internal Combustion of Hydrogen and Stoichiometric Oxygen for Turbines and Piston Engines, CIMAC Conference Paper No. T13, Oslo, Norway.

36 H. Jericha, E. Gottlich, 2002, Conceptual Design for an Industrial Prototype Graz Cycle Power Plant, ASME Paper GT2002-30118.

37 H. Jericha, E. Gottlich, W. Sanz, F. Heitmeir, 2004, Design Optimization of the Graz Cycle Prototype Plant, *Journal of Engineering for Gas Turbines and Power*, 126: 733–740.

38 W. Sanz, H. Jericha, M. Moser, F. Heitmeir, 2005, Thermodynamic and Economic Investigation of an Improved Graz Cycle Power Plant for CO_2 Capture, *Journal of Engineering for Gas Turbines and Power*, 127: 765–772.

39 W. Sanz, H. Jericha, F. Luckel, F. Heitmeir, 2005, A Further Step Towards a Graz Cycle Power Plant for CO_2 Capture, ASME Paper GT2005-68456.

40 H. Jericha, W. Sanz, E. Göttlich, 2008, Design Concept for Large Output Graz Cycle Gas Turbines, *Journal of Engineering for Gas Turbines and Power*, 130: 011701-1–10.

41 R. Anderson, H. Brandt, S. Doyle, K. Pronske, F. Viteri, 2003, Power Generation with 100% Carbon Capture and Sequestration, *Proceedings of the 2nd Annual Conference on Carbon Sequestration*, Alexandria, VA, USA, 5–8 May, 2003.

42 K. Pronske, L. Trowsdale, S. Macadam, F. Viteri, F. Bevc, D. Horazak, 2006, An Overview of Turbine and Combustor Development for Coal-Based Oxy-Syngas Systems, ASME Paper GT2006-90816.

43 R. Anderson, S. MacAdam, F. Viteri, D. Davies, J. Downs, A. Paliszewski, 2008, Adapting Gas Turbine to Zero Emission Oxy-fuel Power Plants, ASME Paper GT2008-51377.

44 R. Anderson, F. Viteri, R. Hollis, M. Hebbar, J. Downs, D. Davies, M. Harris, 2009, Application of Existing Turbomachinery for Zero Emissions Oxy-Fuel Power Systems, ASME Paper GT2009-59995.

45 S. MacAdam, R. Anderson, L. Trowsdale, K. Pronske, Kimberlina – A Zero-Emission Multi-Fuel Power Plant and Demonstration Facility, www.cleanenergysystems.com, May 2006, CES 2006 Clean Water Paper.

46 C. Gou, R. Cai, H. Hong, 2006, An Advanced Oxy-Fuel Power Cycle with High Efficiency, *Proceedings of the Institution of Mechanical Engineers*, Part A: *Journal of Power and Energy*, 220: 315–325.

47 C. Gou, R. Cai, G. Zhang, 2006, An Advanced Zero Emission Power Cycle with Integrated Low Temperature Thermal Energy, *Applied Thermal Engineering*, 26: 2228–2235.

48 C. Gou, R. Cai, H. Hong, 2007, A Novel Hybrid Oxy-fuel Power Cycle Utilizing Solar Thermal Energy, *Energy*, 32: 1707–1714.

49 N. Zhang, N. Lior, M. Liu, W. Han, 2010, COOLCEP (Cool Clean Efficient Power): A Novel CO_2-Capturing Oxy-Fuel Power System With LNG (Liquefied Natural Gas) Coldness Energy Utilization, *Energy*, 35: 1200–1210.

50 M. Liu, N. Lior, N. Zhang, W. Han, 2009, Thermo-Economic Analysis of A Novel Zero-CO_2-Emission High-Efficiency Power Cycle Using LNG Coldness, *Energy Conversion and Management*, 50: 2768–2781.

51 S. Deng, H. Jin, R. Cai, R. Lin, 2004, Novel Cogeneration Power System with Liquefied Natural Gas (LNG) Cryogenic Exergy Utilization, *Energy*, 29: 497–512.

52 P. N. Dyer, R. E. Richards, S. L. Russek, D. M. Taylor. Ion Transport Membrane Technology for Oxygen Separation and Syngas Production. *Solid State Ionics*, 2000, 134: 21–33.

53 R. Bredesen, K. Jordal, O. Bolland. High-temperature Membranes in Power Generation with CO_2 Capture. *Chemical Engineering and Processing*, 2004, 43: 1129–1158.

54 U. Linder, E. H. Eriksen, K. I. Asen, 2000, Method of Operating a Combustion Plant and a Combustion Plant. SE Patent Application 0002037, 2000.
55 T. Bruun, B. Werswick, L. Gronstad, K. Kristiansen, U. Linder, A Device and a Method for Operating Said Device. NO Patent Application 20006690, 2000.
56 K. I. Asen, S. Julsrud, Method for Performing Catalytic or Non-catalytic Processes, Wherein Oxygen is One of the Reactants. NO Patent Application 19972630, 1997.
57 T. Griffin, S. G. Sundkvist, K. Asen, T. Bruun, Advanced Zero Emissions Gas Turbine Power Plant. *Transactions of the ASME, Journal of Engineering for Gas Turbines and Power*, 2005, 127: 81–85.
58 E. Yantovski, J. Gorski, B. Smyth, J. E. ten Elshof, ZEITMOP Cycle (Zero Emission Ion Transport Membrane Oxygen Power). *Proceedings of the International Conference ECOS 2002*, 3–5 July, 2002, Berlin. pp. 1153–1160.
59 E. Yantovski, J. Gorski, B. Smyth, J. ten Elshof. Zero-emission Fuel-fired Power Plants with Ion Transport Membrane. *Energy*, 2004, 29: 2077–2088.
60 E. J. Granite, T. O'Brien, Review of Novel Methods for Carbon Dioxide Separation from Flue and Fuel Gases, *Fuel Processing Technology*, 2005, 1423–1434.
61 P. Chiesa, G. Lozza. CO_2 Emission Abatement in IGCC Power Plants by Semiclosed Cycle: Part A – With Oxygen-blown Combustion. *Transactions of the ASME, Journal of Engineering for Gas Turbines and Power*, 1999, 121: 635–641.
62 http://www.total.com/static/en/medias/topic2627/lacq-pilot-information-dossier.pdf (accessed November 2009).
63 D. Cieutat, I. Sanchez-Molinero, R. Tsiava, P. Recourt, N. Aimard, C. Prebende, 2009, The Oxy-Combustion Burner Development for the CO_2 Pilot at Lacq, *Energy Procedia*, 1: 519–526.
64 C. Hustad, I. Tronstad, R. Anderson, K. Pronske, F. Viteri, 2005, Optimization of Thermodynamically Efficient Nominal 40 MW Zero Emission Pilot and Demonstration Power Plant in Norway, ASME Paper GT2005-68640.
65 CO_2-Global Power Ventures AS, Company Information Memo, *Zero Emission Natural Gas (ZENG) Power Modules for the Industrial-Size Market, ZENG Power Module™*, May 2006.

Index

FSC
www.fsc.org
MIX
Papier | Fördert gute Waldnutzung
FSC® C083411

Kolibri 360 GmbH
Ferdinand-Jühlke-Straße 7
99095 Erfurt, Deutschland
produktsicherheit@kolibri360.de